AF676111

Trends and Prospects in Geotechnics

Trends and Prospects in Geotechnics

Editors

Paulo José da Venda Oliveira
António Alberto Santos Correia

MDPI • Basel • Beijing • Wuhan • Barcelona • Belgrade • Manchester • Tokyo • Cluj • Tianjin

Editors
Paulo José da Venda Oliveira
University of Coimbra
Portugal

António Alberto Santos Correia
University of Coimbra
Portugal

Editorial Office
MDPI
St. Alban-Anlage 66
4052 Basel, Switzerland

This is a reprint of articles from the Special Issue published online in the open access journal *Applied Sciences* (ISSN 2076-3417) (available at: https://www.mdpi.com/journal/applsci/special_issues/Trends_and_Prospects_in_Geotechnics).

For citation purposes, cite each article independently as indicated on the article page online and as indicated below:

LastName, A.A.; LastName, B.B.; LastName, C.C. Article Title. *Journal Name* **Year**, *Volume Number*, Page Range.

ISBN 978-3-0365-3819-8 (Hbk)
ISBN 978-3-0365-3820-4 (PDF)

© 2022 by the authors. Articles in this book are Open Access and distributed under the Creative Commons Attribution (CC BY) license, which allows users to download, copy and build upon published articles, as long as the author and publisher are properly credited, which ensures maximum dissemination and a wider impact of our publications.
The book as a whole is distributed by MDPI under the terms and conditions of the Creative Commons license CC BY-NC-ND.

Contents

About the Editors

Paulo José da Venda Oliveira graduated in Civil Engineering at the University of Coimbra (UC) in 1988. He obtained his PhD degree in Engineering Science in 2001, with specialty in Foundations and Geotechnics. From 1988, he has taught Geotechnics courses at the UC, where he currently holds the position of Associate Professor. His teaching career has covers degree, post-graduate, master and doctorate study courses. Additionally, he has served on more than 90 master's degrees and PhD boards. In terms of the transference of knowledge, he has performed numerous laboratory tests of soil mechanics and many technical studies concerning problems of slope instability, retaining walls, dams, foundations, embankments, etc. At present, he is a member of the Institute for Sustainability and Innovation in Structural Engineering (ISISE), which is a research unit funded by the Portuguese Science Foundation (FCT) that is nationally rated as excellent. He has authored more than 130 scientific and technical papers, 38 of them published in international, peer-reviewed journals (ISI/Scopus). He has supervised more than 40 master's degree and PhD dissertations. He is a frequent reviewer of more than 12 geotechnical journals, such as *Computers and Geotechnics, Geosynthetics International, Journal of Materials in Civil Engineering, Journal of Geotechnical and Geoenvironmental Engineering,* and *Géotechnique,* among others. He has supervised various research grants in the field of chemically stabilized soils and the biocementation of soils. His research activity is focused on geotechnics, from soil characterization to numerical modeling. His research interests over the last five years include numerical modeling of the behavior of stabilized soft soils, laboratory characterization of monotonic and cyclic behavior of stabilized soft soils with binders (cement, lime, fly ash, slags, etc.), biopolymers (xanthan gum), and the use of enzymes (urease) to promote the biocementation process.

António Alberto Santos Correia holds a PhD in Civil Engineering and is Assistant Professor at the University of Coimbra, a member of the Research Centre on Chemical Processes and Forest Products Engineering (CIEPQPF), and collaborates with the Institute for Sustainability and Innovation in Structural Engineering (ISISE); https://orcid.org/0000-0002-3260-8729. He is a specialist in sustainable materials and solutions for engineering problems (binders, nanotechnology), ground improvement, and environmental problems (remediation of contaminated soils). He has worked as a geotechnical expert in 18 national projects ranging from dams, slopes, embankments to roads and industrial infrastructures. He has supervised 8 students and individual grants, 4 of which were PhD grants, in addition to 25 MSc theses; he currently supervises 2 PhD and 3 MSc students. António Correia has published more than 90 scientific publications, 28 of which were published in international peer-reviewed journals; has authored 1 book chapter; served as editor of 8 books and author of 2 theses; is Associate Editor and member of the Editorial Board of the international journal Soils and Rocks; is a frequent reviewer for international journals such as *Geotextiles & Geomembranes, Construction & Building Materials, Geotechnical & Geoenvironmental Engineering,* and *Soils & Rocks,* among others. António Correia is General-Secretary of the Board of the Portuguese Commission of Ground Improvement and Portuguese Delegate on the International Technical Committee TC211—Ground Improvement, a member of the European Committee for Standardization (CEN) TC 189 and International Organization for Standardization (ISO) TC 221, UC representative of the cluster Smart Waste Portugal, ambassador of the Alliance Sustainable Development Goals (SDG) Portugal–United Nations Global Compact Network, and Geotechnical Specialist and Senior Fellow of the Portuguese Institution of Engineers.

António Correia has also taken up a central role in science dissemination, in which he organizes symposia, workshops, and conferences at various levels.

Preface to "Trends and Prospects in Geotechnics"

The world is constantly changing, and so is geotechnics. The challenges posed by society are increasingly demanding in terms of the safety and stability of geotechnical works as well as in terms of sustainability. The geotechnical community has been facing up to these challenges by presenting innovative solutions. Advances in constitutive models, sustainable materials, biobased methods, nanotechnology, energy, artificial intelligence, and climate change, among other topics, are examples of the most recent advances, trends, and prospects in soil mechanics and geotechnical engineering. All types of geotechnical works, such as embankments, deep and shallow foundations, retaining structures, tunnels, dams, and slopes, among others, benefit from such advances.

The aim of this Special Issue book is not to provide an exhaustive summary of the latest advances in the field of geotechnics but, rather, to present some works considered innovative and whose practical application may occur in the near future. Therefore, this Special Issue "Trends and Prospects in Geotechnics" brings together twelve papers (from among twenty-six submitted) that, in addition to their scientific merit, address some of the current and future challenges in geotechnics. The published papers cover a wide range of these emerging topics with a specific focus on the research, design, construction, and performance of geotechnical works. These works are expected to inspire the development of geotechnics, contributing to the future construction of more resilient and sustainable geotechnical structures.

In the end, we would like to take this opportunity to acknowledge all the authors who contributed to this Special Issue "Trends and Prospects in Geotechnics". Special thanks are also addressed to all the reviewers for their valuable contributions and to the *Applied Sciences* MDPI editorial team for the invitation and for all the support given during the production of this Special Issue book, especially Ms. Sara Zhan.

Paulo José da Venda Oliveira and António Alberto Santos Correia

Editors

applied sciences

MDPI

Editorial

Trends and Prospects in Geotechnics

Paulo J. Venda Oliveira [1,*] and António Alberto S. Correia [2,*]

1 Institute for Sustainability and Innovation in Structural Engineering (ISISE), Department of Civil Engineering, University of Coimbra, 3030-788 Coimbra, Portugal
2 Chemical Process Engineering and Forest Products Research Centre (CIEPQPF), Department of Civil Engineering, University of Coimbra, 3030-790 Coimbra, Portugal
* Correspondence: pjvo@dec.uc.pt (P.J.V.O.); aalberto@dec.uc.pt (A.A.S.C.)

The Special Issue, titled "Trends and Prospects in Geotechnics", was launched with an invitation to authors from all over the world to address current and future challenges in geotechnics. As is well known, the world is constantly changing, and so are geotechnics. Advances in constitutive models, sustainable materials, biobased methods, nanotechnology, energy, artificial intelligence, and climate change, among other topics, are examples of the most recent advances, trends, and prospects in soil mechanics and geotechnical engineering. Twenty-six manuscripts were submitted to this Special Issue, and twelve were accepted for publication. Contributions were received from 13 countries (Australia, Brazil, China, Germany, Iraq, Japan, Malaysia, Pakistan, Poland, Portugal, South Korea, Sweden, and Thailand), representing 4 continents (America, Asia, Europe, and Oceania), which addressed some of these emerging topics, with a specific focus on the research, design, construction, and performance of geotechnical works.

In the first paper, Sukmak et al. [1] study the effects of several types of cohesive-frictional soil and geotextile reinforcement configurations on the bearing capacity via laboratory tests and numerical simulation. Several factors were studied, which included embedment depth of the top reinforcement layer, width of horizontal planar form of the reinforcement, spacing between geotextile reinforcement layers, and number of reinforcement layers. The outcome of this research provides a preliminary guideline in a design of reinforced soil foundation structures with different ground soils and other reinforced soil foundation structures with different geosynthetic types.

In the second paper, Yan et al. [2] present a novel study that integrates the spatial time domain reflectometry technique, high precision tensiometer, and consistent outflow logging to investigate the dynamic response of moisture distribution, soil suction, and seepage flux during a transient drainage process. After performing experimental validation tests, the authors concluded that the spatial time domain reflectometry technique offers the researcher a higher resolution of moisture distribution variation with time.

The third [3] and twelfth [4] papers deal with tailing dams. In the third paper, Do et al. [3] investigate the effects of the pond filling rates on excess pore water pressure and the stability of an upstream tailings dam by a numerical study. The approach presented in the paper can be a guide for dam owners to keep a sufficiently high pond filling rate, while still ensuring the desirable stability of an upstream tailings dam. On the other hand, in the twelfth paper, Consoli et al. [4] propose a new approach to deal with the stability problems of tailing dams. In this work, a new approach is proposed for tailings disposal: stacking compacted filtered ore tailings–Portland cement blends. The influence of compaction, as well as the amount of Portland cement, on strength and stiffness properties was evaluated. Although there are parameters that require further studies (moisture content, sustainable binders, and confining pressure), the authors concluded that the addition of a binder to the compacted filtered tailings reduces the volume of hydraulically carried out sediments, thus allowing smaller sedimentation structures downstream of the disposal structure.

Citation: Oliveira, P.J.V.; Correia, A.A.S. Trends and Prospects in Geotechnics. *Appl. Sci.* **2022**, *12*, 3347. https://doi.org/10.3390/app12073347

Received: 16 March 2022
Accepted: 24 March 2022
Published: 25 March 2022

Publisher's Note: MDPI stays neutral with regard to jurisdictional claims in published maps and institutional affiliations.

Copyright: © 2022 by the authors. Licensee MDPI, Basel, Switzerland. This article is an open access article distributed under the terms and conditions of the Creative Commons Attribution (CC BY) license (https://creativecommons.org/licenses/by/4.0/).

The fourth [5] and ninth [6] papers are focussed on the study of machine learning algorithms as a tool to accurately predict the geomechanical properties of rock or soils, thus mimimizing the costs associated with the pre-design and design stages of geotechnical structures. In the fourth paper, Ahmad et al. [5] investigate supervised machine learning algorithms (support vector machine, random forest, AdaBoost, and k-nearest neighbour) to predict the rockfill material shear strength. The performance of the supervised machine learning models are assessed using statistical parameters. The SVM model results in the best and highest performance algorithm, which suggests that this algorithm is more robust in comparison with others in rockfill material shear strength prediction. On the other hand, in the ninth paper, Tinoco et al. [6] study the performance of four machine learning algorithms (artificial neural networks, support vector machines, random forest, and multiple regression) to predict the unconfined compressive strength and the tensile strength of soil–binder–water mixtures reinforced with short fibres. Exploring global sensitivity analysis ensured a deeper understanding around the proposed algorithms. The authors concluded that the proposed models were able to catch both mechanical properties behaviour with a promising performance (R^2 higher than 0.95), particularly those based on artificial neural networks.

The fifth [7] and tenth [8] papers study the effects of applying nanomaterials to chemically stabilised soils, i.e., innovative materials with a promising future. In the fifth paper, Takahashi et al. [7] studied the effects of using cellulose nanofibre (CNF) as an additive in Portland cement in the treatment for soft soil. Authors have concluded that CNF can mix the Portland cement evenly, hardly change the permeability, and reduce the variation in the strength, while at the same time, promoting an increase in the initial age and a reduction in the long-term strength development. In the tenth paper, Correia et al. [8] identify the key parameters in the chemical stabilization of soils with multiwall carbon nanotubes (MWCNT). The characteristics of the surfactant (with an impact on MWCNT dispersion) and time (the importance of MWCNT diminishes as the stabilized matrix becomes denser and stronger) seem to be fundamental parameters which affect the geomechanical behaviour of the stabilized soil enriched with MWCNTs. From the study, the authors concluded that MWCNTs applied in a proper concentration and enriched with a specific surfactant type may be a short-term valid alternative to the partial replacement of traditional additives.

The sixth [9] and eleventh [10] papers deal also with chemically stabilised soils, but now focusing on durability, quality control, and quality assurance issues. In the sixth paper, Ehwailat et al. [9] investigate the use of different materials (nano-magnesium oxide, ground granulated blast furnace slag, and rice husk ash) in soil stabilization to prevent ettringite formation (associated to volume increase), thus avoiding the deterioration of civil engineering structures. The results proved the potential of the nano-magnesium oxide-based binders (incorporating ground granulated blast-furnace slag and rice husk ash) as effective soil stabilizers, showing them to be valid alternatives to traditional binders. In the eleventh paper, Kitazume [10] studies the importance of quality control and quality assurance (QC/QA)-related activities along the workflow of soil stabilisation projects. Based on the Japanese experience/results with mechanical mixing technology by vertical shaft mixing tools with horizontal rotating circular mixing blade, the current and recent developments of QC/QA are presented, which demonstrates their importance to clients and engineers.

In the seventh paper, Mendonça et al. [11] present an interesting study where a more sustainable alternative (xanthan-like biopolymer) is proposed to replace the use of Portland cement in soil stabilisation problems. The authors showed that a treatment with xanthan-like biopolymers or with commercial xanthan gum can be used to replace the Portland cement over the short term (curing time less than 14 days), although a greater level of effectiveness is obtained with the use of the commercial xanthan gum, due to its higher level of purity. The soil treatment with xanthan-like biopolymer creates a network of fibres that link the soil particles, while the commercial xanthan gum fills the voids with a homogeneous paste. Although the results are promising, future studies are still needed to ensure that this bio-material is fully safe.

In the eighth paper, Abdulhamid et al. [12] present a soil contaminated with crude oil in order to reduce the level of soil pollution and then reuse it as a construction material. The contaminated soil is chemically stabilised with two types of Portland cement. The geomechanical properties of the contaminated and stabilised soil have shown that the treatment with Portland cement is an effective remediation method for processing waste to produce a safe, dry material acceptable for onsite application. Moreover, the remediation of contaminated soil with crude oil can utilize Portland cement type II, resulting in a more significant improvement compared to ordinary Portland cement.

Funding: This research received no external funding.

Acknowledgments: The Guest Editors finish by acknowledging all the authors who contributed to this special issue, "Trends and Prospects in Geotechnics". A special mention of thanks is also addressed to the Applied Sciences/MDPI Editorial team for the invitation and for all the support given during the production of this special issue.

Conflicts of Interest: The authors declare no conflict of interest.

References

1. Sukmak, G.; Sukmak, P.; Horpibulsuk, S.; Hoy, M.; Arulrajah, A. Load Bearing Capacity of Cohesive-Frictional Soils Reinforced with Full-Wraparound Geotextiles: Experimental and Numerical Investigation. *Appl. Sci.* **2021**, *11*, 2973. [CrossRef]
2. Yan, G.; Bore, T.; Li, Z.; Schlaeger, S.; Scheuermann, A.; Li, L. Application of Spatial Time Domain Reflectometry for Investigating Moisture Content Dynamics in Unsaturated Loamy Sand for Gravitational Drainage. *Appl. Sci.* **2021**, *11*, 2994. [CrossRef]
3. Do, T.; Laue, J.; Mattsson, H.; Jia, Q. Numerical Analysis of an Upstream Tailings Dam Subjected to Pond Filling Rates. *Appl. Sci.* **2021**, *11*, 6044. [CrossRef]
4. Consoli, N.; Vogt, J.; Silva, J.; Chaves, H.; Scheuermann Filho, H.; Moreira, E.; Lotero, A. Behaviour of Compacted Filtered Iron Ore Tailings–Portland Cement Blends: New Brazilian Trend for Tailings Disposal by Stacking. *Appl. Sci.* **2022**, *12*, 836. [CrossRef]
5. Ahmad, M.; Kamiński, P.; Olczak, P.; Alam, M.; Iqbal, M.; Ahmad, F.; Sasui, S.; Khan, B. Development of Prediction Models for Shear Strength of Rockfill Material Using Machine Learning Techniques. *Appl. Sci.* **2021**, *11*, 6167. [CrossRef]
6. Tinoco, J.; Correia, A.; Venda Oliveira, P. Soil-Cement Mixtures Reinforced with Fibers: A Data-Driven Approach for Mechanical Properties Prediction. *Appl. Sci.* **2021**, *11*, 8099. [CrossRef]
7. Takahashi, H.; Omori, S.; Asada, H.; Fukawa, H.; Gotoh, Y.; Morikawa, Y. Mechanical Properties of Cement-Treated Soil Mixed with Cellulose Nanofibre. *Appl. Sci.* **2021**, *11*, 6425. [CrossRef]
8. Correia, A.; Casaleiro, P.; Figueiredo, D.; Moura, M.; Rasteiro, M. Key-Parameters in Chemical Stabilization of Soils with Multiwall Carbon Nanotubes. *Appl. Sci.* **2021**, *11*, 8754. [CrossRef]
9. Ehwailat, K.; Ismail, M.; Ezreig, A. Novel Approach for Suppression of Ettringite Formation in Sulfate-Bearing Soil Using Blends of Nano-Magnesium Oxide, Ground Granulated Blast-Furnace Slag and Rice Husk Ash. *Appl. Sci.* **2021**, *11*, 6618. [CrossRef]
10. Kitazume, M. Recent Development and Future Perspectives of Quality Control and Assurance for the Deep Mixing Method. *Appl. Sci.* **2021**, *11*, 9155. [CrossRef]
11. Mendonça, A.; Morais, P.; Pires, A.; Chung, A.; Oliveira, P. Reducing Soil Permeability Using Bacteria-Produced Biopolymer. *Appl. Sci.* **2021**, *11*, 7278. [CrossRef]
12. Abdulhamid, S.; Hasan, A.; Aziz, S. Solidification/Stabilization of Contaminated Soil in a South Station of the Khurmala Oil Field in Kurdistan Region, Iraq. *Appl. Sci.* **2021**, *11*, 7474. [CrossRef]

applied sciences

MDPI

Article

Load Bearing Capacity of Cohesive-Frictional Soils Reinforced with Full-Wraparound Geotextiles: Experimental and Numerical Investigation

Gampanart Sukmak [1,*], Patimapon Sukmak [1], Suksun Horpibulsuk [2,3,*], Menglim Hoy [2] and Arul Arulrajah [4,*]

1 School of Engineering and Technology, Center of Excellence in Sustainable Disaster Management, Walailak University, 222 Thaiburi, Thasala District, Nakhon Si Thammarat 80161, Thailand; patimapon.su@wu.ac.th
2 School of Civil Engineering, Undergraduate Program in Civil and Infrastructure Engineering, Center of Excellence in Innovation for Sustainable Infrastructure Development, Suranaree University of Technology, Nakhon Ratchasima 30000, Thailand; menglim@g.sut.ac.th
3 The Royal Society of Thailand, Bangkok 10300, Thailand
4 Department of Civil and Construction Engineering, Swinburne University of Technology, Melbourne, VIC 3122, Australia
* Correspondence: gampanart.su@wu.ac.th (G.S.); suksun@g.sut.ac.th (S.H.); aarulrajah@swin.edu.au (A.A.); Tel.: +66-0-7567-2383 (G.S.); +66-0-4422-4322 (S.H.); +61-3-9214-5741 (A.A.)

Citation: Sukmak, G.; Sukmak, P.; Horpibulsuk, S.; Hoy, M.; Arulrajah, A. Load Bearing Capacity of Cohesive-Frictional Soils Reinforced with Full-Wraparound Geotextiles: Experimental and Numerical Investigation. *Appl. Sci.* **2021**, *11*, 2973. https://doi.org/10.3390/app11072973

Academic Editors: Paulo José da Venda Oliveira and António Alberto Santos Correia

Received: 17 February 2021
Accepted: 23 March 2021
Published: 26 March 2021

Publisher's Note: MDPI stays neutral with regard to jurisdictional claims in published maps and institutional affiliations.

Copyright: © 2021 by the authors. Licensee MDPI, Basel, Switzerland. This article is an open access article distributed under the terms and conditions of the Creative Commons Attribution (CC BY) license (https://creativecommons.org/licenses/by/4.0/).

Abstract: This research investigated the effects of types of cohesive-frictional soil and geotextile reinforcement configurations on the bearing capacity of reinforced soil foundation (RSF) structures, via laboratory test and numerical simulation. The four reinforcement configurations studied for the RSF included: (i) horizontal planar form of geotextile, (ii) full-wraparound ends of geotextile, (iii) full-wraparound ends of geotextile with filled-in sand, and (iv) full-wraparound ends of geotextile with filled-in sand and sand backfill. The foundation soils studied were mixtures of fine sand and sodium bentonite at replacement ratios of 0, 20, 40, 60, 80, and 100% by dry weight of sand to have various values of plasticity index (PI). The numerical analysis of RSF structures was performed using PLAXIS 2D software. Several factors were studied, which included: embedment depth of the top reinforcement layer (U), width of horizontal planar form of the reinforcement (W), and spacing between geotextile reinforcement layers (H). Number of reinforcement layers (N) was varied to determine the optimum parameters of U/B, W/B, H/B, and N, where B is the footing width. The most effective improvement technique was found for the full wraparound ends of geotextile with filled-in sand and sand backfill. The outcome of this research will provide a preliminary guideline in a design of RSF structure with different ground soils and other RSF structures with different geosynthetic types.

Keywords: ground improvement; earth reinforcement; geotextiles; numerical analysis

1. Introduction

The Reinforced Soil Foundation (RSF) structure system using geosynthetic reinforcement has grown significantly over the past four decades. Geosynthetics, such as geotextiles, geogrids, and geocells, are widely used in ground improvement and transportation geotechnics projects, to improve the load bearing capacity and decrease the settlement of weak soil foundation [1–9]. In addition, the embedded geosynthetic reinforcement can inhibit the development of stress bulbs, which induces the progressive failure of soil underneath the footing [10–12].

Recently, a wraparound ends of geotextile technique was introduced to increase the load-bearing capacity of soils [6,7,13,14]. The bearing capacity of the wraparound ends of geotextile reinforced sand increased by approximately twice, as compared with the unreinforced sand [6]. Moreover, this wraparound technique is more suitable for constructing in a confined area, unlike the conventional techniques that incorporate the horizontal planar form of geotextile reinforcements.

Previous research on the bearing capacity of RSF structures [6,13–15] was mainly conducted via experimental tests and numerical simulations using either cohesionless (sand) or cohesive (clay) soils, which were reinforced with the horizontal planar form and wraparound of geotextile reinforcement. Numerical simulations were undertaken to compute the deformation, pore pressure, stress, and strain at various locations underneath the RSF structures [15–18].

Natural soils are commonly cohesive-frictional by nature and they are composed of coarse and fine particles [19]. The shear strength parameters of soil are essential for predicting the load bearing capacity of RSF structures. Several researchers established correlations between plasticity index (PI) and the strength parameters for predicting the behavior of cohesive-frictional soil [20–24]. However, the investigation of RSF using the full-wraparound technique and cohesive-frictional soil is limited. Therefore, the physical model tests and numerical simulation on RSF structures were carried out in this research. Four reinforcement layout configurations studied were (i) horizontal planar form of geotextile, (ii) full-wraparound ends of geotextile with filled-in foundation soils, (iii) full-wraparound ends of geotextile with filled-in sand, and (iv) full-wraparound ends of geotextile with filled-in sand and sand backfill. To obtain various PIs, the foundation soils studied were mixtures of sand and sodium bentonite at different ratios of 0, 20, 40, 60, 80 and 100% bentonite.

Large-scale load-bearing tests were performed to simulate the RSF structures with varying the embedment depth of top reinforcement layer (U), width of horizontal planar reinforcement (W), the spacing between geotextile reinforcement layers (H), and the number of reinforcement layers (N) to determine the optimum parameters of U/B, W/B, H/B and N, where B is the width of footing. The physical model test results of load-bearing capacities of RSF were compared with the numerical simulation results to calibrate the numerical model. The PLAXIS 2D program, which has been proved to be a robust and accurate tool [25–28], was used to investigate the performance of RSF.

A relationship between PI and ultimate bearing capacity of unreinforced cohesive-frictional soils was proposed. The numerical simulations were then carried out to evaluate the bearing capacity improvement of various types of soil reinforced with different configurations of reinforcement by varying the N layers. The simulation technique performed in this research can explain the stress distribution and soil movement underneath the model rigid footing, and it can be applied to the real practice. Based on the analysis of the numerical result, the appropriate reinforcement configurations were recommended for the foundation soils at different PIs. The outcome of this study will provide a preliminary guideline in predicting the bearing capacity improvement of cohesive-frictional soils that are reinforced with geotextile.

2. Theoretical Background

The ultimate bearing capacity of a foundation is defined as the maximum magnitude of the applied load for which failure plane extends to a considerable volume of soils. Prandtl [29] first proposed the bearing capacity equation, by assuming that the soil is weightless. Terzaghi [30] proposed the ultimate bearing capacity equation of shallow foundations on soils by extending the equation proposed by Prandtl [29].

For strip footing, the ultimate bearing capacity ($\sigma_{b(ult)}$) of cohesive-frictional soils is calculated, as follows [30]:

$$\sigma_{b(ult)} = cN_c + \gamma D_f N_q + 0.5\gamma B N_\gamma \quad (1)$$

where c is cohesion of soil, γ is unit weight of soil, B is width of footing, D_f is depth of foundation, N_c, N_q and N_γ are the bearing capacity factors, which are functions of the internal friction angle, ϕ. The bearing capacity factors proposed by Meyerhof [31] are expressed by the following equations:

$$N_q = e^{\pi \tan\phi} \tan^2\left(45 + \frac{\phi}{2}\right) \tag{2}$$

$$N_c = (N_q - 1)\cot\phi \tag{3}$$

$$N_\gamma = (N_q - 1)\tan(1.4\phi) \tag{4}$$

Common techniques for improving soil-bearing capacity are the soil replacement by high quality material [32–36] and geosynthetic reinforcement. For the soil replacement technique, Chen and Abu-Farsakh [11] suggested the thickness of sand soil over clayey soil of more than four times the footing dimension.

The RSF technique has considerable potential as an alternative cost-effective method. Several investigators [1,5,8,14,37–39] have experimentally and numerically studied the behaviors of the footing on the geosynthetic reinforced cohesionless and cohesive soils. All previous investigators indicated that the RSF technique can increase the ultimate bearing capacity and decrease the settlement of footing. Previous studies revealed that the width of the footing (B), embedment depth of top reinforcement layer (U), width of horizontal planar form of reinforcement (W), vertical spacing between reinforcement layers (H), and number of reinforcement layers (N) were dominant factors controlling the soil-bearing capacity. The optimum values of U/B, W/B and H/B have been recommended: U/B = 0.25–0.5, W/B = 6, H/B = 0.2–0.4, and N = 4 [39,40].

3. Experimental and Numerical Procedures

3.1. Soil Samples

In order to obtain various PIs, a fine sand was mixed with bentonite (montmorillonite) at various ratios of 0, 20, 40, 60, 80 and 100% bentonite, and labeled as B0, B20, B40, B60, B80 and B100, respectively. The bentonite powder was a commercial product, which is mainly used as drilling mud for geotechnical construction. The commercial bentonite powder was purchased from Geomechanical Service Co., Ltd., Bangkok, Thailand. The grain size distribution of bentonite powder was investigated by laser diffraction grain size analyzer. The bentonite consisted of 98% clay and 2% silt. The liquid limit (LL) and plastic limit (PL) of bentonite obtained from the Casagrande's and rolling thread methods, according to ASTM-D4318, were 347% and 33%, respectively.

The fine sand was air-dried and passed through a 2.00 mm sieve (No.10). The tested sand (B0) consisted of 98% sand and 2% silt, as shown in Figure 1a. Its average grain size, D50 was 0.5 mm. The bentonite and sand had specific gravity values of 2.35 and 2.80, respectively, and they were classified as a high plasticity clay (CH) and poorly graded sand (SP), respectively, according to the Unified Soil Classification System (USCS).

Figure 1a also shows the grain size distribution curves of the sand-bentonite mixtures. Figure 1b shows the relationships between clay fraction content (<0.002 mm) and LL, PL and PI. The LL significantly increased from 49 to 347% with the increased clay fraction content from 19 to 98% due to the increase in water holding capacity of bentonite. The linear increase of PL, LL and PI with the increased bentonite content (clay fraction) is in agreement with the previous researches [41,42]. The slope of the relationship between PI and clay content, defined as activity [43], was equal to 3.61 and, therefore, classified as active soil. Table 1 summarizes the optimum water content and maximum dry unit weight of these mixtures under the standard Proctor energy (ASTM-D698 [44]).

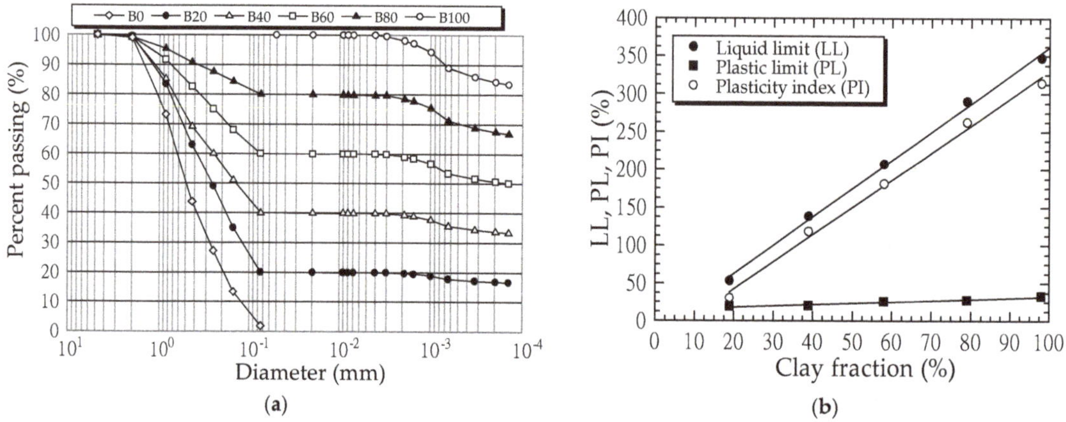

Figure 1. (**a**) Grain size distribution curves of all soil mixtures; (**b**) liquid limit (LL), plastic limit (PL), and plasticity index (PI) of all soil mixtures.

Table 1. Physical and engineering properties of soils.

Soil Mixtures	Total Unit Weight (kN/m^3)	Dry Unit Weight (kN/m^3)	Optimum Water Content (%)	Liquid Limit LL (%)	Plastic Limit PL (%)	Plasticity Index (PI) (%)	Internal Friction Angle (Degree)	Cohesion (kPa)
B0	17.66	17.14	3.0	-	-	-	40	2.8
B20	18.69	16.87	10.8	49	19	30	30	7
B40	17.40	15.01	16.0	138	20	118	12	21
B60	16.17	13.32	21.4	207	26	181	9	23
B80	15.81	12.65	25.0	290	28	262	5	25
B100	14.17	10.99	29.0	347	33	314	0	33

The dried sand-bentonite mixtures were mixed with water by spraying water droplets to attain their optimum water content (obtained from laboratory compaction tests). The soil preparation and direct shear test details can be referenced to Sukmak et al. [19]. The values of cohesion and internal friction angle were obtained from a 65 mm in diameter of a direct shear device at a constant shearing rate of 0.5 mm/min.

Figure 2 shows the internal friction angle and cohesion versus clay fraction relationship. The cohesion linearly increased, while the internal friction angle decreased in a polynomial manner as the clay fraction increased. For clay fraction < 60% (B0, B20, B40 and B60), the internal friction angle significantly decreased with increasing the clay content. Previous researchers [45–47] reported that the ϕ value decreased, as the clay fraction content increased, due to increased lubrication from the clay paste filling the void spaces between sand. The rate of reduction in value was relatively low when clay fraction > 60% (B80 and B100). The highest cohesion of 33 kPa and lowest friction angle of 0 were found at B100. Table 1 summarizes the physical and engineering properties of tested soils.

3.2. Geotextile Reinforcement

A woven geotextile that was manufactured from the high strength polyester (PET) was used as the reinforcement in this study. The material properties of geotextile were: mass per unit area = 445 g/m^2, nominal thickness = 0.9 mm, tensile strength = 200 kN/m, and 50 kN/m in machine and cross-machine directions, respectively, and the tensile strain at failure in machine direction = 10%.

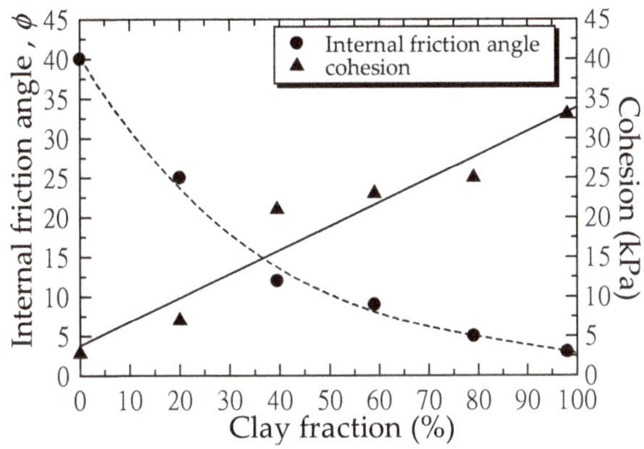

Figure 2. Relationship between soil strength parameters (internal friction angle and cohesion) and clay fraction content.

3.3. Load-Bearing Test Apparatus

Figure 3 illustrates the load-bearing model test apparatus. The rigid reaction frame was fabricated from a 200 mm × 150 mm wide steel flange beam. The 10-mm thick steel container had internal dimensions of 1000 mm length, 600 mm width, and 700 mm depth. The side of the container was strongly braced to prevent its lateral deformation from the soil compaction and load-bearing testing. The friction between the tested soils and the model container was minimized while using a smooth plate.

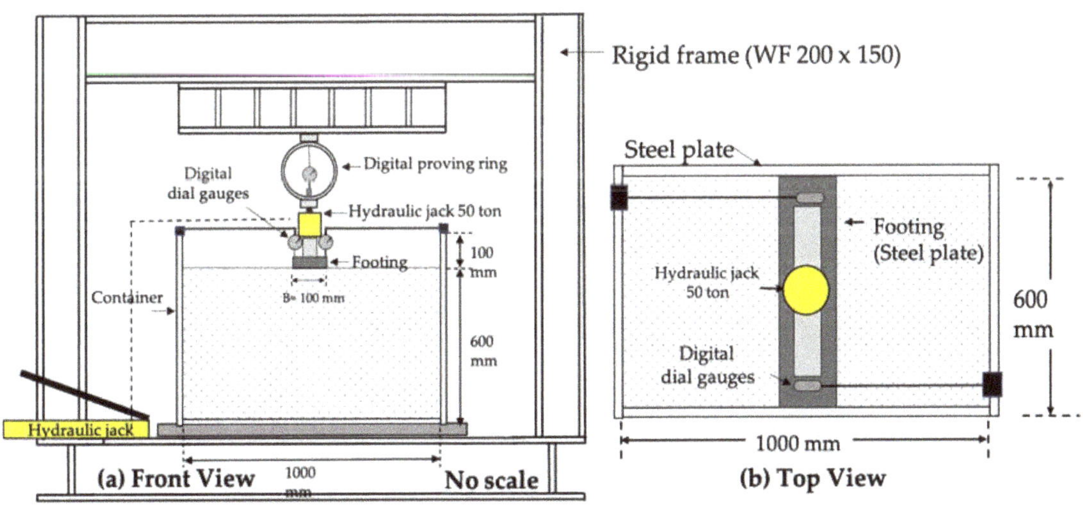

Figure 3. A schematic view of the tests set up for load-bearing tests.

The model rigid footing was a strip footing made of 30 mm thick rigid plate with dimensions of 596 mm length and 100 mm width. The width of footing, B = 100 mm was fixed for all the tests. The base of footing was attached to the rough sandpaper by glue. The horizontal and vertical boundaries of model container were 10 times and six times of the footing width, respectively, in order to minimize boundary effects. The model tests were conducted under the plane-strain condition.

3.4. Experimental and Numerical Programs

This study was undertaken to investigate the relationship between bearing capacity of unreinforced soil and PI and obtain the most effective values of influence parameters for the studied RSF: the embedment depth of top reinforcement layer (U), width of horizontal planar reinforcement (W), the vertical spacing between reinforcement layers (H), and number of reinforcement layers (N).

To evaluate the development in bearing capacity of RSF with different geotextile reinforcement layout configurations, the number of reinforcement layers (N) were varied from 1 to 4. The bearing capacity development was analyzed in terms of a bearing improvement factor, a ratio of the bearing capacity of reinforced soils to the bearing capacity of unreinforced soil (q_b/q_{b0}). The numerical simulations were carried out to evaluate the performance of studied RSF. The simulated result was the stress distribution and soil movement underneath the rigid footing.

The numerical analysis was performed on the RSF with various reinforcement layout configurations, as shown in Figure 4, including Series A to F. The simulated results were then compared with the laboratory model test results. Figure 4a indicates the experimental model test of the strip footing on different unreinforced soil mixtures (Series A) to evaluate the effect of PI on its bearing capacities. The techniques of reinforcement were the horizontal planar reinforced soils (Figure 4b), the full-wraparound ends of geotextile with filled-in foundation soils (Figure 4d), the full wrapround ends of geotextile with filled-in sand (Figure 4e), and the full wraparound ends of geotextile with filled-in sand and sand backfill (Figure 4f). Table 2 provides the details of experimental and numerical programs.

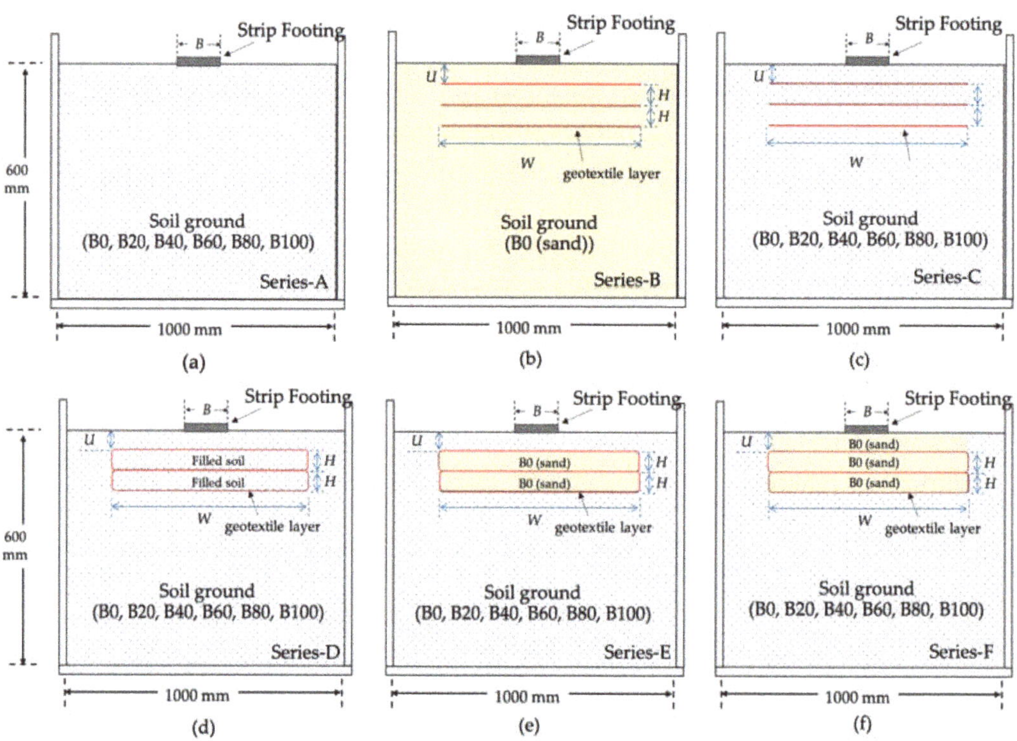

Figure 4. Reinforcement layout configurations; **(a–c)** for experimental and numerical investigations and **(d–f)** for numerical investigation.

Table 2. Details of experimental and numerical programs.

Investigation Cases	Reinforcement Layout Configurations	Test Series	Soil Ground	Filled Soil	Reinforcement Layout Configurations	Test Series
Experimental and numerical investigations	Unreinforced soil	A	B0, B20, B40, B60, B80, B100	-	-	-
	Horizontal planar reinforced soils	B-1	B0	-	N = 1, W/B = 6	U/B = 0.1, 0.3, 0.5, 1 and 2
		B-2		-	N = 1, U/B = 0.3	W/B = 2, 3, 4, 5 and 6
		B-3		-	N = 2, U/B = 0.3, W/B = 3	H/B = 0.1, 0.25, 0.5, 0.7
		B-4		-	U/B = 0.3, W/B = 3, H/B = 0.5	N = 1, 2, 3 and 4
	Horizontal planar reinforced soils	C-1 C-2 C-3 C-4 C-5 C-6	B0 B20 B40 B60 B80 B100	-	U/B = 0.3, W/B = 3, N = 1, 2, 3 and 4	R_{inter} = 0 to 1.0
Numerical investigations	Full wraparound ends of geotextile	D-1 D-2 D-3 D-4 D-5 D-6	B0 B20 B40 B60 B80 B100	B0 B20 B40 B60 B80 B100	U/B = 0.3, W/B = 3, H/B = 0.5	N = 2, 3 and 4
	Full wraparound ends of geotextile with filled-in sand	E-1 E-2 E-3 E-4 E-5	B20 B40 B60 B80 B100	B0	U/B = 0.3, W/B = 3, H/B = 0.5	N = 2, 3 and 4
	Full wraparound ends of geotextile reinforcement with filled-in sand and sand backfill	F-1 F-2 F-3 F-4 F-5	B20 B40 B60 B80 B100	B0	U/B = 0.3, W/B = 3, H/B = 0.5	N = 2, 3 and 4

Series B was conducted to determine the optimal U/B, W/B and H/B of the strip footing on sand (B0) reinforced with different number of horizontal layers of geotextile (N) (see Figure 4b). For a single horizontal reinforcement layer (N = 1), the U/B ratios were varied from 0.1 to 2.0 with a constant W/B ratio of 6.0. Subsequently, the optimum W/B ratio was obtained by varying W/B ratios from 2 to 6 with a constant optimal ratio of U/B. For N = 2, the optimum spacing value between reinforcement layers was obtained by varying H/B ratios from 0.1 to 0.7 at both optimum ratios of U/B and W/B. The optimum values of U/B, W/B and H/B ratios were selected for modeling Series C, D, E and F.

In the numerical analysis of Series C to F, the interaction factors (R_{inter}) between soil and geotextile reinforcement were required and they were determined from the back-analysis of test results of Series C. The R_{inter} value was varied until the simulated curves were in good agreement with the experimental curves. The R_{inter} relates the interface strength (friction resistance on the surface between geotextile and soil) to the soil strength (frictional angle and cohesion). The R_{inter} value is intermediate between smooth (R_{inter} = 0) and fully rough (R_{inter} = 1.0). Kazi et al. [6,7] and Shadi et al. [48,49] reported that the R_{inter} value of 0.66 can be used to simulate the interaction behavior of sandy soil that was reinforced with geotextile with different reinforcement configurations (with and without partial and full-wraparound geotextile).

The numerical results of Series C, D, E and F would illustrate the improvement of bearing capacity of RSF with different reinforcement configurations.

3.5. Construction and Test Methods

For series A, the soil was compacted in 50 mm height layers with a vibratory compactor until the maximum dry unit weight was attained. The total compacted soil thickness was 600 mm. The density tests were conducted after the end of construction to control the compaction quality and repeatability of the experiments. The results revealed that all test sections were reproducible.

After the soil foundation was compacted for the full depth, the compacted soil was excavated and the reinforced soil foundation (RSF) with various configurations (Series B to C) was constructed, as shown in Figure 4. The geotextile layers in the machine direction were placed parallel to the width of the footing at the desired magnitude of U, W, H and N according with the reinforcement layout configuration that is given in Table 2.

The footing was then placed on the RSF. The vertical loading was applied on the footing by a 50 kN capacity hydraulic jack that was supported by a digital proving ring and a rigid reaction frame. The bearing force was measured using a digital proving ring, which was installed between the rigid reaction frame and hydraulic jack (see Figure 3). The bearing force was applied at a constant rate of 0.5 mm/min. The vertical movement of the footing was monitored with two digital dial-gauges installed at the edge of each side of footing. The bearing force and vertical movement were recorded until the end of test at 15 mm settlement.

3.6. Numerical Modeling

The numerical simulation was conducted using a finite element program, PLAXIS 2D (Figure 5). The bottom boundary of the model was fixed against movement in all directions, while the vertical boundaries were free to move in the vertical direction but restricted in the horizontal direction. The soil elements were modeled using 15-noded triangular elements. Very fine meshes were used around the geometry line of footing and reinforcements to minimize the mesh size effect. The linear-elastic perfectly plastic Mohr-Coulomb (MC) model was utilized for soil and interface elements. Five basic parameters for MC-model, Young's modulus (E), Poisson's ratio (υ), cohesion (c), internal friction angle (ϕ), and dilation angle (ψ), were obtained from conventional laboratory tests. The coefficient of earth pressure at rest, K_0 was calculated from ($1-\sin\phi$). Table 3 summarizes the numerical soil parameters.

Table 3. The soil parameters used in numerical analysis.

Soil Mixtures	E (kN/m^2)	υ	ϕ (Degree)	c (kPa)	ψ (Degree)	R_{inter}
B0	20,000	0.3	40	2	10	0.66
B20	17,000	0.3	30	9	0	0.59
B40	15,000	0.35	12	18	0	0.54
B60	13,000	0.4	9	20	0	0.48
B80	8000	0.4	5	25	0	0.45
B100	5000	0.4	0	33	0	0.42

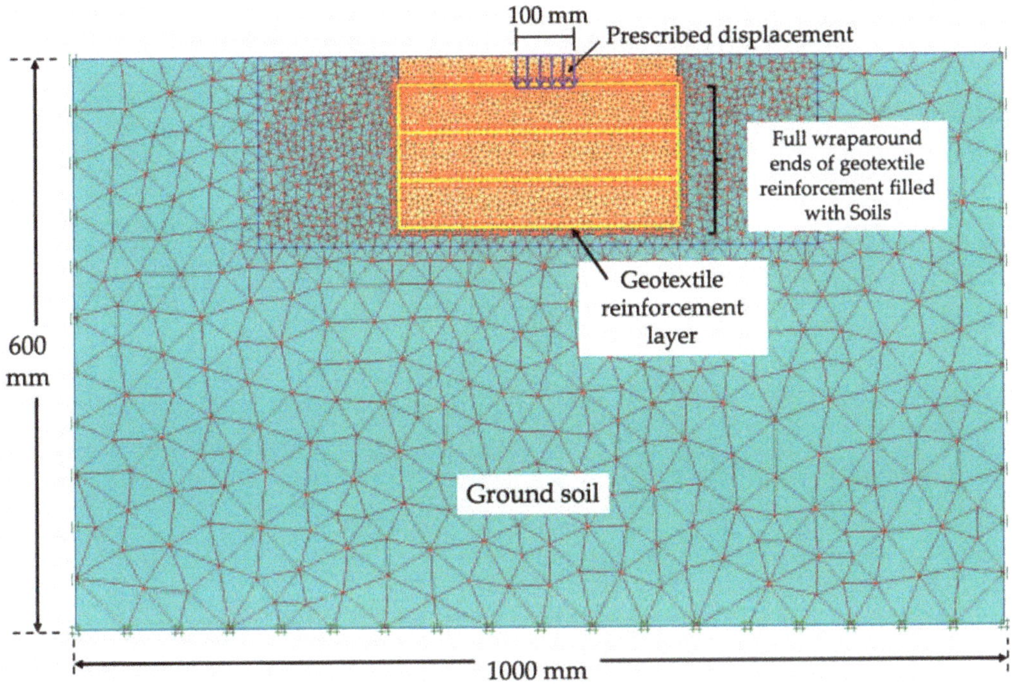

Figure 5. Finite-element model for load-bearing tests.

The geotextile material was modeled using a geogrid element (line elements), which is generally used to model geosynthetics reinforcement in the PLAXIS program. The geotextile was assumed as slender structures with a nonzero value of normal stiffness and zero bending stiffness, since it can only sustain tensile force, but not compression forces [50]. The elastic axial stiffness (EA) of geotextile material was equal to 200 kN/m obtained from the tensile strength test according to ASTM-D4595 [51]. The layout configuration of the geotextile reinforcement (Figure 4 and Table 2) and the material properties (Table 3) were simulated.

The interaction parameter (R_{inter}) was simulated with the interface elements that were located between the geotextile reinforcement and surrounding soils. The R_{inter} values for different configurations of geotextile reinforcement were summarized in Table 3.

The load was applied on the footing using the prescribed displacement method. The calculation was terminated when the final settlement reached 15 mm. The pressure (stress point; Sig-yy) and vertical settlement (displacement node; U_y) values under the center of strip footing were compared with the measured results. The bearing capacity ratio (q_b/q_{b0}) at the S/B ratio of 15% was used to illustrate the bearing capacity improvement due to the geotextile reinforcement, where q_b = bearing capacity of RSF and q_{b0} = the bearing capacity of unreinforced soil. Table 2 summarizes the conditions of experimental model tests and numerical simulations.

4. Test Results and Discussion

4.1. Load-Bearing Capacity of Unreinforced Soils (Series A)

Figure 6 shows the relationship between bearing stress (q) and settlement ratio of unreinforced soils with different PI: dense sand (B0) and stiff cohesive-frictional soils (B20, B40, B60, B80 and B100). The shear stress continued to increase as the soil deformed. For B0, q sharply increased with increasing the S/B ratio until the q was obtained at S/B ratio approximately equal to 5%, and then gradually decreased. For B20, B40, B60, B80 and B100, the q gradually increased with increasing the S/B ratio. After S/B > 10%, the q was almost constant without peak. The q_{b0} decreased as the bentonite content increased i.e., q_{b0} of B100 was the lowest, while q_{b0} of B0 was the highest. In other words, the q_{b0} decreased with the increase of PI. The soil surface on both sides of the foundation noticeably heaved, indicating the general shear failure, which is the common failure mode for a shallow foundation on the dense sand and stiff cohesive soils [4].

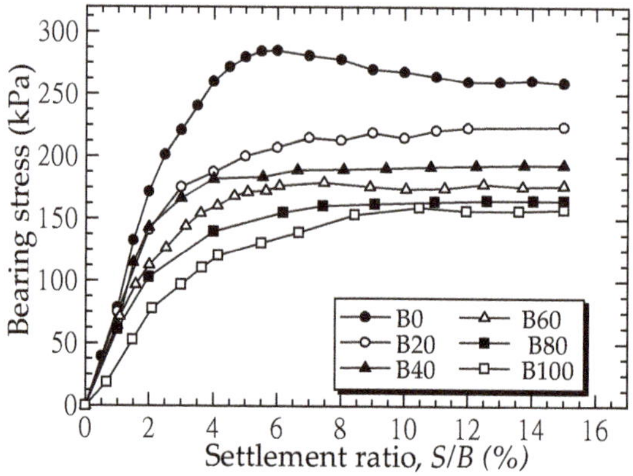

Figure 6. Relationship between the bearing stress and settlement ration (S/B) of different unreinforced soils.

Figure 7 shows the relationship between q_{b0} and PI at S/B = 15%. The measured q_{b0} was compared with the calculated bearing capacity using Equations (1)–(5). The q_{b0} with various PI was satisfactorily predicted by Meyerhof (1963)'s equation. The q_{b0} decreased in a polynomial function as the PI increased:

$$q_{b0} = 0.0011\,(\mathrm{PI})^2 - 0.658\,(\mathrm{PI}) + 255.28 \text{ for } 0\% < \mathrm{PI} < 314\% \quad (5)$$

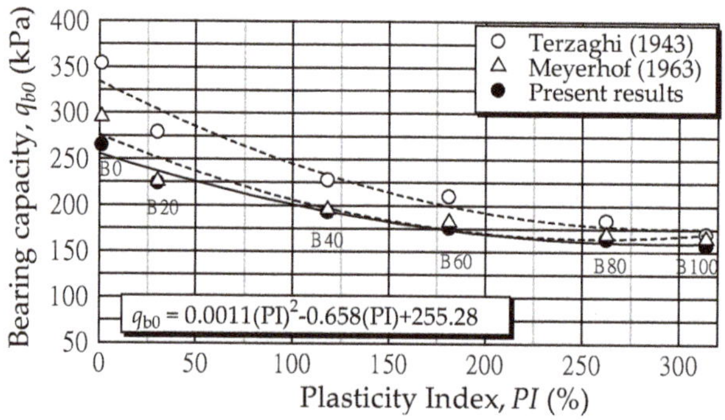

Figure 7. Relationship between bearing capacity and plasticity index (PI) of tested soils.

4.2. Load-Bearing Capacity of B0 Mixture Reinforced with Horizontal Planar Geotextile Reinforcement with N Layers (Series B)

Series B-1 to B-3 were carried out on the strip foundation on the B0 mixture reinforced with single horizontal planar layer (Series B-1 and B-2) and two horizontal planar layers (Series B-3) of geotextile reinforcement. The ratios of U/B, W/B and H/B were varied to obtain the optimum values. Series B-4 was for the B0 mixture reinforced with various layers of horizontal planar geotextile (N = 1 to 5) at constant ratio of U/B, W/B and H/B.

Figure 8 shows the relationships between the q_b and settlement ratio with various values of U/B, W/B, H/B and N. Figure 9 shows the variations of the q_b/q_{b0}.

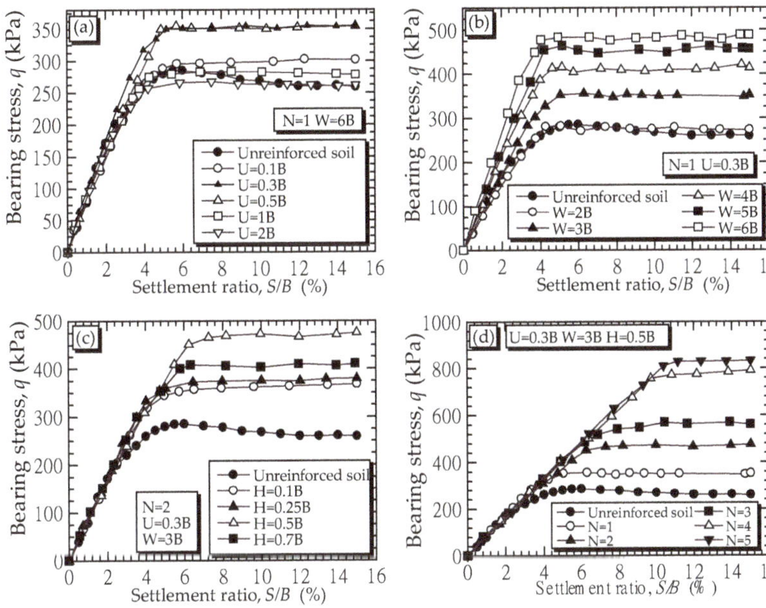

Figure 8. Relationship between bearing stress and S/B ratio with various parameters of (**a**) U/B; (**b**) W/B; (**c**) H/B; and (**d**) N.

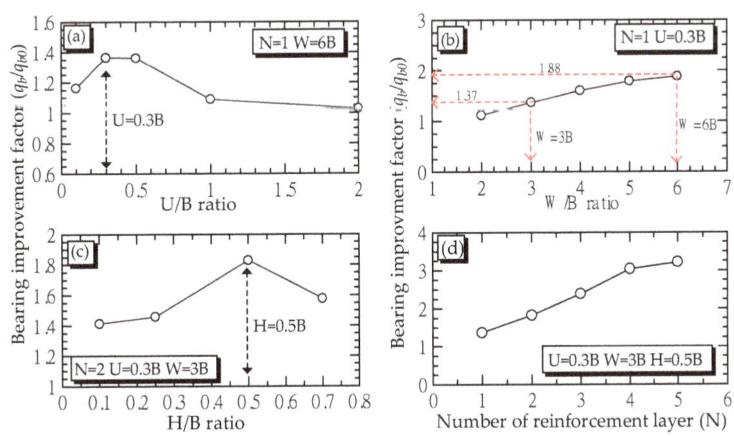

Figure 9. Relationship between bearing stress and S/B ratio with various parameters of (**a**) U/B; (**b**) W/B; (**c**) H/B; and (**d**) N.

For Series B-1, the measured q_b of B0 reinforced with the single-layer planar geotextile at various U/B ratios and the W/B ratio of 6.0 is shown in Figure 8a. The bearing stress, q, increased when increasing the S/B ratio up to the S/B ratio of approximately 5% and then the q_b was essentially constant with the increased S/B ratio. Figure 9a shows the q_b/q_{b0} and U/B relationship. The maximum q_b/q_{b0} ratio was 1.39 at the U/B = 0.30. When the U/B > 1, the q_b/q_{b0} ratios were approximately equal to 1.0. In other words, the installation of reinforcement deeper than B did not improve the q_b. Das et al. [2] stated that the decrease in q_b when U/B < 0.3 was due to the lack of confinement for the top reinforcing layer. While, when U/B > 1, the failure surface in the soil is located above the top of reinforcement layer.

For Series B-2, Figures 8b and 9b show the measured q_b and q_b/q_{b0} of B0 reinforced with N = 1 at a constant U/B ratio of 0.3 with various W/B ratios, respectively. For S/B = 0 to 5%, the slope of q versus S/B plots increased with the increased W/B (Figure 8b). The q of all the W/B ratios became approximately constant when S/B > 5%. Figure 9b revealed that the q_b/q_{b0} increased as the W/B ratio increased. Khing et al. [40] suggested that the W/B ratio of 6.0 was optimal for a strip footing on sand supported by horizontal planar geosynthtic reinforcement, which is not applicable for the field projects, which with have a limited construction area. In this study, the full-wraparound geotextile reinforcement technique was investigated to reduce the reinforcement width. The W/B = 3.0 was therefore selected for Series C, D, E and F.

The q_b and q_b/q_{b0} of B0 reinforced with N = 2 at a constant ratio of U/B = 0.3 and W/B = 3.0 with various H/B ratios are shown in Figures 8c and 9c (Series B-3). The q_b/q_{b0} increased with the H/B increased from 0.1 to 0.5. Subsequently, the q_b/q_{b0} decreased when H/B > 0.5.

Figures 8d and 9d show the simulated result for Series B-4 with N =1 to 5, U/B = 0.3, W/B = 3.0, and H/B = 0.5. The S/B ratios at the failure increased from 4 to 10% for N increasing from 1 to 5, while the same gradient of q versus S/B relationship is noted (Figure 8d). Figure 9d shows that the q_b value significantly increased with increasing the N value when N < 4. In other words, the most effective N was 4.

Based on Series B test results, the values of U/B = 0.3, W/B = 3.0, H/B = 0.5, and N = 4 were used for numerical studies of Series C, D, E and F.

4.3. Interface Factor

Figure 10 shows the laboratory (solid line) q of the RSF (Series C) on B0, B20, B40, B60, B80 and B100 reinforced with N = 1, U/B = 0.3, and W/B = 3.0. The R_{inter} values were determined by a trial and error method until the simulated q versus S/B curves were comparable with the measured q versus S/B curves.

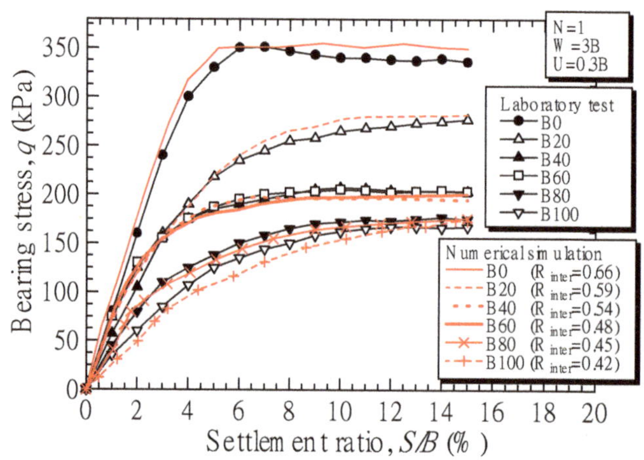

Figure 10. Measured and simulated bearing stress results of strip footing resting on different soils reinforced with single horizontal planar geotextile layer with U/B = 0.3 and W/B = 3.0.

Figure 11 shows the relationship between R_{inter} and PI. The R_{inter} value linearly decreased with the increase in PI due to the reduction in interface skin friction between soil and geotextile reinforcement:

$$R_{inter} = -0.0007\ (PI) + 0.63 \text{ for } 0\% < PI < 314\% \tag{6}$$

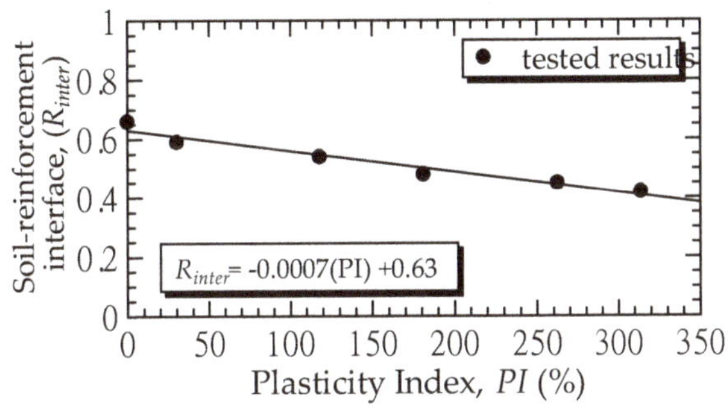

Figure 11. Relationship between the soil-reinforcement interface and plasticity index (PI).

4.4. Performance of RSF with Different Reinforcing Techniques

Figure 12 shows the relationship between the simulated q and S/B of B20, B60 and B100 with the four configurations (Series C, D, E and F). For all cases, the q initially increased with the increase of S/B ratio up to a particular S/B_f. When $S/B > S/B_f$, the q value was slightly increased, and it can be assumed as achieving the failure state. For Series C, D and E, the S/B_f = 12, 8, 6, 5, and 5% for B20, B40, B60, B80 and B100, respectively, for N = 1–4. The q for Series F was fully developed at the S/B_f = 14% for N = 1 to 4. It was evident that the development of q in Series F was the highest when compared with other Series. The development of q_b depended on not only the types of tested soil, but also the reinforcing techniques (planar and wraparound ends), which will be discussed in the next section.

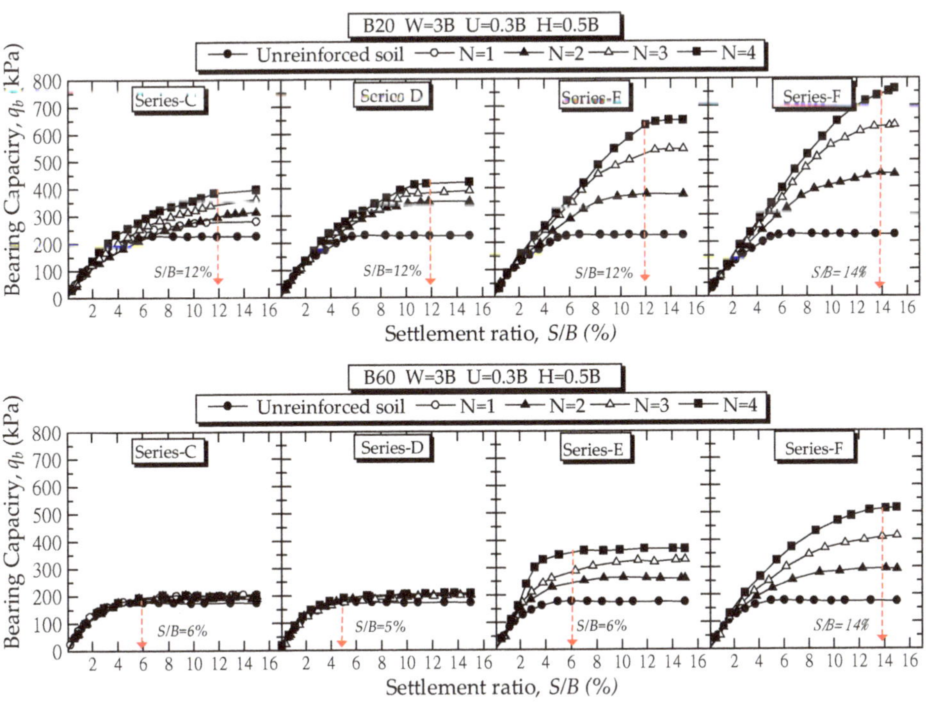

Figure 12. *Cont.*

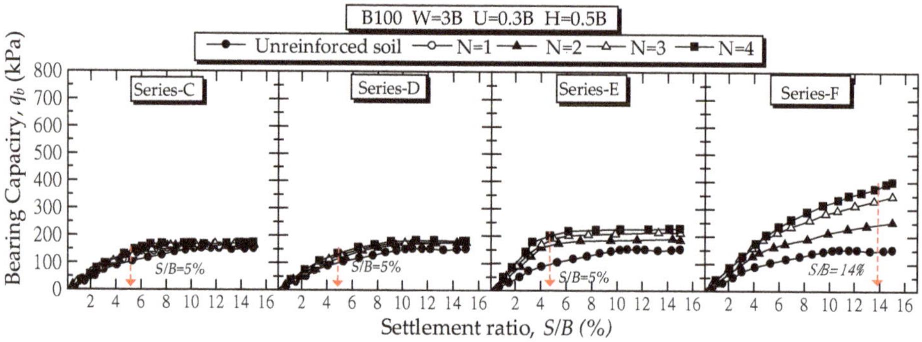

Figure 12. Simulated bearing stress results of strip footing resting on different soils reinforced with different reinforcing techniques.

Figure 13 shows the relationship between q_b/q_0 versus N for Series C, D, E and F. Figures 14–19 show the simulated results of the total displacement vectors and stress distribution (Sig-yy) of B0, B20 and B100 at the S/B ratio of 15% for Series C, D, E and F with N = 4, respectively. When the unreinforced soil (Series A) was subjected to the vertical applied stresses, the soil particles beneath the footing moved downward to horizontal direction on both sides with large areas, as shown in Figures 14a, 15a and 16a for B0, B20 and B100, respectively. The triangular shaped wedge under the footing was clearly observed, as shown Figures 17a, 18a and 19a for B0, B20 and B100, respectively.

When PI < 30% (B0 and B20), the q_b/q_0 increased from 1.4 to 2.4 for B0 and 1.24 to 1.76 for B20 due to the increase in N from 1 to 4, as shown in Figure 13 (for Series C). On the other hand, when PI > 30% (B40, B60, B80 and B100), the q_b/q_0 development was comparatively low with the additional N. The q_b/q_0 ratio of B40, B60, B80 and B100 was approximately from 1.03 to 1.22 for N = 1 to 4. These results can be explained by Figures 14b and 15b, in that the amount of horizontal displacement vectors at soil-geotextile interfaces of B0 and B20 densely occurred on all layers of geotextile reinforcement. However, the vectors of B100 mostly occurred only on the first layer while it lightly occurred on other layers, as shown in Figure 16b.

In general, the geotextile reinforcement can significantly increase the q_b by its high tensile resistance and the soil-reinforcement interface shear resistance [3,40]. The soil particles under the footing moved along the geotextile in lateral direction, as shown in Figures 14b, 15b and 16b, while the high interface strength tried to prevent the movement of soil particles and, therefore, improved the q_b. Moreover, the applied stress to the soil was lower when compared to unreinforced soil as some stresses were transferred to the geotextile. The deformed reinforcement can support the part of applied stress to the soil by developing upward force, as shown in Figures 17b, 18b and 19b.

The full-wraparound ends of geotextile filled with the foundation soil (Series D) can significantly improve the q_b of B0, B20 and B40 (PI < 118%) with increasing the N layers, as shown in Figure 13. The q_b/q_0 was improved in a range of 1.77 to 3.30 for B0, 1.45 to 1.90 for B20, and 1.44 to 1.75 for B40 when the N value increased from 2 to 4. Most of the displacement vectors of soils appeared within the full-wraparound geotextile ends, as illustrated in Figures 14c and 15c. When comparing Series D to Series C, the q_b development in Series D was higher than that for Series C for the same foundation soils. Nevertheless, when PI > 118% (B60, B80 and B100), the q_b/q_0 was lower than 1.25 for all N values. The large soil movement between the footing and first layer of geotextile was observed (Figure 16c) when compared with soil with PI < 118% (Figures 14c and 15c for B0 and B20, respectively). As such, this technique was recommended for foundation soil with PI < 118%.

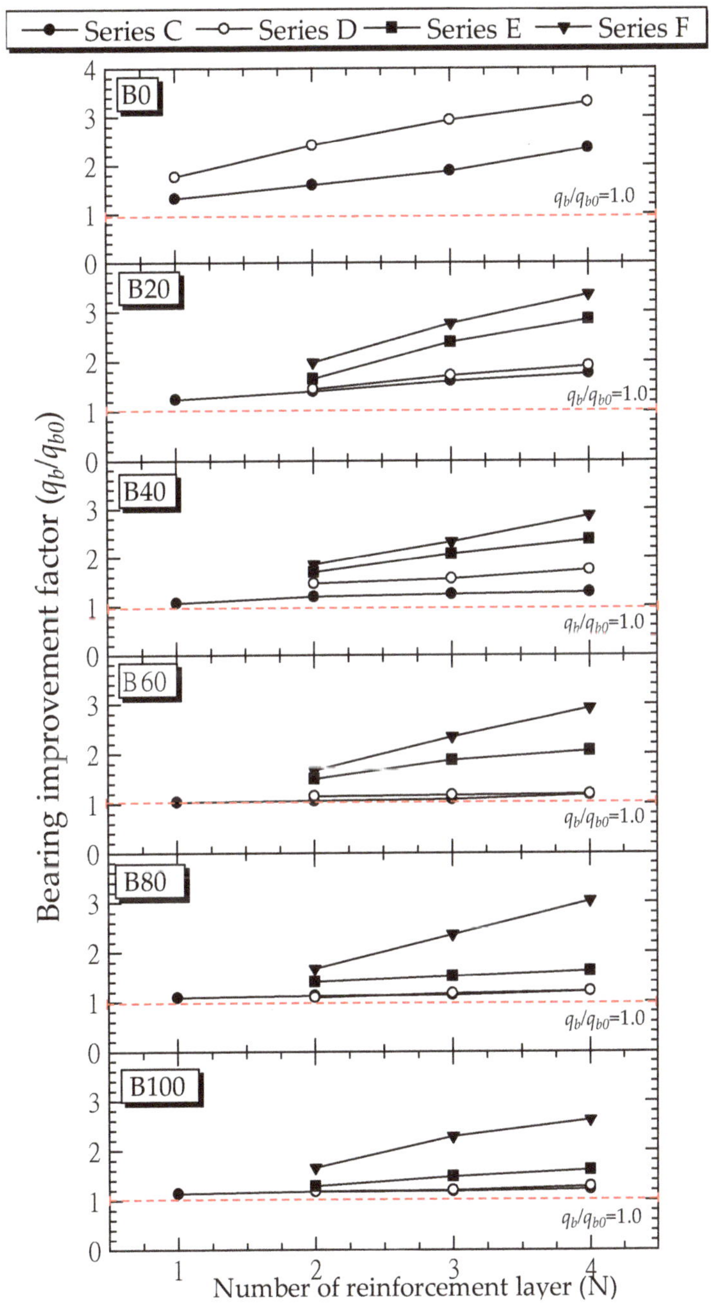

Figure 13. Relationship between bearing improvement factor and number of reinforcement layer of soils reinforced with different Reinforcement layout configurations (series C, D, E and F).

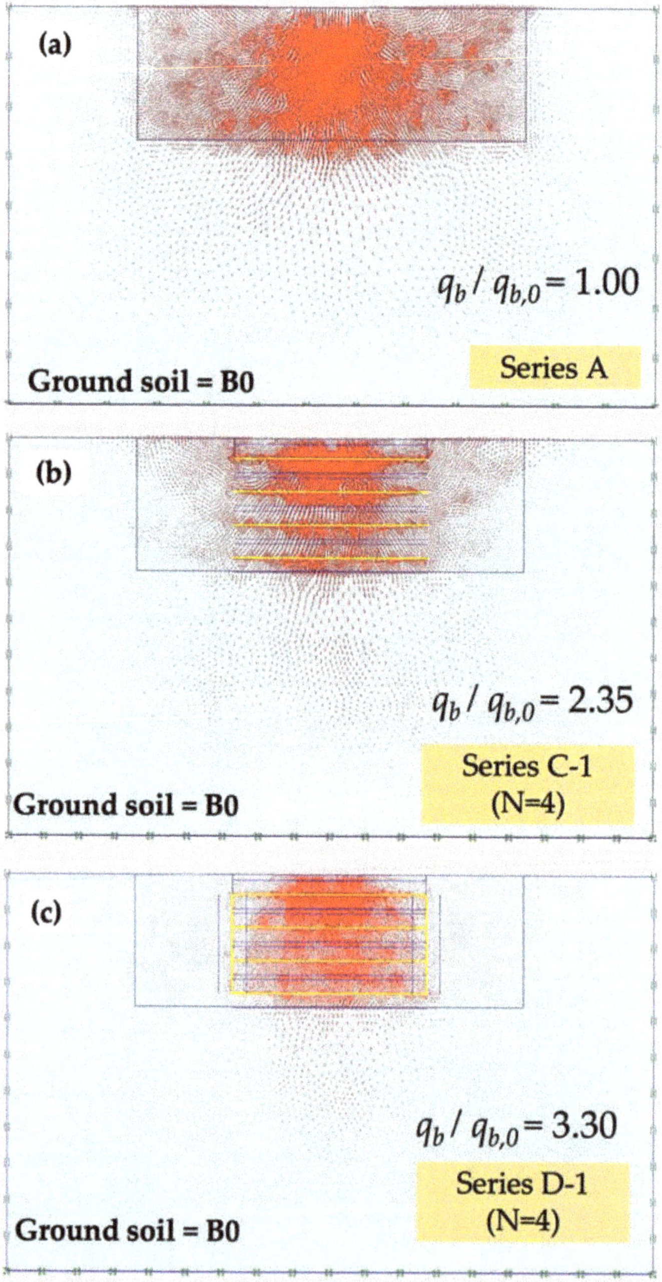

Figure 14. Simulated results of total displacement vectors of: (**a**) unreinforced B0 Soil; (**b**) B0 reinforced by series C; and, (**c**) B0 reinforced by series D for N = 4 at the S/B ratio of 15%.

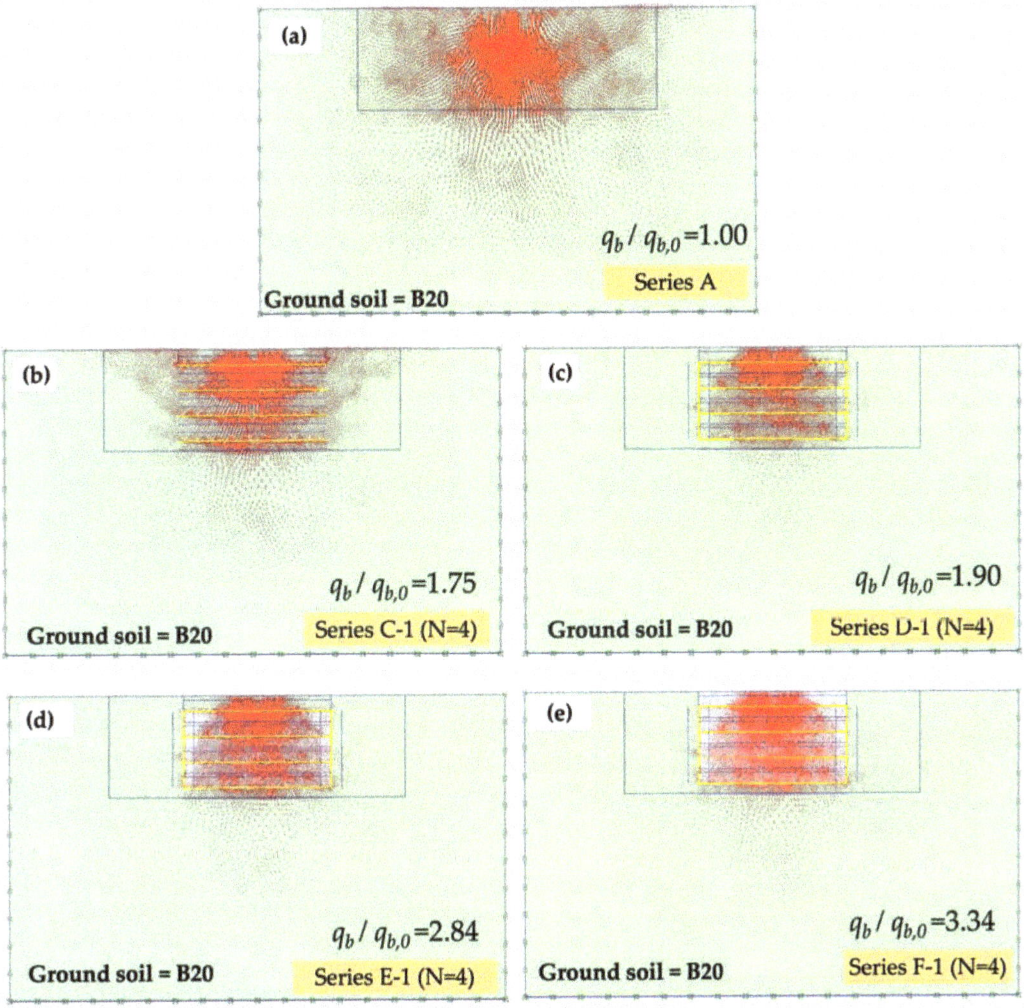

Figure 15. Simulated results of total displacement vectors of: (**a**) unreinforced B20; (**b**) B20 reinforced by series C; (**c**) B20 reinforced by series D; (**d**) B20 reinforced by series E; and (**e**) B20 reinforced by series F for N = 4 at the S/B ratio of 15%.

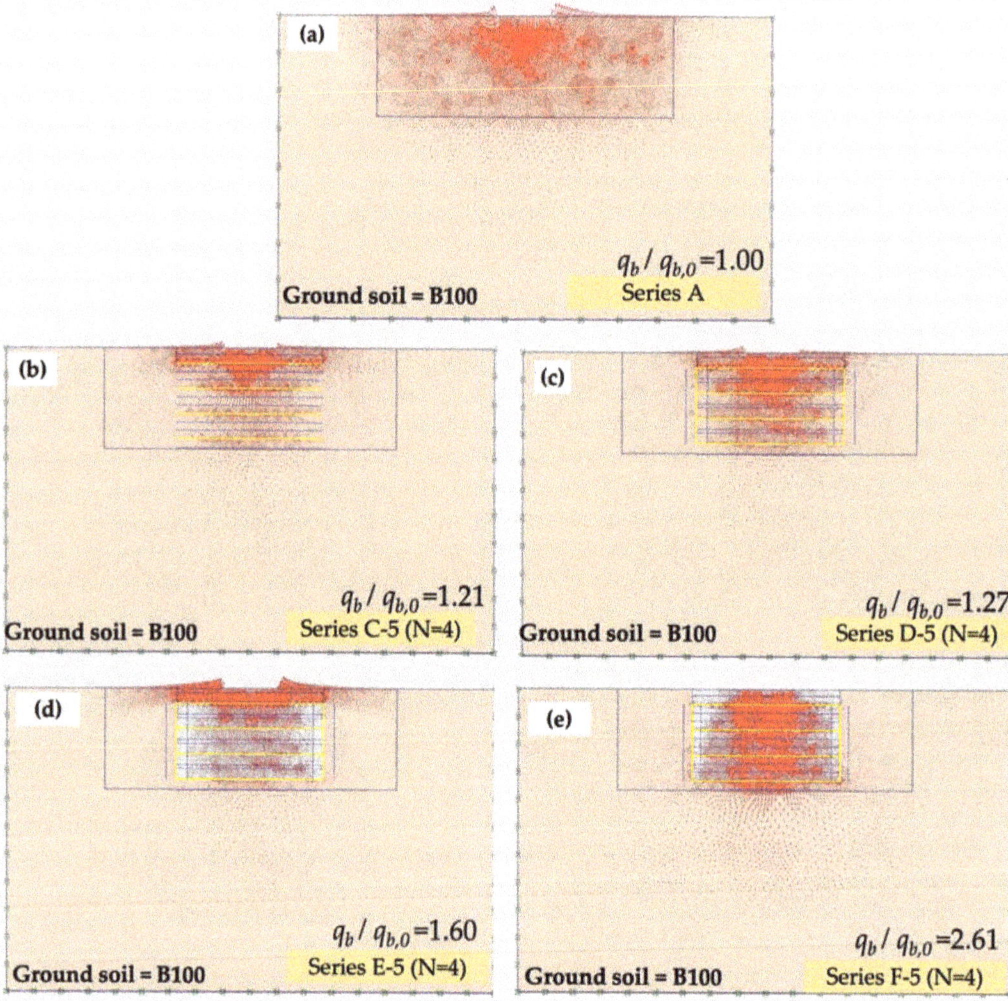

Figure 16. Simulated results of total displacement vectors of: (**a**) unreinforced B100; (**b**) B100 reinforced by series C; (**c**) B100 reinforced by series D; (**d**) B100 reinforced by series E; and (**e**) B100 reinforced by series F for N = 4 at the S/B ratio of 15%.

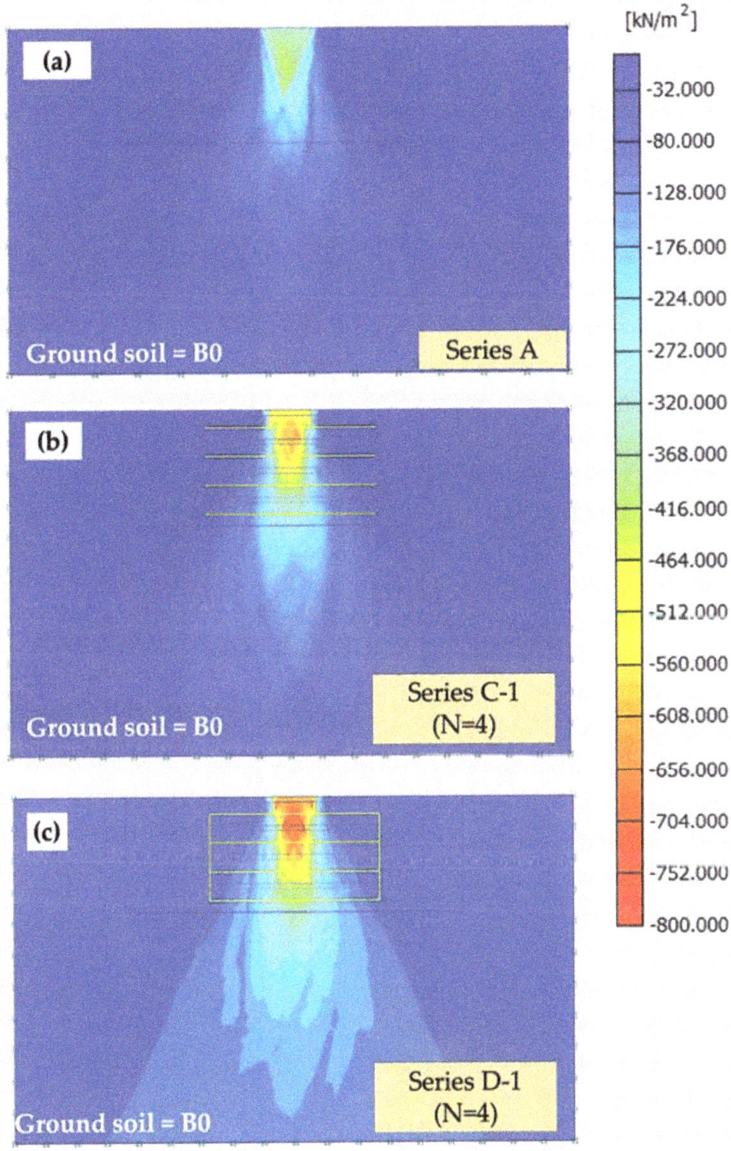

Figure 17. Simulated results of stress distribution (sig-yy) of: (**a**) unreinforced B0 Soil; (**b**) B0 reinforced by series C; and (**c**) B0 reinforced by series-D for N = 4 at the S/B ratio of 15%.

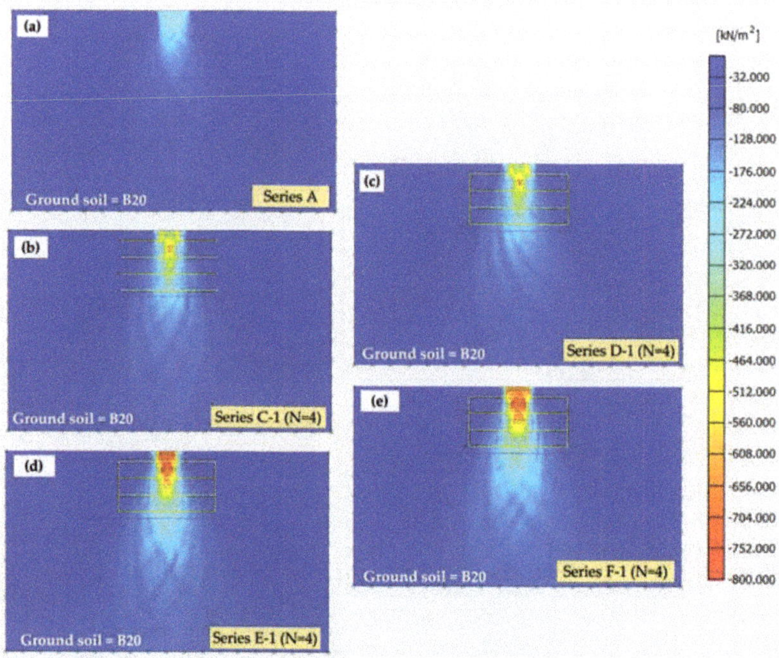

Figure 18. Simulated results of stress distribution (sig-yy) of: (**a**) unreinforced B20; (**b**) B20 reinforced by series C; (**c**) B20 reinforced by series D; (**d**) B20 reinforced by series E; and (**e**) B20 reinforced by series-F for N = 4 at the S/B ratio of 15%.

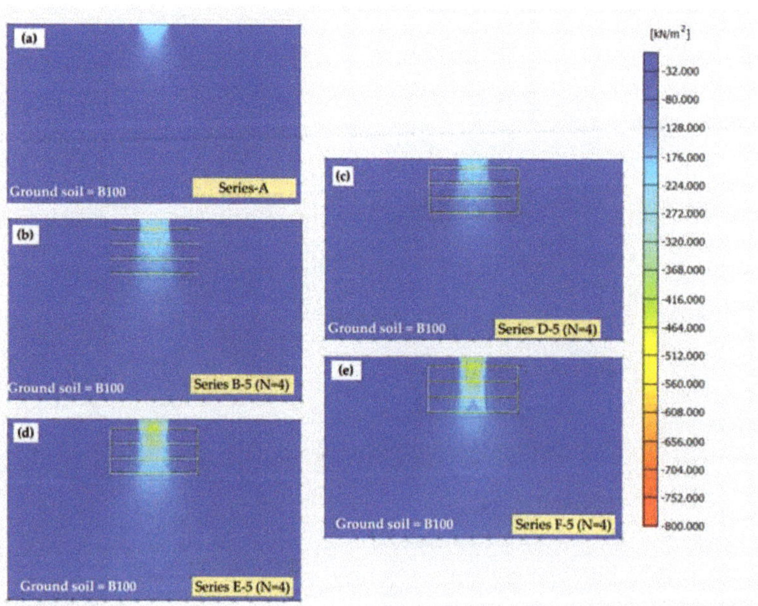

Figure 19. Simulated results of stress distribution (sig-yy) of: (**a**) unreinforced B100; (**b**) B100 reinforced by series C; (**c**) B100 reinforced by series D; (**d**) B100 reinforced by series E; and (**e**) B100 reinforced by series-F for N = 4 at the S/B ratio of 15%.

The q_b/q_0 development for Series E (filled sand in wraparound geotextile) decreased with the increase in PI of foundation soils. The q_b/q_0 value increased from 1.70 to 2.84 for B20, 1.69 to 2.36 for B40, and 1.50 to 2.06 for B60, 1.40 to 1.62 for B80, and 1.2 to 1.60 for B100 when N increased from 2 to 4. For foundation soil with PI < 181% (B0, B20, B40 and B60), the applied stresses were transferred through foundation soil and distributed to wraparound geotextiles. The displacement vectors under the footing moved downward, as shown in Figures 14c and 15d. The usage of high-quality soil (B0) as fill material can improve the stiffness of full-wraparound ends of geotextile. The applied stress distributed through the foundation soil under the footing to the top of geotextile layer and then the applied stress transferred to the bottom of geotextile layer, as shown in Figures 17c and 18d. However, the stress distribution of foundation soil depended on its soil shear strength and stiffness.

With the low stiffness of soil with PI > 181% (B80 and B100) on the comparably stiffer RSF, the soil above the RSF largely deformed, as shown in Figure 16d. Therefore, the applied stresses could not be perfectly transferred to the other layers of geotextile; hence, the development of q_b with the addition of N was insignificant (see Figure 13). In addition, the distributed stress in B0 inside full-wraparound layers was lower when compared to the case of foundation soil with PI < 181%, as illustrated in Figures 17c, 18d and 19d.

To improve the performance of q_b for the wraparound ends technique with filled sand, the replacement of foundation soil above the RSF by B0 was studied in Series F. It was evident that the q_b in Series F was the highest when compared with other Series (Figure 13). The q_b/q_0 increased linearly with the increased N = 2 to 4 and varied from 1.35 to 3.34 for B20, 1.86 to 2.87 for B40, 1.66 to 2.91 for B60, 1.67 to 3.03 for B80, and 1.15 to 2.61 for B100. Figures 14c, 15e and 16e show that, for all of the foundation soils, the displacement vectors were within the full-wraparound ends of geotextile. The applied stresses can distribute from footing to bottom geotextile reinforcement, as shown in Figures 17c, 18d and 19d. Accordingly, the q_b development depended on the soil underneath the bottom layer. The q_b development decreased with the increasing of PI. This technique changed the shallow foundation to deep foundation.

Based on the analysis of numerical results, the q_b improvement depended on both the reinforcement configurations and the PI of foundation soil. The horizontal planar reinforced soils (Series C), the full-wraparound ends with filled-in foundation soil (Series D), and the full wrapround ends with filled-in sand (Series E) were recommended for foundation soil with PI < 30%, PI < 118%, and PI < 181%, respectively. The highest of q_b/q_0 ratio was achieved when using the full wrapround ends of geotextile with filled-in sand and replaced sand under footing (Series F).

The replacement method with cohesionless soil (high-quality soil), such as sand and gravel, is a classical technique for improving the bearing capacity of footing on cohesive-frictional soils (low-quality soil). The q_b improvement is significantly dependent upon the depth of soil replacement. Numerical simulations were performed on the studied soils (B0 and B100) to compare the q_b/q_0 value from the replacement method with that from Series F of B100 with the same boundary condition (Figure 5). Based on the simulation, when the B0 with the replacement width of 3B was used to improve bearing capacity of footing on B100 at various replacement depth, the q_b/q_0 was equal to 1.05, 1.27, 1.54, 1.61 and 1.62 for the replacement depth of 0.5B, 1B, 1.5B, 2B and 3B, respectively. This implied that the suitable replacement depth was approximately 2B. When compared with results of Series F (see Figure 12c), the same q_b/q_0 can be achieved with only 0.8B replacement depth when N = 2 was implemented. Moreover, the greater q_b improvement (q_b/q_0 > 1.62) was possible for Series F with the increased N (see Figure 13).

5. Conclusions

This paper investigated the effects of several types of cohesive-frictional soil and geotextile reinforcement configurations on the bearing capacity via a laboratory test and numerical simulation. The reinforcement used in this study was a woven geotextile that was manufactured from the high strength polyester (PET). The four reinforcement configurations studied were: (i) horizontal planar form of geotextile, (ii) full-wraparound ends of geotextile with foundation soils, (iii) full-wraparound ends of geotextile with filled-in sand, and (iv) full-wraparound ends of geotextile with filled-in sand and sand replaced foundation soil. The studied foundation soils were mixtures of fine sand and sodium bentonite at different replacement ratios of 0%, 20%, 40%, 60%, 80% and 100% by the dry weight of sand. The bearing capacity ratio (q_b/q_0) was used to illustrate the bearing capacity improvement due to the geotextile reinforcement, where q_b = bearing capacity of RSF and q_{b0} = the bearing capacity of unreinforced soil. The following conclusions can be drawn from this study.

1. For unreinforced soil, the q_{b0} decreased as the PI increased because of the reduction of soil shear strength. The q_{b0} can be predicted in terms of PI using a polynomial function: $q_{b0} = 0.0011(\mathrm{PI})^2 - 0.658(\mathrm{PI}) + 255.28$ when $0\% < \mathrm{PI} < 314\%$. The q_{b0} with various PI is satisfactorily predicted by Meyerhof (1963)'s equation. The laboratory investigation showed that the effective values of influence parameters for planar geotextile reinforcement were U/B = 0.3, W/B = 3.0, H/B = 0.5, and N = 4.
2. For the horizontal planar form of geotextile reinforcement, when PI < 30, the q_b/q_0 increased with the increase in N from 1 to 4. The high soil-geotextile interface shear resistance tended to prevent the movement of soil particles and increased the bearing capacity. On the other hand, when PI > 30% (B40, B60, B80 and B100), the q_b/q_0 development was comparatively low with the additional N, because the soil movement highly occurred only on the first layer.
3. For the full-wraparound ends of geotextile with filled-in foundation soils, when PI < 118%, the q_b/q_0 was improved in a range of 1.77 to 3.30 for B0, 1.45 to 1.90 for B20, and 1.44 to 1.75 for B40 when the N value increased from 2 to 4. Nevertheless, when PI > 118%, the q_b/q_0 was lower than 1.25 for all N values. When compared to the horizontal planar form of geotextile, the q_b development in full-wraparound ends of geotextile was higher for the same foundation soils.
4. For the full-wraparound ends of geotextile with filled-in sand, the q_b/q_0 value increased when the N increased from 2 to 4. For foundation soil with PI < 181%, the applied stress distributed through the foundation soil under the footing to the geotextiles effectively. However, the soil above the RSF was largely deformed with the low stiffness of soil with PI > 181% on the comparably stiffer RSF. With sand backfill under the footing, the q_b/q_0 increased linearly with the increase of N from 2 to 4. The q_b/q_0 varied from 1.35 to 3.34 for B20, 1.86 to 2.87 for B40, 1.66 to 2.91 for B60, 1.67 to 3.03 for B80, and 1.15 to 2.61 for B100.
5. The horizontal planar reinforced soils, the full-wraparound ends of geotextile with filled-in foundation soil, and the full wraparound ends of geotextile with filled-in sand, were recommended for foundation soils with PI < 30%, PI < 118%, and PI < 181%, respectively. The sand backfill above the full wraparound ends of geotextile with filled-in sand can help effectively transfer the applied stress to RSF, hence the significant improvement in q_b.
6. It is kept in mind that the studied results are only valid for a woven geotextile. However, the knowledge gained from this research can be used as a guideline for further studies on different types of geosynthestics.

Author Contributions: Conceptualization and methodology, G.S. and P.S.; software and validation, G.S. and S.H.; writing—original draft preparation, G.S.; writing—review and editing, S.H., M.H. and A.A.; supervision, G.S. All authors have read and agreed to the published version of the manuscript.

Funding: This work is financially supported by the Walailak University, No. WU6223. The authors also acknowledge the support from the National Science and Technology Development Agency under the Chair Professor program [grant no P-19-52303].

Institutional Review Board Statement: Not applicable.

Informed Consent Statement: Not applicable.

Conflicts of Interest: The authors declare no competing interest.

References

1. Chen, R.-H.; Wu, C.-P.; Huang, F.-C.; Shen, C.-W. Numerical analysis of geocell-reinforced retaining structures. *Geotext. Geomembr.* **2013**, *39*, 51–62. [CrossRef]
2. Das, B.M.; Shin, E.C.; Omar, M.T. The bearing capacity of surface strip foundations on geogrid-reinforced sand and clay—A comparative study. *Geotech. Geol. Eng.* **1994**, *12*, 1–14. [CrossRef]
3. Dash, S.K.; Krishnaswamy, N.R.; Rajagopal, K. Bearing capacity of strip footings supported on geocell-reinforced sand. *Geotext. Geomembr.* **2001**, *19*, 235–256. [CrossRef]
4. Han, J. *Principles and Practice of Ground Improvement*; John Wiley & Sons, Inc: Hoboken, NJ, USA, 2015.
5. Hegde, A.M.; Sitharam, T.G. Three-dimensional numerical analysis of geocell-reinforced soft clay beds by considering the actual geometry of geocell pockets. *Can. Geotech. J.* **2015**, *52*, 1396–1407. [CrossRef]
6. Kazi, M.; Shukla, S.K.; Habibi, D. Behavior of embedded strip footing on sand bed reinforced with multilayer geotextile with wraparound ends. *Int. J. Geotech. Eng.* **2015**, *9*, 437–452. [CrossRef]
7. Kazi, M.; Shukla, S.K.; Habibi, D. An Improved Method to Increase the Load-Bearing Capacity of Strip Footing Resting on Geotextile-Reinforced Sand Bed. *Indian Geotech. J.* **2015**, *45*, 98–109. [CrossRef]
8. Shirazi, M.G.; Rashid, A.S.A.; Nazir, R.B.; Rashid, R.H.B.A.; Moayedi, H.; Horpibulsuk, S.; Samingthong, W. Sustainable soil bearing capacity improvement using natural limited life geotextile reinforcement—A review. *Minerals* **2020**, *10*, 479. [CrossRef]
9. Rashid, A.S.A.; Shirazi, M.G.; Nazir, R.; Mohamad, H.; Sahdi, F.; Horpibulsuk, S. Bearing capacity performance of soft cohesive soil treated by kenaf limited life geotextile. *Mar. Georesour. Geotechnol.* **2020**, *38*, 755–760. [CrossRef]
10. Chakraborty, M.; Kumar, J. Bearing capacity of circular foundations reinforced with geogrid sheets. *Soils Found.* **2014**, *54*, 820–832. [CrossRef]
11. Chen, Q.; Abu-Farsakh, M. Ultimate bearing capacity analysis of strip footings on reinforced soil foundation. *Soils Found.* **2015**, *55*, 74–85. [CrossRef]
12. Shahin, H.M.; Nakai, T.; Morikawa, Y.; Masuda, S.; Mio, S. Effective use of geosynthetics to increase bearing capacity of shallow foundations. *Can. Geotech. J.* **2017**, *54*, 1647–1658. [CrossRef]
13. Benmebarek, S.; Djeridi, S.; Benmebarek, N.; Belounar, L. Improvement of bearing capacity of strip footing on reinforced sand. *Int. J. Geotech. Eng.* **2018**, *12*, 537–545. [CrossRef]
14. Djeridi, S.; Benmebarek, N.; Benmebarek, S. Effect of Full Wrap-Around Ends of Geotextile on the Bearing Capacity of Sand. In *Advances in Science, Technology & Innovation*; IEREK Interdisciplinary Series for Sustainable Development; Springer: Berlin/Heidelberg, Germany, 2019; pp. 291–294. [CrossRef]
15. Rao, P.; Liu, Y.; Cui, J. Bearing capacity of strip footings on two-layered clay under combined loading. *Comput. Geotech.* **2015**, *69*, 210–218. [CrossRef]
16. Benmebarek, S.; Benmoussa, S.; Belounar, L.; Benmebarek, N. Bearing Capacity of Shallow Foundation on Two Clay Layers by Numerical Approach. *Geotech. Geol. Eng.* **2012**, *30*, 907–923. [CrossRef]
17. Benmebarek, S.; Remadna, M.S.; Benmebarek, N.; Belounar, L. Numerical evaluation of the bearing capacity factor $N\gamma'$ of ring footings. *Comput. Geotech.* **2012**, *44*, 132–138. [CrossRef]
18. Conte, E.; Donato, A.; Troncone, A. Progressive failure analysis of shallow foundations on soils with strain-softening behaviour. *Comput. Geotech.* **2013**, *54*, 117–124. [CrossRef]
19. Sukmak, K.; Sukmak, P.; Horpibulsuk, S.; Han, J.; Shen, S.-L.; Arulrajah, A. Effect of fine content on the pullout resistance mechanism of bearing reinforcement embedded in cohesive–frictional soils. *Geotext. Geomembr.* **2015**, *43*, 107–117. [CrossRef]
20. Lupini, J.F.; Skinner, A.E.; Vaughan, P.R. The drained residual strength of cohesive soils. *Géotechnique* **1981**, *31*, 181–213. [CrossRef]
21. Maio, C.D.; Fenellif, G.B. Residual strength of kaolin and bentonite: The influence of their constituent pore fluid. *Géotechnique* **1994**, *44*, 217–226. [CrossRef]
22. Mesri, G.; Cepeda-Diaz, A.F. Residual shear strength of clays and shales. *Géotechnique* **1986**, *36*, 269–274. [CrossRef]
23. Sridharan, A.; Rao, S.; Murthy, N. Liquid Limit of Montmorillonite Soils. *Geotech. Test. J.* **1986**, *9*, 156–159. [CrossRef]
24. Tan, T.; Goh, T.; Karunaratne, G.; Lee, S. Shear Strength of Very Soft Clay-Sand Mixtures. *Geotech. Test. J.* **1994**, *17*, 27–34. [CrossRef]
25. Bergado, D.T.; Youwai, S.; Teerawattanasuk, C.; Visudmedanukul, P. The interaction mechanism and behavior of hexagonal wire mesh reinforced embankment with silty sand backfill on soft clay. *Comput. Geotech.* **2003**, *30*, 517–534. [CrossRef]
26. Khedkar, M.S.; Mandal, J.N. Pullout behaviour of cellular reinforcements. *Geotext. Geomembr.* **2009**, *27*, 262–271. [CrossRef]
27. Suksiripattanapong, C.; Chinkulkijniwat, A.; Horpibulsuk, S.; Rujikiatkamjorn, C.; Tangsuttinon, T. Numerical Analysis of Bearing Reinforcement Earth (BRE) Wall. *Geotext. Geomembr.* **2012**, *32*, 28–37. [CrossRef]
28. Sukmak, K.; Han, J.; Sukmak, P.; Horpibulsuk, S. Numerical parametric study on behavior of bearing reinforcement earth walls with different backfill material properties. *Geosynth. Int.* **2016**, *23*, 435–451. [CrossRef]
29. Prandtl, L. Uber die eindringungstestigkeit plastisher baustoffe und die festigkeit von schneiden [in German]. *Z. Fur Angew. Math. Mech.* **1921**, *1*, 15–30. [CrossRef]
30. Terzaghi, K. *Theoretical Soil Mechanics*; John Wiley & Sons, Inc.: Hoboken, NJ, USA, 1943.

31. Meyerhof, G.G. Some recent research on the bearing capacity of foundations. *Can. Geotech. J.* **1963**, *1*, 16–26. [CrossRef]
32. Burd, H.J.; Frydman, S. Bearing capacity of plane-strain footings on layered soils. *Can. Geotech. J.* **1997**, *34*, 241–253. [CrossRef]
33. Hanna, A.M.; Meyerhof, G.G. Design charts for ultimate bearing capacity of foundations on sand overlying soft clay. *Can. Geotech. J.* **1980**, *17*, 300–303. [CrossRef]
34. Ismail Ibrahim, K.M.H. Bearing capacity of circular footing resting on granular soil overlying soft clay. *HBRC J.* **2016**, *12*, 71–77. [CrossRef]
35. Kumar, J.; Chakraborty, M. Bearing capacity of a circular foundation on layered sand–clay media. *Soils Found.* **2015**, *55*, 1058–1068. [CrossRef]
36. Madhav Madhira, R.; Sharma, J.S.N. Bearing Capacity of Clay Overlain by Stiff Soil. *J. Geotech. Eng.* **1991**, *117*, 1941–1948. [CrossRef]
37. Huang, C.-C.; Tatsuoka, F. Bearing capacity of reinforced horizontal sandy ground. *Geotext. Geomembr.* **1990**, *9*, 51–82. [CrossRef]
38. Lovisa, J.; Shukla, S.K.; Sivakugan, N. Behaviour of prestressed geotextile-reinforced sand bed supporting a loaded circular footing. *Geotext. Geomembr.* **2010**, *28*, 23–32. [CrossRef]
39. Yetimoglu, T.; Wu Jonathan, T.H.; Saglamer, A. Bearing Capacity of Rectangular Footings on Geogrid-Reinforced Sand. *J. Geotech. Eng.* **1994**, *120*, 2083–2099. [CrossRef]
40. Khing, K.H.; Das, B.M.; Puri, V.K.; Cook, E.E.; Yen, S.C. The bearing-capacity of a strip foundation on geogrid-reinforced sand. *Geotext. Geomembr.* **1993**, *12*, 351–361. [CrossRef]
41. Kumar, G.V.; Muir Wood, D. Fall cone and compression tests on clay-gravel mixtures. *Geotechnique* **1999**, *49*, 727–739. [CrossRef]
42. Polidori, E. Relationship between the atterberg limits and clay content. *Soils Found.* **2007**, *47*, 887–896. [CrossRef]
43. Skempton, A.W. The colloidal activity of clays. In Proceedings of the 3rd International Conference on Soil Mechanics and Foundation Engineering, Zurich, Switzerland, 16–27 August 1953; Volume 1, pp. 57–61.
44. *ASTM-D698 Standard Test. Methods for Laboratory Compaction Characteristics of Soil Using Standard Effort (12,400 ft-lbf/ft3 (600 kN-m/m3))*; ASTM International: West Conshohocken, PA, USA, 2012.
45. Al-Shayea, N.A. The combined effect of clay and moisture content on the behavior of remolded unsaturated soils. *Eng. Geol.* **2001**, *62*, 319–342. [CrossRef]
46. Tiwari, B.; Marui, H. A New Method for the Correlation of Residual Shear Strength of the Soil with Mineralogical Composition. *Geotech. Geol. Eng.* **2005**, *131*, 1139–1150. [CrossRef]
47. Wang, S.; Chan, D.; Lam, K.C. Experimental study of the effect of fines content on dynamic compaction grouting in completely decomposed granite of Hong Kong. *Constr Build. Mater.* **2009**, *23*, 1249–1264. [CrossRef]
48. Shadi, A.; Sanjay, K.S.; Alireza, M. Behaviour of sandy soil reinforced with geotextile having partially and fully wrapped ends. *Ground Improvement.* **2021**, *174*, 29–41. [CrossRef]
49. Shadi, A.; Sanjay, K.S.; Alireza, M. Numerical Investigation of Wraparound Geotextile Reinforcement Technique for Strengthening Foundation Soil. *Int. J. Geomech.* **2019**, *19*. [CrossRef]
50. Brinkgreve, R.B.J.; Broere, W.; Waterman, D. *PLAXIS 2D—Version 8 Tutorial Manual*; Plaxisbv: Delft, The Netherlands, 2006.
51. *ASTM-D4595 Standard Test. Method for Tensile Properties of Geotextiles by the Wide-Width Strip Method*; ASTM International: West Conshohocken, PA, USA, 2017.

applied sciences

MDPI

Article

Application of Spatial Time Domain Reflectometry for Investigating Moisture Content Dynamics in Unsaturated Loamy Sand for Gravitational Drainage

Guanxi Yan [1,*], Thierry Bore [1], Zi Li [1,2,3], Stefan Schlaeger [4], Alexander Scheuermann [1] and Ling Li [1,2,3]

1 Geotechnical Engineering Centre, School of Civil Engineering, University of Queensland, St. Lucia, Brisbane, QLD 4072, Australia; t.bore@uq.edu.au (T.B.); lizi@westlake.edu.cn (Z.L.); a.scheuermann@uq.edu.au (A.S.); liling@westlake.edu.cn (L.L.)
2 School of Engineering, Westlake University, Hangzhou 310024, China
3 Institute of Advanced Technology, Westlake Institute for Advanced Study, Hangzhou 310024, China
4 Science-Engineering-Measurement, Sceme.de GmbH, Horn-Bad Meinberg, HRB 7181 Amtsgericht Lemgo, Germany; stefan.schlaeger@sceme.de
* Correspondence: g.yan@uq.edu.au

Abstract: The strength of unsaturated soil is defined by the soil water retention behavior and soil suction acting inside the soil matrix. In order to obtain the suction and moisture profile in the vadose zone, specific measuring techniques are needed. Time domain reflectometry (TDR) conventionally measures moisture at individual points only. Therefore, spatial time domain reflectometry (spatial TDR) was developed for characterizing the moisture content profile along the unsaturated soil strata. This paper introduces an experimental set-up used for measuring dynamic moisture profiles with high spatial and temporal resolution. The moisture measurement method is based on inverse modeling the telegraph equation with a capacitance model of soil/sensor environment using an optimization technique. With the addition of point-wise soil suction measurement using tensiometers, the soil water retention curve (SWRC) can be derived in the transient flow condition instead of the static or steady-state condition usually applied for conventional testing methodologies. The experiment was successfully set up and conducted with thorough validations to demonstrate the functionalities in terms of detecting dynamic moisture profiles, dynamic soil suction, and outflow seepage flux under transient flow condition. Furthermore, some TDR measurements are presented with a discussion referring to the inverse analysis of TDR traces for extracting the dielectric properties of soil. The detected static SWRC is finally compared to the static SWRC measured by the conventional method. The preliminary outcomes underpin the success of applying the spatial TDR technique and also demonstrate several advantages of this platform for investigating the unsaturated soil seepage issue under transient flow conditions.

Keywords: spatial TDR; moisture content; soil suction; SWRC; transient flow

Citation: Yan, G.; Bore, T.; Li, Z.; Schlaeger, S.; Scheuermann, A.; Li, L. Application of Spatial Time Domain Reflectometry for Investigating Moisture Content Dynamics in Unsaturated Loamy Sand for Gravitational Drainage. *Appl. Sci.* **2021**, *11*, 2994. https://doi.org/10.3390/app11072994

Academic Editor: Paulo José da Venda Oliveira

Received: 28 January 2021
Accepted: 23 March 2021
Published: 26 March 2021

Publisher's Note: MDPI stays neutral with regard to jurisdictional claims in published maps and institutional affiliations.

Copyright: © 2021 by the authors. Licensee MDPI, Basel, Switzerland. This article is an open access article distributed under the terms and conditions of the Creative Commons Attribution (CC BY) license (https://creativecommons.org/licenses/by/4.0/).

1. Introduction

In nature, the vadose zone above the water table can be several meters high. In this zone, the moisture stored in the soil matrix does not fill the pores completely. This form of water is no longer in positive potential but negative, as it presents a capillary tensile interaction that binds the soil particles [1]. This binding interaction thus enhances the mechanical properties of unsaturated soil, resulting in less deformation [1,2] and higher shear strength [3,4] compared to saturated soil, where water molecules encounter a compressive interaction [5]. Unsaturated soil mechanical behavior is, therefore, dependent on environmental conditions, which enhance or alleviate this binding effect, such as evapotranspiration and draining or flooding and intensive rainfall [1,2,6]. On the other hand, due to the air invasion into the soil matrix to form such a capillary water meniscus, the amount of effective hydraulic conductive channels is reduced as well and this degradation of geo-gas and groundwater flow conduits in unsaturated soil are highly concerning for the investigation of solute transport in the vadose zone [6]. For quantifying the contribution of capillary effects as water protentional in energy form and moisture content to the hydro-mechanical properties of unsaturated soil, the soil suction profile and soil moisture content profile need to be determined in order to characterize the soil water retention curve (SWRC), which is the core unsaturated soil constitutive relationship between the soil suction and moisture content (gravimetric/volumetric) [2].

The measurement of SWRC can be conducted either in the laboratory or in the field. In the laboratory, the standard axis translation technique (ATT) is often implemented [1]. Briefly, there is a one-dimensional representative elementary volume (REV) scaled soil specimen placed in a small fluid leakage-free chamber; through adding air pressure on top of the specimen, soil moisture can be expelled out of the soil matrix; the suction is the difference between applied air on top and water pressure at the bottom; the water moisture variation can be determined by measuring the mass of specimen or volume of water expelled out. Although ATT can be implemented in the laboratory due to the small set-up, this technique can only measure the SWRC under the static condition, for which there is no further moisture redistribution and capillary non-equilibrium. In the field, SWRC is usually given by separately measuring the suction profile using tensiometers and the moisture profile using in-situ sampling soil for the oven-drying method [7]. Compared to the ATT technique on a laboratory scale, this approach is more labor-expensive. In addition, sampling of soil disturbs the original configuration of the soil strata further, inducing moisture reconfiguration. This phenomenon is inevitable for in situ measurement and leads to potential falsification. Due to these limitations, there is always disagreement between the SWRC measured using ATT and in situ methods. Moreover, neither approach is able to measure the SWRC under the transient flow condition due to the requirement of equilibrium achievement and sampling procedure. Therefore, even for a homogeneous soil specimen, the conventional methods constrain the understanding of the spatial effect on soil suction determination [8,9], the temporal effect on moisture reconfiguration [10] and the dynamic effect in SWRC [11–13]. To overcome these limitations, there is a strong need for a laboratory experiment maintaining the homogeneity of a specimen in a full-scale soil profile without uncontrollable environmental conditions.

For measuring the SWRC under the transient flow condition, the spatially distributed moisture content needs to be measured with high temporal resolution. Some previous literature has implemented the soil column experiment using point-wise moisture sensors and tensiometers for logging the transient response of soil moisture and suction [12,14–16]. Nevertheless, these series of experiments only studied the dynamic effect of moisture and suction in a short sand column of less than 1 m, and this did not cover the entire suction profile of sandy soil (approximately 0–2 m). Even though these dynamic SWRC studies allow the quantification of suction and moisture under the transient flow condition, they are not able to provide information on global suction/moisture profiles due to the smaller experimental scale. To collect information on both global suction and moisture profiles, the experimental set-up should be large enough to cover the sandy soil suction profile, and the moisture content profile should also be logged within a smaller time step as well. Therefore, the spatial time domain reflectometry (spatial TDR) developed by the Soil Moisture Group (SMG) at the University of Karlsruhe [17–19] was applied to the experimental studies of the non-uniqueness of the SWRC on a sand column experiment and levee model [13,20]. However, in these previous experimental investigations, a high air entry (HAE) ceramic disk, having higher hydraulic resistivity than sand, was placed under the specimen to build up the hydraulic connectivity between the hanging column reservoir and specimen. This might also result in alleviation of moisture seepage velocity and, additionally, reduce the significance of the dynamic response of suction and moisture redistribution.

With the aim of mitigating the shortcomings of these previous soil column tests, a novel experimental platform is developed in this study. In this experimental platform, the spatial TDR sensors are installed in sand columns to measure the TDR waveforms consistently during one-step gravitational drainage experiments. Taking advantage of this valuable moisture measuring technique, the moisture profile along a 240-cm sand column can be calculated using an inverse modeling method developed by SMG [17–19]. With this powerful inverse technique, and by consistently logging TDR traces along the soil strata, a dynamic moisture profile with high spatial and temporal resolution can be achieved. With the addition of temporal logging of soil suction using a tensiometer along different elevations of the soil column, it is able to determine not only the point-wise SWRC at varied elevations but also the global suction/moisture profile under transient flow conditions. In this study, this novel laboratory-scale experimental set-up is presented. Furthermore, some preliminary outcomes of the gravitational drainage test are presented to demonstrate the capabilities of this experimental platform and eventually compared to the conventional hanging column method with discussions.

2. Spatial Time Domain Reflectometry

2.1. Basic Principles of Time Domain Reflectometry

Time domain reflectometry (TDR) has become an important monitoring technique for various civil engineering problems [21–24]; the main application is the monitoring of moisture content for porous media [25–28]. A TDR measurement system is composed of a pulse generator combined with an oscilloscope and a transmission line system (see Figure 1a). The transmission line includes a leading coaxial cable and a sensing waveguide, called the TDR probe, which is embedded in the medium for characterization. The characteristics (length and geometry) of the probe depend on the application. The classical geometry remains in the rod configuration. The pulse generator delivers an electromagnetic pulse along the transmission line, and the oscilloscope records returning reflection. Reflection will occur at any impedance discontinuities along the transmission line. In particular, the pulse will be reflected at the beginning and end of the TDR probes. The travel time analysis between these two reflections can be used to determine the flight time Δt of the pulse to travel along the probe. This flight time is generally computed based on the dual tangent method [29], where the reflection arrival is located at the intersection of the two tangents to the TDR curve (see Figure 1b). Finally, the apparent permittivity ε_{app} of the medium can be determined with the well-known relation:

$$\varepsilon_{app} = \left(\frac{c_0 \Delta_t}{2L_p}\right)^2 \tag{1}$$

where c_0 is the speed of the light and L_p is the length of the probe. Please note that for most porous media, the permittivity is complex and frequency-dependent (dispersion). The term apparent permittivity is used here because the time-domain method does not intend to compute the complete frequency-dependent complex permittivity.

In a second step, the apparent permittivity is linked to the parameter of interest—in most cases, water content, θ. The most frequent method is to use the empirical equation. The classical example is the famous Topp equation [30]:

$$\theta_{Topp} = -5.3 \times 10^{-2} + 2.92 \times 10^{-2}\varepsilon_{app} - 5.5 \times 10^{-4}{\varepsilon_{app}}^2 + 4.3 \times 10^{-6}{\varepsilon_{app}}^3 \tag{2}$$

The accuracy of such an empirical model highly depends on the nature of the soil related to density, the salinity of soil water, mineral composition, etc. For example, Topp's equation is not suitable for high swelling clay mineral [31]. An alternative to this solution is to derive a material-specific calibration functions [32].

Another approach relies on a physical model based on mixture equations [33,34]. Mixing models consider the soil as a medium that is composed of different phases. The relative permittivity of the mixture ε_m is the result of the sum of the dielectric properties of each phase multiplied by its volume fraction:

$$\varepsilon_m^a = \sum_{k=1}^{N} V_k.\varepsilon_{r,k}^a \tag{3}$$

where N is the number of phases, a is a structure parameter, V_k is the volume fraction and $\varepsilon_{r,k}$ is the relative permittivity of the solid phase. The classical configuration for the mixture equation is the three-phase mixtures (air–solid–liquid) with a structure index equal to 0.5, called the Complex Reflective Index Model (CRIM) [35]. In this configuration, ε_m can be expressed directly as a function of water content [36]. The main source of interest of such models is the versatility: they can take into account the temperature and frequency dependence of the single components. The main disadvantage of such a method is the lack of knowledge and understanding of the interaction between single components.

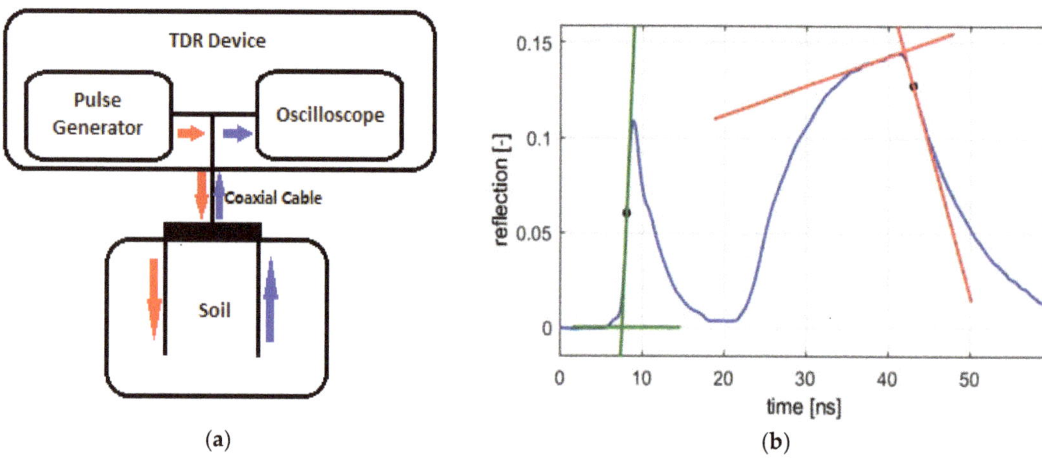

Figure 1. (**a**) Schematic description of Time Domain Reflectometry (TDR) device (pulse generator for sending a signal and oscilloscope for receiving) and TDR sensor plugged into a bulk of soil; (**b**) TDR waveform for saturated sand with the travel time determined by tangent method (reflection coefficient is dimensionless).

2.2. Spatial Time Domain Reflectometry Sensor Development

For the purpose of measuring natural porous media on a large scale, conventional TDR has its limitations, especially concerning the measurement of spatially distributed moisture, as probes are usually constructed as small three-pin or two-pin probes, which can only give a point measurement. Spatial TDR was developed by SMG at the Karlsruher Institute of Technology (KIT), Karlsruhe, Germany, to enable moisture profile measurements [17–19,37]. A flat ribbon cable consisting of three-line copper wire covered with polyethylene insulation is usually used for this purpose. An example of the spatial TDR sensor is given in Figure 2a. The corresponding conceptual model of the electrical circuit of a TDR transmission line and capacitance model are separately shown in Figure 2b,c. The electrical circuit of an infinitesimal section along the cable can be treated as the electrical circuit plotted in Figure 2b. The electrical design of the sensor was developed by Huebner, C. et al. [18].

For such a infinitesimal section, information on the surrounding soil and coating material properties can be extracted from the conductance (G) and capacitance (C) using inverse modeling of the telegraph equation [19,38–41]. For instance, one version developed by Schlaeger [5] is selected to conduct the inverse analysis in this study. Before inverse modeling, some assumptions and set-up conditions have to be achieved:

1. The resistance is assumed to be constant at a value of zero for lower frequencies (<10^4 Hz), because, in practical cases, dielectric losses are much higher than resistance losses, with the exception of long sensors buried into a nearly lossless material such as snow [18];
2. The inductance is assumed to be constant (L_0) for lower frequencies (<10^4 Hz), because only the external inductance remains at the highest frequency (10^6~10^9 Hz), and a transition frequency around 100 kHz ensures the insignificant influence of the inductance increase at a low-frequency range within the time window of TDR measurement [18];
3. The conductance and capacitance depend on the surrounding moist sandy soil and are assumed to be independent of frequency for lower frequencies (<10^5 Hz) [17];
4. The performance of the flat ribbon cable sensor is very sensitive to the installation in accordance with the 3-D electromagnetic modeling analysis [37] because the air-filled gap of 0.25 mm on both sides of the flat ribbon cable causes significant underestimation of moisture content, while a water-filled gap leads to overestimation [17,36].

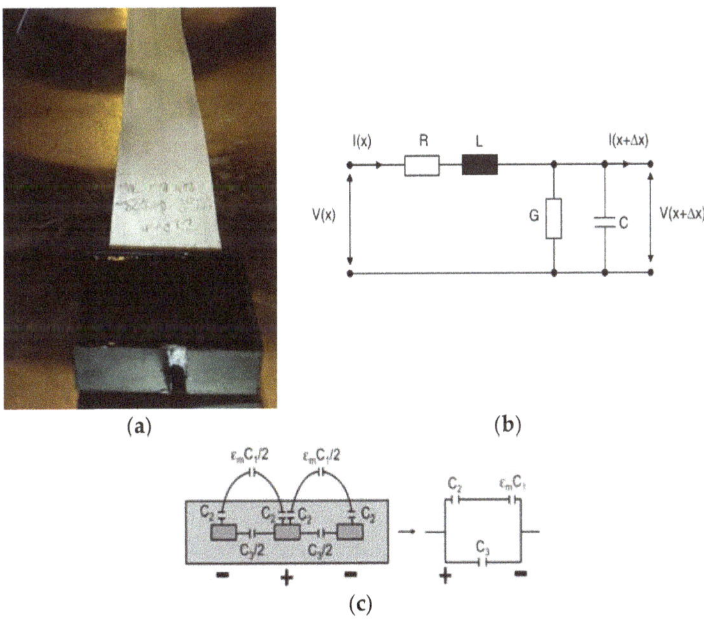

Figure 2. (**a**) Flat ribbon cable used for spatial TDR measurements; (**b**) equivalent circuit of an infinitesimal section of cable sensor [18]; (**c**) capacitance model of the flat ribbon cable [17].

2.3. Spatial Time Domain Reflectometry Forward Modeling

Based on the above assumptions and ideal sensor installation conditions, the inverse technique on two-way TDR tracing to extract C and G profiles can be executed. According to the electrical circuit simplified in Figure 3a, the forward modeling telegraph equation can predict the TDR trace along a flat ribbon cable. Therefore, the finite difference method is applied to numerically solve these equations with boundary conditions exactly matching the physical sensor design to predict the TDR trace [38,40–42]. To simplify the original telegraph equations for solving, Schlaeger transformed two first-order governing partial differential equations (PDE) into a single PDE in second-order as Equation (4):

$$\left[LC(x)\frac{\partial^2}{\partial t^2} + LG(x)\frac{\partial}{\partial t} + \frac{\partial L/\partial x}{L}\cdot\frac{\partial}{\partial x} - \frac{\partial^2}{\partial x^2}\right]U_i(x,t) = 0 \tag{4}$$

where L is inductance, assumed to be constant because of the second assumption; C and G are, separately, capacitance and conductance in functions of cable length x because of the third assumption; U is the voltage predicted for plotting the TDR trace (i = 1,2 represents first and second TDR tracing); the resistance R is assumed to be zero because of the first assumption depicted in Figure 3a. The appropriate initial and boundary conditions for solving Equation (4) to replicate the TDR trace have been discussed in Schlaeger's forward modeling work in detail [42]. Due to the requirement of two unknown input state variables (C and G profiles), two-way TDR tracing has to be implemented using a TDR device combined with a multiplexer, as described in Figure 3b [18,19]. The TDR device should send voltage steps through one channel of the multiplexer and then should be immediately switched to the second channel to collect the TDR waveform from the other terminal. Thus, through solving Equation (4) twice to match the two TDR waveforms collected from both terminals, the two unknown input profiles can be determined. Moreover, the forward modeling domain is constrained within two sensor terminals corresponding to two inflection points along a single TDR waveform [17].

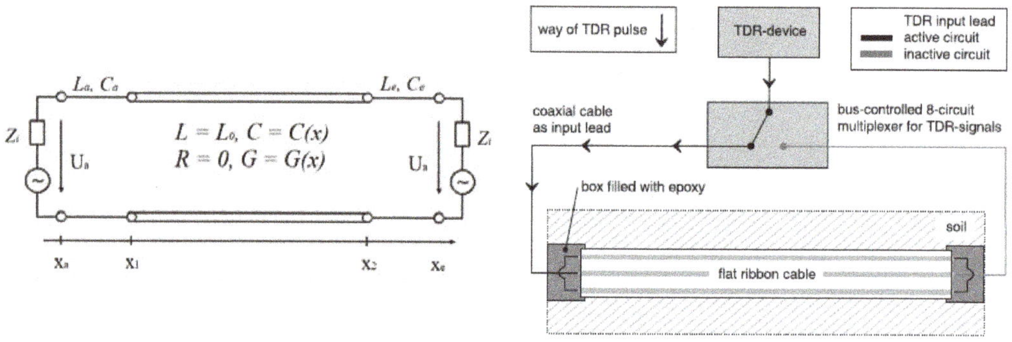

(a) (b)

Figure 3. (**a**) The schematic plot of spatial TDR measuring electrical circuit with previous assumptions; the flat ribbon cable is located between x_1 and x_2, coaxial cable is between x_a and x_1 or x_2 and x_e [19]; (**b**) the implementation of two-way TDR tracing using TDR device and multiplexer [17].

2.4. Spatial Time Domain Reflectometry Two-Way Inverse Analysis

To extract the C and G profiles, the last step is the application of the optimization technique for minimizing the difference between forward modeling and TDR measurement results. Generally, for an optimization problem, the objective function is usually the square difference between prediction and measurement, starting from an initial estimation and finally determining with the global minimum. Schlaeger [19] derived the objective function as Equation (5):

$$F(C,G) = \sum_{i=1}^{2} \int_0^T \left[U_i^M(x_i,t,C,G) - U_i^m(x_i,t)\right]^2 dt \tag{5}$$

where F is the objective function of C and G; U_i^M and U_i^m are, individually, the predicted and measured voltage (i = 1,2 corresponds to two-way TDR tracing). This function has been successfully optimized using the conjugate gradient method with the reconstruction of C and G by Schlaeger [19]. The optimization procedure is to input one initial estimation of the C and G profiles into forward modeling to calculate the cost function and update the input by previous minimization outcomes until the cost function achieves the global minimum. The derivation of the Jacobin gradient of the objective function using the adjoint PDE of Equation (4) can be sourced from the mathematical study [19].

2.5. Spatial Time Domain Reflectometry One-Way Inverse Analysis

On the other hand, for the only reconstruction of the capacitance profile using one-way TDR tracing, Becker and Schlaeger [43] applied an empirical equation between C and G proposed by Håkansson [44]:

$$G(C) = \begin{cases} G_\infty \cdot (1 - e^{\frac{C_0 - C}{C_d}}) & \text{if } C \geq C_0 \\ 0 & \text{if } 0 \leq C \leq C_0 \end{cases} \quad (6)$$

where G and C are the conductance and the capacitance of soil and sensor isolation, C_0 and C_d are the fitting parameters for capacitance; and G_∞ is the fitting parameter for the conductance. Using Equation (6), the inverse analysis only solves one TDR trace measured from each terminal, because the capacitance profile is the only unknown input in need of optimization. This relationship was successfully used for the three-pin probe sensor in the previous study of technique development [43]. In this study, the values of fitting parameters are C_0 = 50 pF/m, C_d = 40 pF/m and G_∞ = 90 mS/m.

2.6. Spatial Time Domain Reflectometry Post-Analysis

Once the capacitance profile can be inversely determined using this algorithm, the permittivity profile of the surrounding soil along the sensor can be calculated in accordance with the capacitance model characterizing the cable sensor, as shown in Figure 2c. Before calculating the permittivity of the soil (ε_m), the three capacitance values C_1, C_2 and C_3 ought to be experimentally calibrated. This calibration has been completed previously (see [17,18]), and the permittivity is calculated based on the capacitance model shown in Figure 2c as:

$$C(\varepsilon_m) = \frac{\varepsilon_m C_1 \cdot C_2}{\varepsilon_m \cdot C_1 + C_2} + C_3 \quad (7)$$

where C_1 (14.8 pF/m), C_2 (323 pF/m) and C_3 (3.4 pF/m) are parameters of the capacitance model, C is the total capacitance of the soil and coating isolation, ε_m is relative permittivity [17]. Eventually, the soil moisture profile can be calculated by the empirical models (Equation (2)) or the phase-mixing model (Equation (3)), which needs specific calibration using a Vector Network Analyzer (VNA) with the open-ended coaxial cell [45–47].

Instead of the reconstruction of the capacitance profile along the sensor, the travel time determined using the tangent method along one TDR trace can also be used to calculate the mean capacitance using:

$$C(\varepsilon_m) = \frac{1}{L} \cdot \left(\frac{t}{2L_p}\right)^2 \quad (8)$$

where C is the total capacitance for the entire measuring zone, ε_m is relative permittivity, L is inductance (constant L_0 = 756 nH/m), t is travel time and L_p is the length of the sensor. Equation (8) is useful to validate the mean moisture content for the total measuring volume. The mean moisture content later can be determined using Equation (8) for soil permittivity with Topp's model for the moisture content calculation.

3. Experimental Set-Up for Investigation of Soil Water Retention Behavior

The experimental set-up mainly consists of three logging systems separately recording point-wise soil suction profile, moisture content profile and accumulative outflow. An overview of the experimental platform is shown in Figure 4a, which demonstrates the view of the experimental set-up in the Geomechanics Laboratory at Geotechnical Engineering Centre, School of Civil Engineering, University of Queensland, St. Lucia QLD, Australia.

3.1. Soil Sample Specification

Two types of sandy soil were involved in this experimental set-up. One is the beach sand collected from Bribe Island, and the other one is the Budget Brickies loamy sand collected from an unknown construction site, Queensland, Australia. The particle size of the two types of sand ranges from 0.075 to 2 mm. The D_{50} of beach sand is 0.35 mm, which is slightly larger than the mean size of loam, 0.27 mm. The sieving analysis was implemented for both types of sand in accordance with ASTM standards of particle analysis for coarser soil [48].

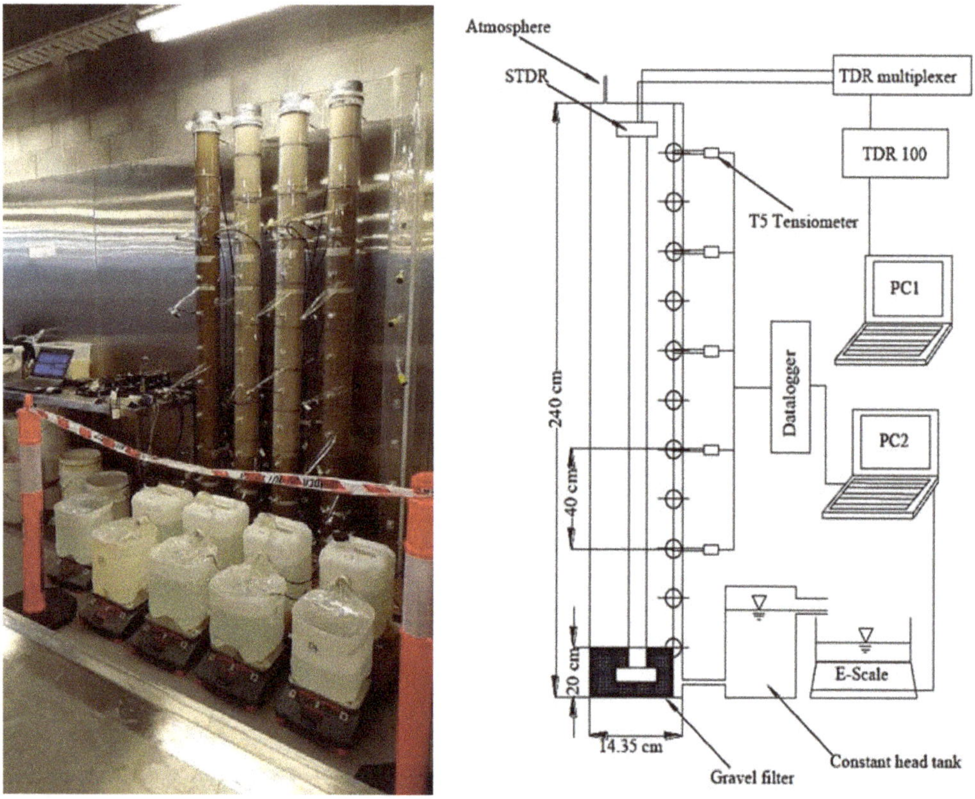

(a) (b)

Figure 4. (**a**) The overview of the physical experimental set-up for gravitational drainage and spontaneous imbibition experiment; (**b**) the specification of sensor configuration and hydraulic boundary condition for gravitational drainage experiment.

The mineralogy of the two samples comprised of quartz at the specific gravity of 2.65. The coefficients of uniformity (C_u) for beach sand and loamy sand are 1.71 and 2.39 separately, and the coefficients of gradation (C_c) are 0.99 and 1.11. Based on the Unified Soil Classification System (USCS), both of them are poorly graded uniform sand (SP). The beach sand is medium sand without any fine content, while the loamy sand is fine sand with a fine content of 5%. The loamy sand is adopted for a demonstration of the results collected by the experimental platform in this study.

3.2. Moisture Profile Logging System Set-Up

As previously mentioned, the spatial TDR technique was applied to soil moisture profile measurement; four flat ribbon cable sensors were individually inserted into the centers of four empty acrylic columns (diameter = 14.35 cm and length = 240 cm). The bottom terminal of spatial TDR is fixed with an artificial gravel filter designed based on published criteria [49]. The falling head test was conducted on the gravel filter to confirm that the hydraulic conductivity of the filter (6.5×10^{-3}~9.5×10^{-3} cm/s) was larger than the hydraulic conductivity of the soil specimen in the column (2.9×10^{-3}~5.6×10^{-3} cm/s). It does not only indicate the end of spatial TDR but is also a highly pervious porous media under the specimen being tested. Moreover, there was no sand or fine gravel flushed out during the soil water drainage process.

Each flat ribbon cable sensor had two terminals, demanding two channels on the multiplexer. Due to the need for four sensors for four columns, a TDR trace logging system consisting of a Campbell Scientific SDM X50 multiplexer© (8-channels) combined with both a TDR100© and a CR1000 data logger© was built to connect to eight connectors of four sensors. The data logging program coded in CRBasic© was then applied for automatically logging eight TDR traces from 8 multiplexer channels to complete one/two-way TDR tracing in each soil column. An overview of the experimental set-up and illustration are separately shown in Figure 4a,b.

This technique and logging system have been successfully applied to soil column tests and embankment water content dynamics measurement [17,20]. According to previous performance measurements [17], spatial TDR can achieve a spatial resolution of 3 cm with an average deviation of around ±2% for moisture content. The TDR100© can only impulse into the single sensor by changing the single switch along eight multiplexer channels. Therefore, for two-way TDR tracing on each sensor, there has to be at least a one-second delay between the two occurrences of TDR pulsing. However, this is much smaller than the moisture redistribution speed in natural sand. Hence, this delay is neglected, and two-way TDR tracing could be seen as simultaneously implemented.

As a gravel filter designed based on filter design criteria [49] guarantees the lowest hydraulic resistivity impacts from the column bottom without internal erosion, the transient flow conditions in the measuring range will not encounter any artificial alleviation of flow velocity. Similar soil column experiments have been implemented by many previous studies [8,9,16,20,50–53]. With the exception of the 2-m soil column test [51], most of them, conducted in shorter soil columns of less than 1.2 m, required a hanging column method with a water table below the bottom of the specimen. In this case, there has to be a high air entry (HAE) porous media under the specimen to build up the hydraulic connectivity between the specimen and water reservoir for suction control. To avoid air penetrating through the HAE disk to subsequently disconnect the suction-applying reservoir from the system, such disks or membranes must possess a smaller pore size distribution (PSD), which is much smaller than the specimen. Usually, the HAE disks are ceramic disks made of clay particles sintered together. Thus, the hydraulic conductivity of the HAE material is much lower than the sandy soil specimen. It is somehow inevitable to generate a quasi-steady-state flow condition using a shorter soil column test with the adoption of an HAE disk. Additionally, such smaller column tests only allow the measurement of half of the suction and moisture profile. In this experiment, due to the use of a 2.4-m-long column with a gravel filter fixing the spatial TDR sensor to the column bottom, there was an opportunity to neglect those artificial constraints in previous experimental set-ups. On the other hand, the previous 2-m soil column test applied the oven-drying method for determining the segments of the soil column only after equilibrium was achieved [51]. In comparison to the previous study, the current experimental set-up in this study can provide moisture measurement with a high temporal and spatial resolution by applying the spatial TDR technique. These advantages demonstrate the novelties of this experimental platform.

3.3. Suction Profile Logging System Set-Up

There were five UMS T5 tensiometers© (Figure 5a) inserted into the unsaturated zone at the elevations of 40, 60, 100, 140, 180 cm, where all were the above constant water pressure head of 36 ± 1 cm, as shown in Figure 4b. Moreover, there is another tensiometer inserted into the saturated zone under the water table at the elevation of 20 cm, which is not specified in Figure 4b. There is a spatial resolution of 20 cm for suction profile logging at the column bottom in order to detect the dynamic pore water pressure around the capillary fringe area for the future investigation of SWRC dynamic effects under variable water pressure boundary conditions. In total, there are twenty-four sensors on four columns (six sensors on each column). These sensors are connected to one Geo-Datalogger (DT85G, Pacific Data©). The temporal resolution of data logging is set to 30 s. As the UMS T5 tensiometers used a Wheatstone bridge circuit (Figure 5b), according to the logger manual [54], the output value of this data logger is:

$$B_{out} = \frac{V_{out}}{V_{ex}} \cdot 10^6 \quad (9)$$

where B_{out} is the dimensionless voltage, V_{out} is the measured bridge output voltage, V_{ex} is the excitation voltage. In addition, the T5 sensor measuring range is from −85 KPa for suction up to +100 KPa for positive water pressure. Thus, calibration between the dimensionless voltage and water pressure was carried out for each sensor within the positive water pressure range. Due to the perfect linear equation fitting into the calibration data points in Figure 5b, the calibration equation can be applicable for the entire measurement range. Twenty-four sensors share the same slope in the linear calibration equation, except for differences in the interception (−0.2~1.5 kPa) due to the varying physical offset on each sensor. It should be noted that the physical offset depends on the maintenance of the T5 sensors. After storing them for a long period, the calibration has to be reconducted to determine the new offset values due to the aging issue of the membrane in the sensor body. When the water table drops to the hydraulic head of the constant head tank, and equilibrium is approached, the suction value can also be double-calibrated and validated by calculating the suction head above the groundwater table. Meanwhile, the trend of the dynamic response of pore pressure can also be collected under the transient flow condition. The specification of the T5 tensiometer is summarized in Table 1.

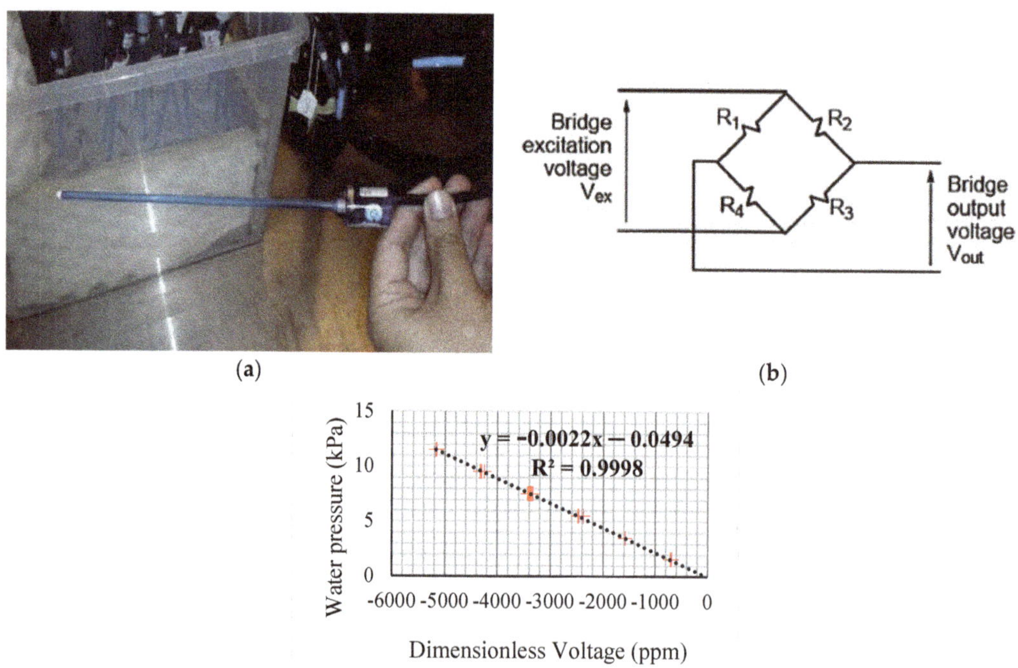

(a) (b) (c)

Figure 5. (**a**) UMS T5 tensiometer with a 20-cm shaft (deionized water fully filled) [53]; (**b**) the electrical circuit of T5 tensiometer [54]; (**c**) the linear calibration between dimensionless voltage bout and water pressure (the red crosses are data measured for calibration using linear regression) [53].

Table 1. Specification of UMS T5 micro-tensiometer [55].

Specification	Range
Measuring range	+100–85 kPa
Precision	±0.5 kPa
Shaft diameter	5 mm
Shaft length	20 cm
Output signal	−100 Mv + 85 mV

For an investigation of the soil suction profile in transient flow conditions, the response delay of the pressure sensor is of particular concern. The response time of tensiometers was fully investigated by Klute et al. [56]. For every tensiometer, there will be a time delay in the pressure response between the specimen and sensor body because of the permeability of the ceramic cup on the shaft tip. However, because the T5 sensor only measures a low suction range and the thickness of the ceramic cup is smaller than the HAE disk, there is only minor hydraulic resistance between the sensor and specimen. According to the T5 tensiometer manual [55], it only takes 5 seconds to approach an accurate value, and this is already the fastest option for suction measurement among other methods requiring moisture redistribution in a long equilibrium period, such as filter paper, dew point meters and ATT [1,2,57]. As the temporal resolution of suction logging is 30 seconds, such a small delay in response can be neglected for this experimental set-up.

3.4. Outflow Logging Set-Up, Initial and Boundary Conditions

There are four bench scales used to log accumulative outflow. As illustrated in Figure 4c, a constant head tank is attached to the output at the bottom of the column. The overflow of the constant head tank is collected by another water tank located on an electrical bench scale. In the case of evaporation leading to underestimation of outflow, every tank is covered by cling wraps with holes only for flow conduits. In Figure 4a,b, four bench scales are all connected to a USB hub with a data transmission cable between RS232 and a USB interface. In Figure 4b, one extra tank on the bench scale is located in the leftmost position with exactly the same ambient conditions to measure the water evaporation for moisture loss compensation. The maximum capacity of the bench scale (Ohaus Ranger 3000 R31P30©) is 30 kg, with a precision of ±1 g. The logging time step is also set at 30 s. On the other end of the column, cling wrap was also used to fully cover the open top of the soil column with a small pinhole to apply atmospheric pressure. The initial condition was to retain the water table. Since the valve between the constant tank and the saturated sand column was opened, the experiment was initiated. The data logging system was turned on only a few minutes before.

3.5. Specimen Installation and Operating Procedure

The density control of sand specimen installation in the column was conducted by tamping on the side of the acrylic column and vertically compacting on top of each layer. The top and bottom of each column were fixed to be exactly vertical to avoid tilting off during compaction and any inclines afterwards. In regard to minimizing the density variation along such a long column, the thickness of each soil layer was controlled to 22 ± 1 cm. With the same inner diameter of the column, it was possible to achieve a mean dry density of 1.61 ± 0.05 g/cm^3 (porosity 39% ± 2%) for beach sand and 1.45 ± 0.1 g/cm^3 (porosity 45% ± 4%) for loamy sand. The water used for fabricating this saturated specimen was tap water from the Geomechanics Laboratory of the University of Queensland, St. Lucia QLD, Australia, due to a lack of access to a large amount of de-aired water produced close by. However, it is non-saline water for general purpose. Under a low suction range for sand soil (0~22 kPa), soil water temperature 22 ± 1 °C and normal room temperature 23 ± 1 °C, there should be no further issue regarding nucleation, such as cavitation and boiling. The segmentation of each layer was carefully controlled to mismatch each pinhole for tensiometer insertion. A previous soil column test mentioned two approaches for tensiometer installation: installing a sensor through a pre-drilled hole or installing a sensor during the compaction process [58]. Here, the T5 sensors were all inserted into the column during the compaction process. Compared to the first method, the second ensures perfect contact between each sensor and ambient soil [58]. The gap between each sensor and pinhole was sealed with both a waterproof rubber O-ring and gas leakage-free thread tap. Each T5 sensor was originally filled with deionized water using an automatically de-aired water refilling kit, manufactured by UMS®. Except for the occurrence of the ceramic tip dried out for soil gas percolation, there will be bubble-free water in the sensor and shaft up to −101 KPa by mechanical vacuuming.

For the gravitational drainage test, the soil was oven-dried before loading into the column to avoid soil segregation carrying bubbles inside the microscale soil structure. Beach sand is completely cohesionless sand. Once the sample was oven-dried, there was no more segregation, while the loamy sand needed some crushing and mechanical shaking because of slightly cohesion. The water table was always above each layer that was being compacted for the preparation of the saturated specimen. The spatial TDR sensor was located in the center of the column, with good contact between soil and sensor, so there should be no more concerns about the failure of the spatial TDR technique because of gaps between the soil and sensor [37] and inductance variation due to distortion of the sensor geometry [17].

4. Results and Discussion

This experiment was originally designed to investigate the dynamic effects in SWRC [11], the hydraulic properties of unsaturated soil and the validity of the theories of two-phase flow in porous media for the transient flow condition [59–61]. Thus, this experimental platform will be used for different hydraulic boundary conditions in a sequence of the constant head, multistep in/outflow by head control. Currently, as the first stage of the constant head boundary, this experiment was conducted to assess gravitational drainage in a fully saturated sample and spontaneous imbibition to fully dry sample, which needs several months to reach equilibrium conditions without further moisture/suction variation.

However, some results regarding gravitational drainage only took a few weeks to approach equilibrium. Therefore, this part of the results is presented as a demonstration of the experimental platform's success in this study. Comparison between this soil column test and the standard hanging column test for loamy sand is also given to depict the difference and discussion is dedicated to identifying the potential problems in the conventional testing technique. Prior to the measurement of SWRC, the spatial TDR measurement and analysis are conducted to highlight the difficulties and problems in the post-analysis of spatial TDR waveforms.

4.1. *Spatial TDR Tracing during Water Table Decreasing*

When this experimental platform was firstly set up, before loading the sample into the column, some previous tests were conducted to validate the signal variation of TDR traces. Thus, a pre-test was completed by dropping the water table for a single column from an originally saturated condition in order to characterize the decrease in the reflection coefficient along TDR waveforms. An illustration of this pre-test is shown in Figure 6a and the TDR traces for different water tables are shown in Figure 6b. With the water table dropping downward, the TDR traces logged from 0 to 3 min clearly show an increase in the reflection coefficient along the flat ribbon cable measured from the top. After four min, the water table is stable since no variation in the TDR waveform is visible.

4.2. *Spatial TDR Waveform Variation along the Sand Column in the Drainage Test*

The TDR traces for the gravitational drainage test are shown in Figure 7. Two TDR pulses are separately sent to the sensor in the column from the top and bottom. A pair of TDR traces measured from both ends of the sensor at the same time can be used to conduct the inverse analysis. According to Figure 7a (Terminal Top) and b (Terminal

Bottom), it is clear that the reflection coefficient of TDR traces increases with the falling water table and water draining out of the soil column. Since air enters the pore space at the top end of the column with the dropping of the water table, TDR traces measured from the top and end show an increase in the reflection coefficient up to 0.3. Meanwhile, the second reflection points in these TDR traces, indicating the transition between the sensor and the second coaxial cable, decrease from 110 to 100 ns. The TDR traces measured from the bottom display significant changes at the end of the TDR trace, showing a more pronounced transition from the sensor to a coaxial cable (drop after the second rise in the signal). The comparison of the signals demonstrates the strong variation in the TDR traces, corresponding to the change in the moisture profile. It further proves the capability of the spatial TDR technique for characterizing information on moisture content dynamics in large-scale unsaturated soil experiments.

4.3. Validation of Spatial TDR by Outflow Logging

The measurement from the spatial TDR sensor was firstly validated by the outflow data logged using an electrical bench scale. First, the tangent method was used to determine the travel time for each TDR trace sent from the column bottom. Second, the previously determined travel time was used to calculate the total mean capacitance using Equation (8). Then, the real part of soil permittivity can be calculated using Equation (7). Finally, the volumetric moisture content is calculated by Topp's model in Figure 8. Based on the outflow and volume of the specimen, the mean volumetric moisture content was eventually calculated and plotted in Figure 8. The moisture prediction from the spatial TDR sensor strongly agrees with the moisture content calculated from the outflow data. This demonstrates the success of electrical design with corresponding parameter calibration of the spatial TDR sensor and the applicability of Topp's model for this loamy sand.

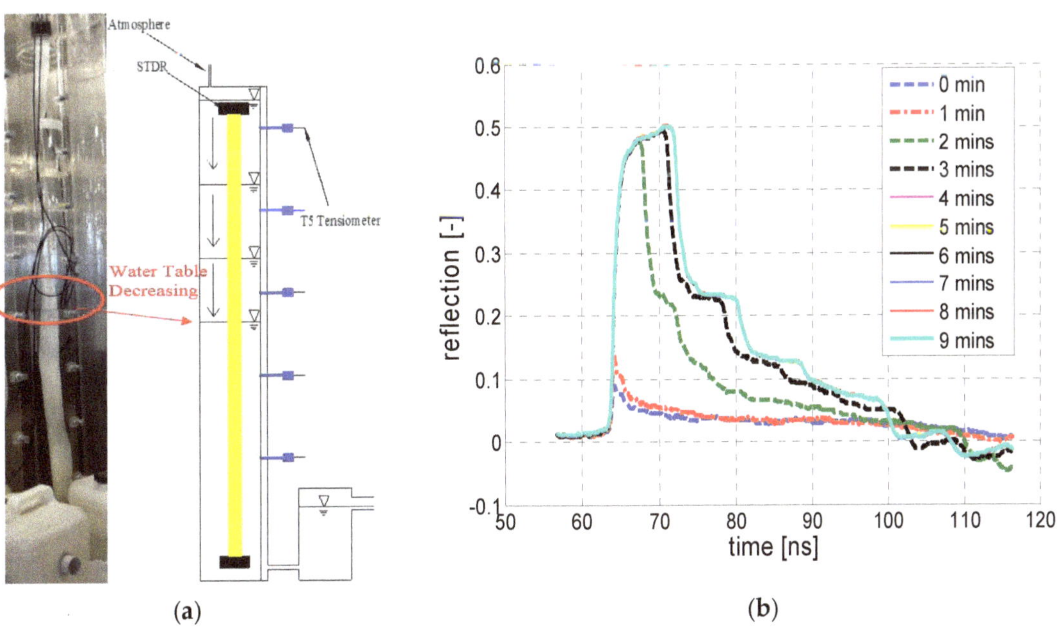

Figure 6. (**a**) The illustration of a single column drainage test without sample filled; (**b**) TDR waveforms indicating a drop in water table [62].

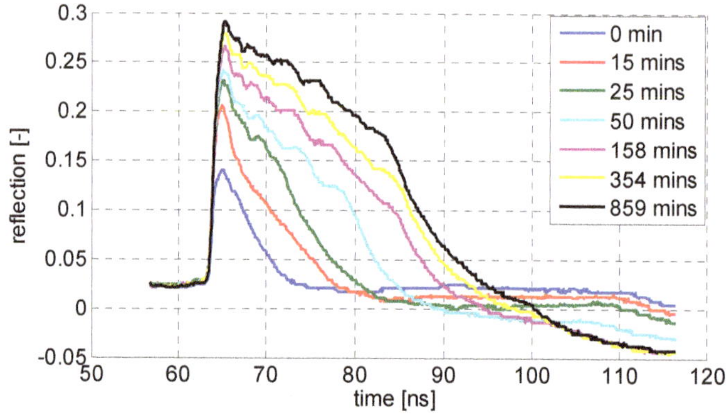

(a)

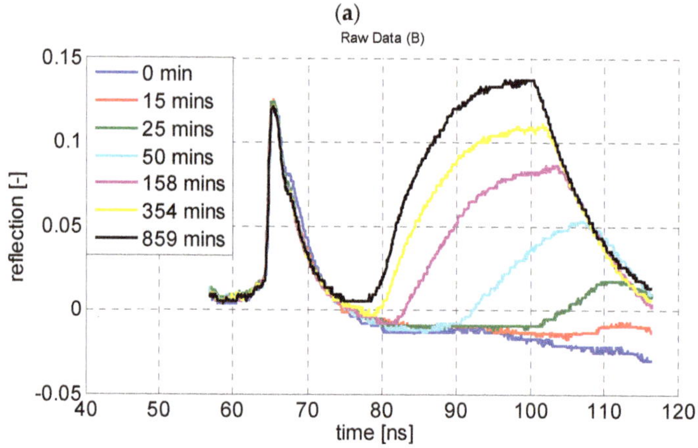

(b)

Figure 7. TDR traces for gravitational drainage test: (**a**) TDR traces measured from the top; (**b**) TDR traces measured from the bottom [62].

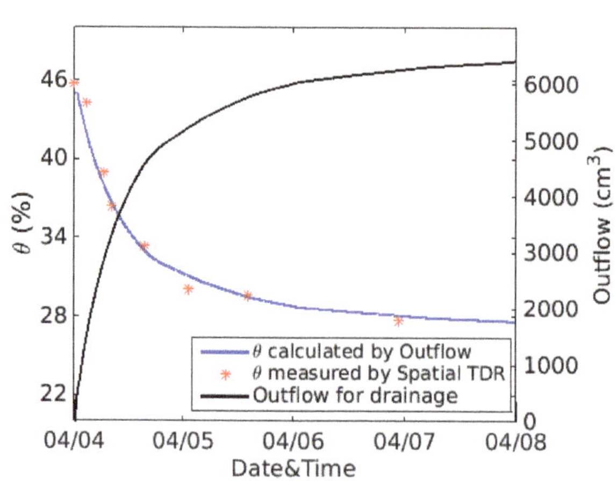

Figure 8. The mean water content dynamics for the loamy sand column in comparison with the mean water content from outflow data.

4.4. Inverse Analysis of Spatial TDR Trace and Dynamic Moisture Profile

Figure 9a shows the performance of optimization using one-way inverse analysis of the spatial TDR technique. By applying the optimization, the TDR traces simulated using the forward modeling telegraph equation can be well-matched with the TDR traces measured using the data logging system. Based on this acceptable fitting performance, the dynamic moisture profile can be finally calculated using Topp's model, as shown in Figure 9b. From the beginning at 12:00 pm on 4th April, the moisture content profile was fully saturated, with a mean moisture content of around 45%. Once the one-step gravitational drainage commenced, the pore water in the upper zone drained fast in the first 8 h. After two days, by 1:56 am on 6th April, it almost achieved the equilibrium condition. The data logging from the two following days shows no significant variation in moisture profiles, indicating the final equilibrium condition achieved.

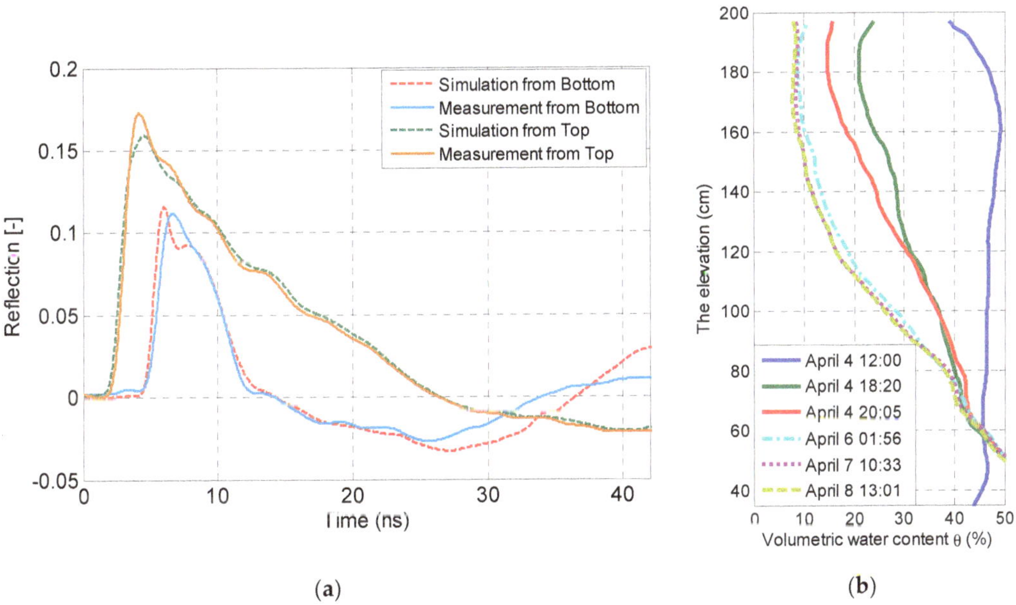

Figure 9. (**a**) The fitting performance of capacitance reconstruction using one-way inversion analysis on both terminals (TDR trace sent from bottom and top); (**b**) the dynamic volumetric moisture content profile measured using the spatial TDR inverse analysis.

It should be noted that there is an overestimation of moisture in the capillary fringe zone because of the transition zone between the sensor and epoxy terminals. Thus, it should not be used for the determination of saturated moisture content. Instead, the saturated moisture content at the initial condition should be used to cut off the irresponsible partition to determine both the air entry value (AEV) and moisture profile in the saturated area. These series of dynamic moisture profiles demonstrate the success of inverse analysis of spatial TDR technique functioning in principle, while more effort might be required to enhance the optimization method and take consideration of frequency dependence for measuring more conductive porous media.

4.5. Validation of Pressure Measurement and Dynamic Response of Water Pressure

Figure 10a shows the good agreement between the measured suction and theoretical water pressure calculated using the hydrostatic concept at the equilibrium stage. It proves the accuracy of the previous linear calibration between value B_{out} and water pressure for the T5 tensiometer with the datalogger. Figure 10b gives the dynamic response of measured suction for one-step drainage for six months. In the first six hours, the water pressure sharply decreased from positive values to the negative, indicating soil suction.

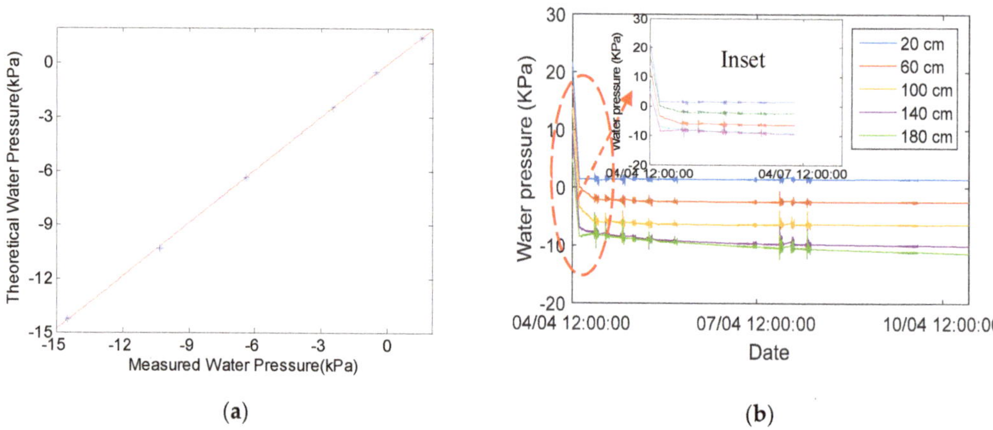

Figure 10. (**a**) The comparison between tensiometer measurement and theoretical water pressure; (**b**) the dynamic response of tensiometer on each point.

A few days later, the pore water pressure still gradually dropped to equilibrium. For the rest of the six-month period, only the suction value from the tensiometer inserted at 180 cm continually decreased. This further reduction for such a long period might be due to the inevitable evaporation on the upper layer, while there were two more soil layers covered by the cling wrap over the measured zone to alleviate the impacts of subsurface evaporation. The suction logging system properly functions for this drainage test, and the dynamic behavior of negative pore pressure can be successfully recorded using this experimental platform.

4.6. Soil Water Retention Curve Measurement Compared to SWRC Using the Standard Method

Finally, the suction and moisture logging collected at equilibrium are used to plot the SWRC shown in Figure 11 in comparison to the SWRC measured by the standard hanging column method. In Figure 11, there is a slight difference in initial density between the two methods of around 2–3%. The zone of residual moisture content is around 8–9% for both methods, whereas the capillary storage (moisture content changing with soil suction) and AEV show significant differences between these two methods. Even the porosity controlled for the standard method is higher, the AEV is 1.5 kPa higher than the AEV by the large column test. In fact, the larger porosity should provide a smaller AEV.

Moreover, due to the smaller AEV achieved in the large column test, the slope of SWRC is extended, so there is rather a gradual reduction in moisture content by increasing the soil suction than the sharper wetting front measured by the standard hanging column method. Previous studies on the spatial variation in suction calculated using the height difference between a ceramic disk and water table have demonstrated the sensitivity of suction determination by varying the specimen thickness [9]. This is reconfirmed again in our experimental exploration as one of the initiative motivations of this experimental study.

However, it is still not possible to thoroughly conclude the failure or inaccuracy of using the standard method due to the insufficient repetition for constructing such a large experimental operation. Furthermore, for a non-deformable soil matrix, the initial density control determines the pore size distribution, which governs the static SWRC and AEV [63,64]. This could be also a reason for the differences. This trial still further upgrades our understanding of SWRC regarding the spatial variation in suction and moisture. There will be a higher expectation that with further repetition of the same test in the laboratory and field, with systematic control of impact factors from environmental conditions and better precision of moisture given by the inverse analysis of the spatial TDR technique, the dynamic and spatial effects of SWRC can be more comprehensively unveiled for better two-phase flow seepage simulation, unsaturated soil strength estimation and deformation prediction.

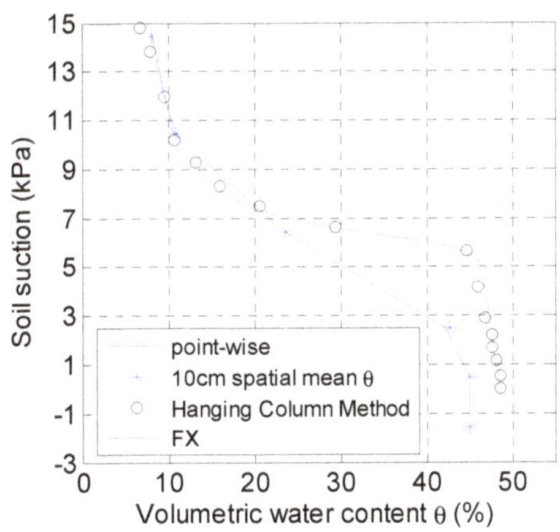

Figure 11. The comparison between SWRC measurements from the large soil column test (the mean θ averaged spatially ± 5 cm) and standard hanging column method (the data points given by the hanging column method are fitted using Fredlund and Xing (FX) model [63]).

5. Summary and Reflection

The soil water retention curve (SWRC) is one of the most important constitutive relationships for the simulation of the hydraulic and mechanical behavior of unsaturated soil. As unsaturated soil effective stress, shear strength, earth pressure, consolidation and swelling are highly governed by both the soil moisture content and suction, the pre-estimation of the air–water seepage process determines the accuracy of estimating unsaturated soil's mechanical behavior. Due to this prior importance, it is critical to comprehensively investigate the dynamic and spatial effects of soil suction and moisture redistribution because many previous studies indicate such a difference between static and dynamic measurement.

This study integrates the spatial time domain reflectometry technique, high-precision tensiometer and consistent outflow logging to investigate the dynamic response of moisture distribution, soil suction and seepage flux during a transient drainage process. As the first stage, the experimental platform has been validated by the comparison between each pair of logging systems. This demonstrates the validity and functionality of this dynamic SWRC testing platform for specific loamy sand. The preliminary result shows that this system can be applied to capture both the dynamic response and the final equilibrium stage. The SWRC measured at the final stage is compared to the SWRC using the standard hanging column method, and there is a significant difference in air entry values and capillary storage between two methods. This finding further motivates the investigation of spatial effects in suction determination regarding the flaws embedded in the conventional testing method.

On the other hand, even though this experimental set-up involves excellent techniques to consistently log three state variables, the complexity of the sensoring technique and logging system leads to difficulty regarding sufficient replication in a short period. The dynamic behavior of SWRC under draining conditions needs further analysis, and the corresponding measuring technique can still be improved in future experimental investigation. In principle, the application of the spatial TDR technique offers the researcher a higher resolution of moisture distribution varying with time, and in principle, it functions for this experimental set-up. With a great appreciation of the TDR technique development, the hydrologist and geotechnical engineer can have a better understanding of the unobservable phenomenon that is not considered in the conventional theoretical framework.

Author Contributions: Conceptualization, G.Y., Z.L. and L.L.; Data curation, T.B.; Formal analysis, G.Y. and T.B.; Investigation, Z.L., T.B. and A.S.; Methodology, T.B., A.S. and S.S.; Project administration, L.L.; Resources, T.B.; Supervision, A.S. and L.L.; Writing—original draft, G.Y.; Writing—review & editing, T.B. and A.S. All authors have read and agreed to the published version of the manuscript.

Funding: This research was funded by Linkage Project (LP120100662), from Australian Research Council. The APC was funded by the University of Queensland International Scholarship.

Institutional Review Board Statement: Not applicable.

Informed Consent Statement: Not applicable.

Data Availability Statement: Not applicable.

Acknowledgments: The first author would like to acknowledge support from the Geomechanics Laboratory of the University of Queensland, Australia and also thanks for anonymous reviewers.

Conflicts of Interest: The authors declare no conflict of interest.

References

1. Fredlund, D.G.; Rahardjo, H. *Soil Mechanics For Unsaturated Soils*; John Wiley: New York, NY, USA, 1993.
2. Lu, N.; Likos, W.J. *Unsaturated Soil Mechanics*; John Wiley: New York, NY, USA, 2004.
3. Khalili, N.; Khabbaz, M. A unique relationship of chi for the determination of the shear strength of unsaturated soils. *Geotechnique* **1998**, *48*, 681–687. [CrossRef]
4. Vanapalli, S.; Fredlund, D. Comparison of different procedures to predict unsaturated soil shear strength. *Geotech. Spec. Publ.* **2000**, 195–209. [CrossRef]
5. Terzaghi, K.; Terzaghi, K.; Engineer, C.; Czechoslowakia, A.; Terzaghi, K.; Civil, I.; Tchécoslovaquie, A.; Unis, E. *Theoretical Soil Mechanics*; John Wiley: New York, NY, USA, 1943; Volume 18th.
6. Bear, J. Dynamics of fluids in porous media. *Soil Sci.* **1975**, *120*, 162–163. [CrossRef]
7. ASTM D2216-10. Standard test methods for laboratory determination of water (moisture) content of soil and rock by mass. *Am. Soc. Test. Mater. (ASTM)* **2010**. [CrossRef]
8. Bottero, S.; Hassanizadeh, S.M.; Kleingeld, P. From local measurements to an upscaled capillary pressure–saturation curve. *Transp. Porous Media* **2011**, *88*, 271–291. [CrossRef]
9. Sakaki, T.; Illangasekare, T.H. Comparison of height-averaged and point-measured capillary pressure–saturation relations for sands using a modified Tempe cell. *Water Resour. Res.* **2007**, *43*. [CrossRef]
10. Barenblatt, G. Filtration of two nonmixing fluids in a homogeneous porous medium. *Fluid Dyn.* **1971**, *6*, 857–864. [CrossRef]
11. Hassanizadeh, S.M.; Celia, M.A.; Dahle, H.K. Dynamic effect in the capillary pressure–saturation relationship and its impacts on unsaturated flow. *Vadose Zone J.* **2002**, *1*, 38–57. [CrossRef]
12. Das, D.B.; Mirzaei, M. Dynamic effects in capillary pressure relationships for two-phase flow in porous media: Experiments and numerical analyses. *AIChE J.* **2012**, *58*, 3891–3903. [CrossRef]
13. Scheuermann, A.; Galindo-Torres, S.; Pedroso, D.; Williams, D.; Li, L. Dynamics of water movements with reversals in unsaturated soils. In Proceedings of the 6th International Conference on Unsaturated Soils, UNSAT, Sydney, Australia, 2–4 July 2014; pp. 1053–1059.
14. Sakaki, T.; O'Carroll, D.M.; Illangasekare, T.H. Direct quantification of dynamic effects in capillary pressure for drainage–wetting cycles. *Vadose Zone J.* **2010**, *9*, 424–437. [CrossRef]
15. O'Carroll, D.M.; Phelan, T.J.; Abriola, L.M. Exploring dynamic effects in capillary pressure in multistep outflow experiments. *Water Resour. Res.* **2005**, *41*. [CrossRef]
16. Mirzaei, M.; Das, D.B. Experimental investigation of hysteretic dynamic effect in capillary pressure–saturation relationship for two-phase flow in porous media. *AIChE J.* **2013**, *59*, 3958–3974. [CrossRef]
17. Scheuermann, A.; Huebner, C.; Schlaeger, S.; Wagner, N.; Becker, R.; Bieberstein, A. Spatial time domain reflectometry and its application for the measurement of water content distributions along flat ribbon cables in a full-scale levee model. *Water Resour. Res.* **2009**, *45*. [CrossRef]
18. Huebner, C.; Schlaeger, S.; Becker, R.; Scheuermann, A.; Brandelik, A.; Schaedel, W.; Schuhmann, R. Advanced measurement methods in time domain reflectometry for soil moisture determination. In *Electromagnetic Aquametry*; Kupfer, K., Ed.; Springer: Berlin/Heidelberg, Germany, 2005; pp. 317–347.

19. Schlaeger, S. A fast TDR-inversion technique for the reconstruction of spatial soil moisture content. *Hydrol. Earth Syst. Sci. Discuss.* **2005**, *9*, 481–492. [CrossRef]
20. Scheuermann, A.; Montenegro, H.; Bieberstein, A. Column test apparatus for the inverse estimation of soil hydraulic parameters under defined stress condition. In *Unsaturated Soils: Experimental Studies*; Schanz, T., Ed.; Springer: Berlin/Heidelberg, Germany, 2005; pp. 33–44.
21. Robinson, D.; Jones, S.B.; Wraith, J.; Or, D.; Friedman, S. A review of advances in dielectric and electrical conductivity measurement in soils using time domain reflectometry. *Vadose Zone J.* **2003**, *2*, 444–475. [CrossRef]
22. Kupfer, K.; Trinks, E.; Wagner, N.; Hübner, C. TDR measurements and simulations in high lossy bentonite materials. *Meas. Sci. Technol.* **2007**, *18*, 1118. [CrossRef]
23. Siggins, A.; Gunning, J.; Josh, M. A hybrid waveguide cell for the dielectric properties of reservoir rocks. *Meas. Sci. Technol.* **2011**, *22*, 025702. [CrossRef]
24. Zhang, J.; Nakhkash, M.; Huang, Y. Electromagnetic imaging of layered building materials. *Meas. Sci. Technol.* **2001**, *12*, 1147. [CrossRef]
25. Roth, C.; Malicki, M.; Plagge, R. Empirical evaluation of the relationship between soil dielectric constant and volumetric water content as the basis for calibrating soil moisture measurements by TDR. *J. Soil Sci.* **1992**, *43*, 1–13. [CrossRef]
26. Skierucha, W.; Wilczek, A.; Szypłowska, A.; Sławiński, C.; Lamorski, K. A TDR-based soil moisture monitoring system with simultaneous measurement of soil temperature and electrical conductivity. *Sensors* **2012**, *12*, 13545–13566. [CrossRef] [PubMed]
27. Kaatze, U.; Hübner, C. Electromagnetic techniques for moisture content determination of materials. *Meas. Sci. Technol.* **2010**, *21*, 082001. [CrossRef]
28. Kaatze, U. Reference liquids for the calibration of dielectric sensors and measurement instruments. *Meas. Sci. Technol.* **2007**, *18*, 967. [CrossRef]
29. Huisman, J.; Weerts, A.; Heimovaara, T.; Bouten, W. Comparison of travel time analysis and inverse modeling for soil water content determination with time domain reflectometry. *Water Resour. Res.* **2002**, *38*, 13-1–13-8. [CrossRef]
30. Topp, G.; Davis, J.; Annan, A.P. Electromagnetic determination of soil water content: Measurements in coaxial transmission lines. *Water Resour. Res.* **1980**, *16*, 574–582. [CrossRef]
31. Kelleners, T.; Robinson, D.; Shouse, P.; Ayars, J.; Skaggs, T. Frequency dependence of the complex permittivity and its impact on dielectric sensor calibration in soils. *Soil Sci. Soc. Am. J.* **2005**, *69*, 67–76. [CrossRef]
32. Dirksen, C.; Dasberg, S. Improved calibration of time domain reflectometry soil wate-r content measurements. *Soil Sci. Soc. Am. J.* **1993**, *57*, 660–667. [CrossRef]
33. Sihvola, A.H. *Electromagnetic Mixing Formulas and Applications*; IET: Stevenage, UK, 1999.
34. Bore, T.; Schwing, M.; Serna, M.L.; Speer, J.; Scheuermann, A.; Wagner, N. A new broadband dielectric model for simultaneous determination of water saturation and porosity. *IEEE Trans. Geosci. Remote Sens.* **2018**, *56*, 4702–4713. [CrossRef]
35. Brovelli, A.; Cassiani, G. Effective permittivity of porous media: A critical analysis of the complex refractive index model. *Geophys. Prospect.* **2008**, *56*, 715–727. [CrossRef]
36. Bore, T.; Wagner, N.; Delepine Lesoille, S.; Taillade, F.; Six, G.; Daout, F.; Placko, D. Error analysis of clay-rock water content estimation with broadband high-frequency electromagnetic sensors—Air gap effect. *Sensors* **2016**, *16*, 554. [CrossRef]
37. Wagner, N.; Trinks, E.; Kupfer, K. Determination of the spatial TDR-sensor characteristics in strong dispersive subsoil using 3D-FEM frequency domain simulations in combination with microwave dielectric spectroscopy. *Meas. Sci. Technol.* **2007**, *18*, 1137. [CrossRef]
38. Leidenberger, P.; Oswald, B.; Roth, K. Efficient reconstruction of dispersive dielectric profiles using time domain reflectometry (TDR). *Hydrol. Earth Syst. Sci. Discuss.* **2006**, *10*, 209–232. [CrossRef]
39. Comandini, F.V.; Bore, T.; Six, G.; Sagnard, F.; Lesoille, S.D.; Moreau, G.; Placko, D.; Taillade, F. FDR for non destructive evaluation: Inspection of external post-tensioned ducts and measurement of water content in concrete. In Proceedings of the 10th International Conference on Nondestructive Evaluation (NDE) in relation to Structure Safety for Nuclear and Pressurized Components, Cannes, France, 1–3 October 2013; p. 8.
40. Norgren, M.; He, S. An optimization approach to the frequency-domain inverse problem for a nonuniform LCRG transmission line. *IEEE Trans. Microw. Theory Tech.* **1996**, *44*, 1503–1507. [CrossRef]
41. Lundstedt, J.; Ström, S. Simultaneous reconstruction of two parameters from the transient response of a nonuniform LCRG transmission line. *J. Electromagn. Waves Appl.* **1996**, *10*, 19–50. [CrossRef]
42. Schlaeger, S. *Inversion von TDR-Messungen zur Rekonstruktion räumlich verteilter bodenphysikalischer Parameter*; Instituts für Bodenmechanik und Felsmechanik der Technischen Hochschule Fridericiana: Karlsruhe, Germany, 2002.
43. Becker, R.; Schlaeger, S. Spatial time domain reflectometry with rod probes. In Proceedings of the 6th Conference on "Electromagnetic Wave Interaction with Water and Moist Substances", ISEMA, Weimar, Germany, 29 May–1 June 2005.
44. Håkansson, G. Reconstruction of Soil Moisture Profile Using Time-Domain Reflectometer Measurements. Ph.D. Thesis, Royal Institute of Technology, Stockholm, Sweden, 1997.
45. Wagner, N.; Bore, T.; Robinet, J.C.; Coelho, D.; Taillade, F.; Delepine-Lesoille, S. Dielectric relaxation behavior of Callovo-Oxfordian clay rock: A hydraulic-mechanical-electromagnetic coupling approach. *J. Geophys. Res. Solid Earth* **2013**, *118*, 4729–4744. [CrossRef]

46. Huang, Y. Design, calibration and data interpretation for a one-port large coaxial dielectric measurement cell. *Meas. Sci. Technol.* **2001**, *12*, 111. [CrossRef]
47. Hübner, C.; Kupfer, K. Modelling of electromagnetic wave propagation along transmission lines in inhomogeneous media. *Meas. Sci. Technol.* **2007**, *18*, 1147. [CrossRef]
48. ASTM D6913-04. Standard test methods for particle-size distribution (gradation) of soils using sieve analysis. *Am. Soc. Test. Mater. (ASTM)* **2009**. [CrossRef]
49. Giroud, J. Development of criteria for geotextile and granular filters. In *Prestressed Geosynthetic Reinforced Soil By Compaction, Proceedings of the 9th International Conference on Geosynthetics, Guaruja, Brazil, 23–28 May 2010*; IGS Brazil: Sao Paulo, Brazil, 2010; p. 4564.
50. Rassam, D.; Williams, D. A dynamic method for determining the soil water characteristic curve for coarse-grained soils. *Geotech. Test. J.* **2000**, *23*, 67–71.
51. Chapuis, R.P.; Masse, I.; Madinier, B.; Aubertin, M. A drainage column test for determining unsaturated properties of coarse materials. *ASTM Geotech. Test J.* **2007**, *30*, 83–89.
52. Lins, Y.; Schanz, T.; Fredlund, D.G. Modified pressure plate apparatus and column testing device for measuring SWCC of sand. *ASTM Geotech. Test J.* **2009**, *32*. [CrossRef]
53. Yan, G. *Dynamic Multiphase Flow in Granular Porous Media*; Faculty of Engineering, Architecture and Information and Technology (EAIT): Queensland, Australia, 2015.
54. ThermoFisher Scientific. *DT80 Range User's Manual*; Thermo Fisher Scientific Australia Pty Ltd.: Scoresby, Australia, 2013; p. 310.
55. UMS. *User Manual of T5/T5x Pressure Transducer Tensiometer*; Meter Group AG: Munchen, Germany, 2009.
56. Klute, A.; Gardner, W. Tensiometer response time. *Soil Sci.* **1962**, *93*, 204–207. [CrossRef]
57. ASTM D6836-02. Test methods for determination of the soil water characteristic curve for desorption using a hanging column, pressure extractor, chilled mirror hygrometer, and/or centrifuge. *Am. Soc. Test. Mater. (ASTM)* **2003**. [CrossRef]
58. Yang, H.; Rahardjo, H.; Wibawa, B.; Leong, E.-C. A soil column apparatus for laboratory infiltration study. *Geotech. Test J.* **2004**, *27*. [CrossRef]
59. Hassanizadeh, S.M.; Gray, W.G. Toward an improved description of the physics of two-phase flow. *Adv. Water Resour.* **1993**, *16*, 53–67. [CrossRef]
60. Richards, L.A. Capillary conduction of liquids through porous mediums. *J. Appl. Phys.* **1931**, *1*, 318–333. [CrossRef]
61. Kalaydjian, F.-M. Dynamic capillary pressure curve for water/oil displacement in porous media: Theory vs. experiment. In Proceedings of the SPE Annual Technical Conference and Exhibition, Washington, DC, USA, 4–7 October 1992.
62. Yan, G.; Scheuermann, A.; Schlaeger, S.; Bore, T.; Bhuyan, H. Application of Spatial Time Domain Reflectometry for investigating moisture content dynamics in unsaturated sand. In Proceedings of the 11th International Conference on Electromagnetic Wave Interaction with Water and Moist Substances, Florence, Italy, 23–27 May 2016; p. 117.
63. Fredlund, D.G.; Xing, A. Equations for the soil-water characteristic curve. *Can. Geotech. J.* **1994**, *31*, 521–532. [CrossRef]
64. Zhou, A.-N.; Sheng, D.; Carter, J. Modelling the effect of initial density on soil-water characteristic curves. *Geotechnique* **2012**, *62*, 669–680. [CrossRef]

applied sciences

MDPI

Article

Numerical Analysis of an Upstream Tailings Dam Subjected to Pond Filling Rates

Tan Manh Do *, Jan Laue, Hans Mattsson and Qi Jia

Department of Civil, Environmental and Natural Resources Engineering, Division of Mining and Geotechnical Engineering, Luleå University of Technology, 97187 Luleå, Sweden; jan.laue@ltu.se (J.L.); Hans.Mattsson@ltu.se (H.M.); qi.jia@ltu.se (Q.J.)
* Correspondence: tandoh@ltu.se

Abstract: One of the challenges in upstream tailings dam projects is to ensure the allowable rate of deposition of tailings in the pond (i.e., pond filling rate) while maintaining the stability of the dam. This is due to the fact that an upstream tailings dam is constructed by placing dikes on top of previously deposited soft tailings, which could lead to a decrease in dam stability because of the build-up of excess pore water pressure. The main purpose of this work is to investigate the effects of pond filling rates on excess pore water pressure and the stability of an upstream tailings dam by a numerical study. A finite element software was used to simulate the time-dependent pond filling process and staged dam construction under various pond filling rates. As a result, excess pore water pressure increased in each raising phase and decreased in the subsequent consolidation phase. However, some of the excess pore water pressure remained after every consolidation phase (i.e., the build-up of excess pore water pressure), which could lead to a potentially critical situation in the stability of the dam. In addition, the remaining excess pore water pressure varied depending on the pond filling rates, being larger for high filling rates and smaller for low filling rates. It is believed that the approach used in this study could be a guide for dam owners to keep a sufficiently high pond filling rate but still ensure the desirable stability of an upstream tailings dam.

Keywords: excess pore water pressure; pond filling rates; upstream tailings dam; numerical analysis

Citation: Do, T.M.; Laue, J.; Mattsson, H.; Jia, Q. Numerical Analysis of an Upstream Tailings Dam Subjected to Pond Filling Rates. *Appl. Sci.* **2021**, *11*, 6044. https://doi.org/10.3390/app11136044

Academic Editors: Paulo José da Venda Oliveira and António Alberto Santos Correia

Received: 5 June 2021
Accepted: 27 June 2021
Published: 29 June 2021

Publisher's Note: MDPI stays neutral with regard to jurisdictional claims in published maps and institutional affiliations.

Copyright: © 2021 by the authors. Licensee MDPI, Basel, Switzerland. This article is an open access article distributed under the terms and conditions of the Creative Commons Attribution (CC BY) license (https://creativecommons.org/licenses/by/4.0/).

1. Introduction

This study focuses on the effects of pond filling rates on excess pore water pressure and the stability of an upstream tailings dam. Owing to a low initial investment (as a small amount of building material is required), the upstream method is the most popular design for a raised tailings embankment, especially in low-risk seismic areas [1]. It should be noted that there are over 3500 tailings dams worldwide, of which 50% are of the upstream design [2]. One of challenges in an upstream tailings dam is to ensure the allowable rates of deposition in the pond (i.e., pond filling rates) and still maintain the stability of a tailings dam [3]. This is due to the fact that an upstream tailings dam is constructed by placing dikes on top of previously deposited soft tailings, which could lead to a decrease in dam stability [4] because of the build-up of excess pore water pressure. During the operation of an upstream tailings dam, a small starter dike is first built at the extreme downstream toe. The dam wall is then progressively raised on the upstream side, mainly founded on the tailings beach. The pond filling operation process is then gradually accomplished as the height of the dam increases during the stage of construction [5,6]. In general, an adequate pond filling rate allows excess pore water pressure to dissipate gradually during the consolidation process. However, if the consolidation process is not completed when a new dike is built, i.e., in case of high pond filling rates, the excess pore water pressure will not have enough time to dissipate. In this sense, excess pore water pressure could build up under the dikes, which could lead to a critical stability situation of the dam [7]. A close relationship can thereby be seen between the consolidation process and the stability of the dam. In the present study, an approach which can simulate both consolidation processes and stability for a tailings dam is needed. Somogyi [8] investigated a slow deposition process in impoundments of slimes and tailings based on a nonlinear one-dimensional numerical approach. Gassner and Fourie [9] used a simple one-dimensional numerical approach for optimizing the allowable rate of deposition on tailings dams.

However, due to the complexities of the consolidation process coupled with the stability analysis as well as the complicated behavior of tailings materials, an advanced numerical approach would be appropriate. Note that the actual consolidation in the facility is two-dimensional under the assumption of plane strain or axisymmetric conditions, at least at the straight portions of the tailings impoundments and dams [10]. With regard to the use of the advanced numerical approach for a tailings dam, there have been a number of studies addressing two-dimensional numerical simulations [7,11–15] and three-dimensional numerical simulations [16–18]. In the study by Psarropoulos and Tsompanakis [12], an investigation on the mechanical behavior and the stability of a tailings dam under static and dynamic loading was performed. Two-dimensional numerical simulations were used in their study for three typical types of tailings dams. Zandarín et al. [13] developed a numerical model to study the stability of tailings dams subjected to the role of capillarity. Their numerical model considered the consolidation processes under self-weight loads, continuous addition of water with the tailings discharge, infiltration of rainwater, and evaporation. Ormann et al. [7,11] and Zardari et al. [15] boused the two-dimensional finite element method for the simulations of both consolidation processes and the stability of an upstream tailings dam. Ormann et al. [7,11] focused on static aspects, i.e., stability of a curved embankment and strengthening by rockfill embankments, whereas Zardari et al. [15] focused on earthquake-induced liquefaction of an upstream tailings dam. However, none of the abovementioned references have closely evaluated the effects of the pond filling rates on excess pore water pressure and the stability of an upstream tailings dam. Vick [3] addressed the need to manage the pond filling rates of an upstream tailings dam to prevent the build-up of excess pore water pressure that can reduce the shear strength of the fill material. Excessive rates of the pond filling would cause a trigger for static liquefaction that has been the underlying cause for many upstream tailings impoundment failures [19].

In this study, a numerical approach (two-dimensional finite element method) was used to investigate the effects of pond filling rates on excess pore water pressure and the stability of an upstream tailings dam. The time-dependent pond filling process and the stage of dam construction under various pond filling rates were considered in the simulations. The UBCSAND model was used for the tailings materials, whereas the Mohr–Coulomb (M-C) model was applied for moraine and rockfill materials. The results show that the finite element method can be a useful tool for studying how an upstream tailings dam should be built in order to be stable enough for different pond filling rates. With the finite element method, a finite model of a tailings dam can be built easily, e.g., a gradual raising of a dam in its geometry.

2. Numerical Modelling and Methods

2.1. General Description

An idealized upstream tailings dam was simulated using the finite element software Plaxis 2D [20]. Note that the plane strain condition is commonly adopted in simulations of tailings dams as it is a long straight section (except for corners) [7,12,21]. A full model was 360 m wide and 55 m high after the last construction stage, as shown in Figure 1a. In general, the model size should be large enough to avoid boundary effects on simulation outputs. In this study, the left vertical boundaries were 205 m away from the dam in all models. The slope inclination of 2H:1V (horizontal to vertical) was adopted as in the previous work of a simplified upstream tailings dam simulation by Psarropoulos and Tsompanakis [12]. The model comprised six soil types: moraine (foundation), rockfill (downstream support), initial dike (starter dike), layered tailings, compacted sand tailings (dikes), and rockfill (Figure 1b). Figure 1b shows the initial pond filling step, which was located at 10 m above the initial ground level. From this level, five construction stages comprising a raising phase (R) and a consolidation phase (C) were simulated in the model. In each raising phase, a new dike was constructed on top of previously layered tailings, assumed to be completed in 10 days and followed by a 10-day raising period for both new layered tailings and rockfill. The consolidation analysis was carried out to investigate the transient system response allowing excess pore water pressure to dissipate gradually during the consolidation process. The time in the consolidation phase (consolidation time) depends on the pond filling rate. In this current work, four case studies corresponding to four pond filling rates of 10.0 m/year, 5.0 m/year, 3.3 m/year, and 2.5 m/year were used to investigate the effects of the pond filling rates on excess pore water pressure and the stability of the upstream tailings dam. Therefore, the consolidation times in each stage were set over 163 days, 345 days, 528 days, and 710 days corresponding to the pond filling rates of 10.0 m/year, 5.0 m/year, 3.3 m/year, and 2.5 m/year, respectively. The phreatic line was assumed to be located at the surface of the tailings impoundment (worst-case scenario). The locations of the phreatic lines were adapted depending on the pond filling steps. In each analysis step, a new set of hydraulic boundary conditions was imposed, i.e., closed boundaries at the bottom (Y_{min}) and the left vertical edge (X_{min}), and open boundaries for others. As for the mechanical boundary conditions, the model was assumed to be fully fixed at its bottom. The horizontal displacements were assumed to be zero along the lateral edges (i.e., both left and right vertical boundaries). These hydraulic and mechanical boundary conditions can be found in many previous works dealing with two-dimensional numerical simulations of upstream tailings dams [7,11–15]. The finite element mesh of the upstream tailings dam after the last construction stage is shown in Figure 2. The finite element mesh in each cluster was composed of 15-node triangular elements. These elements give a fourth-order (quartic) interpolation for displacements [20]. A massive number of elements were generated in the areas of interest (structural zone), providing the finer mesh near the slope with embankments and rockfill. This is due to the fact that these areas would be affected by large strains during the stage of construction. The coarser mesh was then generated at the far-field areas to minimize computation time.

Apart from the fully coupled analysis on deformation and consolidation, the global factor of safety (*FoS*) for slope failure was computed for every stage of construction. The global factor of safety is computed in the finite element software Plaxis 2D using the shear strength reduction method. The idea of this method is that the soil strength is gradually reduced, and when a failure occurs, the corresponding strength reduction factor can be considered as the factor of safety of soil strength [20,22–25]. In this study, the factor of safety should have a value of at least 1.5 under normal operation conditions according to the Swedish dam safety guidelines [26]. In order to maintain the factor of safety of 1.5 at every stage of the construction, a strengthening by rockfill berms was added on the downstream side of the slope. The procedure for the construction of the rockfill berms was simulated in the finite element model. The rockfill berms were placed immediately after each dike construction in the finite element model. The width of the rockfill was then optimized in this study by a parametric sensitivity analysis where the rockfill width was varied until no *FoS* became smaller than approximately 1.5 at every stage of construction. Four rockfill widths of 5.0 m, 10.0 m, 15.0 m, and 20.0 m were applied to reach the target (discussed in Section 3.1, Figure 6).

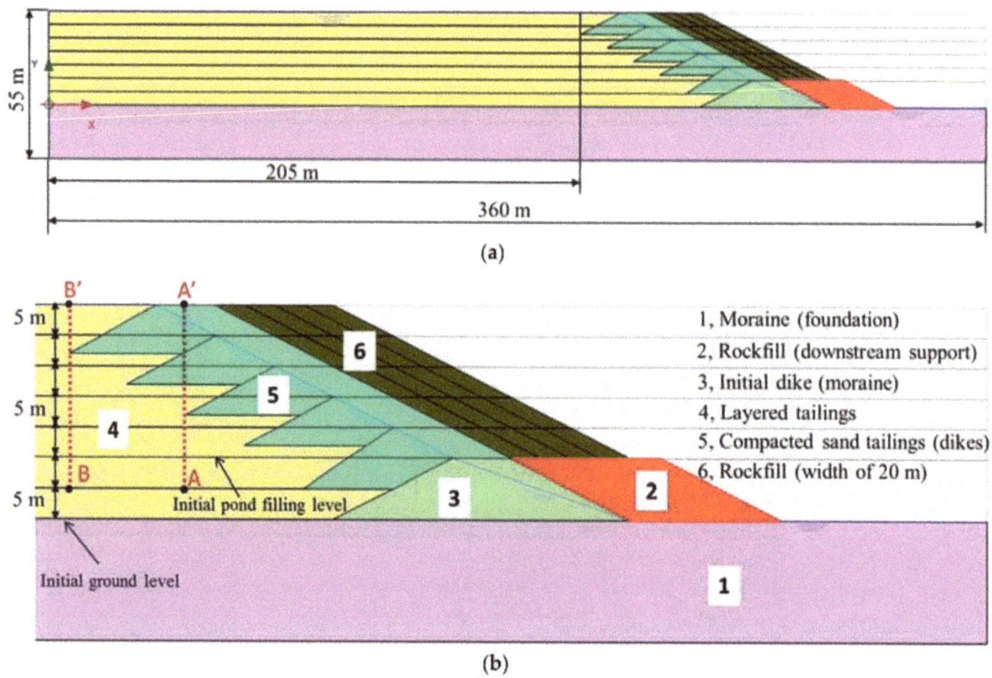

Figure 1. Geometry of the upstream tailings dam (rockfill width of 20 m): (**a**) full model and (**b**) upscaled model (downstream part).

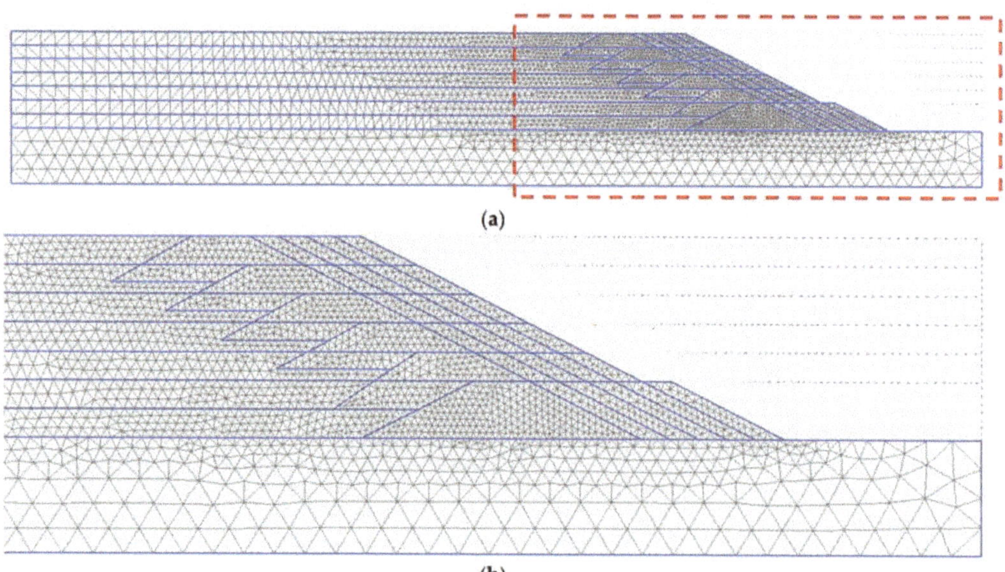

Figure 2. Finite element mesh of the upstream tailings dam: (**a**) full model and (**b**) upscaled model (downstream part).

2.2. Constitutive Models

In this study, the UBCSAND constitutive model was used to simulate the tailings materials, whereas the Mohr–Coulomb (M-C) model was applied for moraine and rockfill materials. The M-C model is a linear elastic–perfectly plastic model. The UBCSAND model is an effective stress plasticity model used in advanced stress deformation analyses of geotechnical materials. A fully coupled analysis (mechanical and groundwater flow calculations) can be performed simultaneously in this model. It can predict the shear stress–strain behavior of soil using an assumed hyperbolic relationship [27]. Therefore, this model can overcome the drawback of the elastic–perfectly plastic M-C model with a possibility to capture the nonlinearity in the elastic part of the soil. In addition, the UBCSAND model is appropriate for tailings materials as excess pore water pressure can build up in tailings during pond filling and dam raising. Figure 3 presents the behavior of tailings materials during drained simple shear tests and simulations. The laboratory tests were performed by Wiklund [28] at the Luleå University of Technology. Several simulation models were performed in this study to evaluate parameters for the UBCSAND model. As can be observed, while the stress–strain curves followed the nonlinear behavior of the experimental data well in the UBCSAND model, they overestimated the shear strength of the tailings materials significantly in the M-C model, regardless of the applied normal stresses (50 kPa, 100 kPa, 150 kPa) or the tailings types (layered tailings, compacted tailings). The UBCSAND model has also been used for materials in the development of excess pore water pressure-induced static liquefaction [29] or seismic liquefaction [15,30].

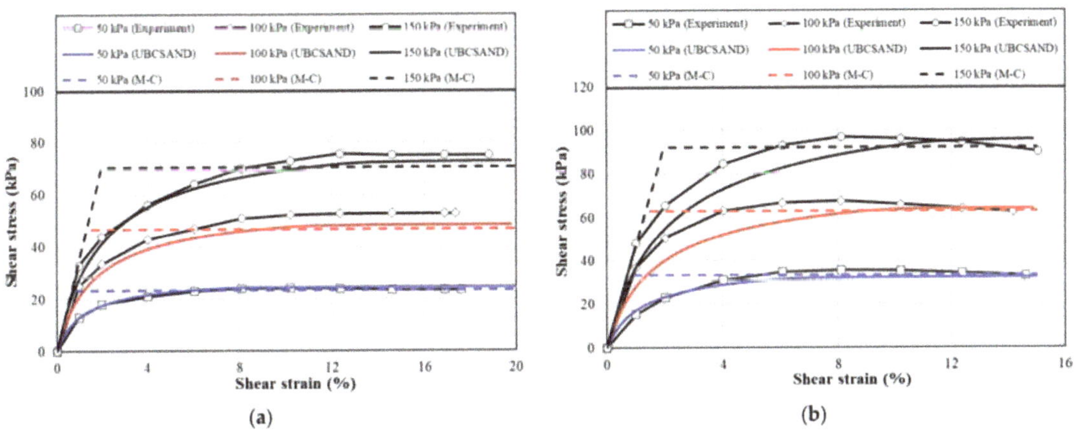

Figure 3. Behavior of tailings materials during drained simple shear tests and simulations: (**a**) layered tailings with dry density of 1.65 t/m^3 and (**b**) compacted tailings with dry density of 1.82 t/m^3.

The M-C input parameter values of moraine and rockfill materials were adopted from previous works [31,32], as summarized in Table 1, except for the Poisson's ratio and the dilatancy angle. In this study, the Poisson´s ratio (υ) and the dilatancy angle (ψ) were assumed to be 0.33 and 0, respectively. These assumed values have been used in many simulation works of tailings dams in Sweden by Ormann et al. [7,11], Knutsson et al. [21], and Zardari et al. [15]. The laboratory tests were performed on tailings materials taken from an iron mine in northern Sweden. In addition, the parametric sensitivity analyses (curve fitting method) for obtaining elastic shear modulus number ($K_G{}^e$), plastic shear modulus number ($K_G{}^p$), and elastic bulk modulus number ($K_B{}^e$) were performed. In the curve fitting method, the friction angles and cohesion, which were evaluated based on the direct shear test results, were fixed. Elastic shear modulus number ($K_G{}^e$) and plastic shear modulus number ($K_G{}^p$) were varied to capture the stress–strain curves from the experiments. Finally, elastic bulk modulus number ($K_B{}^e$) was estimated as 70% of elastic shear modulus number ($K_G{}^e$), as recommended in [20] (material models part). As can be seen in Figure 3, there is a good agreement between the simulations (UBCSAND model) and experimental results, regardless of the tailings types used. The index parameters m_e, n_e, and n_p were assigned the values of 0.5, 0.5, and 0.4, respectively, as suggested in the original UBCSAND report by Beaty and Byrne [27]. The hydraulic conductivity of the tailings was adopted from previous studies [28,31]. In the present study, the hydraulic conductivity values in the horizontal direction were assumed to be 10 times higher than those in the vertical direction due to the layered nature of the tailings [26,31]. All input parameter values of the UBCSAND model used in the numerical analyses are tabulated in Table 2.

Table 1. Parameters of Mohr–Coulomb (M-C) model.

Parameter	Symbols	Moraine (Foundation)	Moraine (Initial Dike)	Rockfill	Unit
Soil unit weight above phreatic level	γ_{unsat}	20	20	18	kN/m^3
Soil unit weight below phreatic level	γ_{sat}	22	22	20	kN/m^3
Young's modulus	E	20,000	20,000	40,000	kN/m^2
Cohesion	c'	1	1	1	kN/m^2
Friction angle	φ'	37	35	42	degree
Horizontal hydraulic conductivity	k_x	4.98×10^{-8}	1.0×10^{-7}	1×10^{-1}	m/s
Vertical hydraulic conductivity	k_y	1.0×10^{-8}	4.98×10^{-8}	1×10^{-1}	m/s

Table 2. Parameters of UBCSAND model.

Parameter	Symbols	Layered Tailings	Compacted Tailings (Dikes)	Unit
Soil unit weight above phreatic level	γ_{unsat}	17	18	kN/m^3
Soil unit weight below phreatic level	γ_{sat}	20	20	kN/m^3
Elastic shear modulus number	$K_G{}^e$	130	130	-
Plastic shear modulus number	$K_G{}^p$	50	75	-
Elastic bulk modulus number	$K_B{}^e$	91	91	-
Elastic bulk modulus index	m_e	0.5	0.5	-
Elastic shear modulus index	n_e	0.5	0.5	-
Plastic shear modulus index	n_p	0.4	0.4	-
Constant volume friction angle	Φ_v	25	28	degree
Peak friction angle	Φ_p	28	36	degree
Cohesion	c'	0	0	kN/m^2
Horizontal hydraulic conductivity	k_x	1.0×10^{-7}	1.0×10^{-6}	m/s
Vertical hydraulic conductivity	k_y	1×10^{-8}	1.0×10^{-7}	m/s

3. Simulation Results

3.1. Stability Analyses

Stability analysis is an important component in the design of any earth structure, including tailings dams. Stability is usually expressed in terms of the factor of safety (*FoS*), which is defined as the ratio between the available shear strength of the soil and the minimum shear strength required against failure [12]. Figure 4 illustrates the factor of safety of the dam (pond filling rate of 5.0 m/year) during the pond filling, comprising raising phases (R) and consolidation phases (C). As shown, the *FoS* decreased gradually and dropped to less than 1.5 after the construction of three dikes of a height of 20 m from the initial pond filling level, i.e., *FoS* was 1.37 for R3 and 1.48 for C3. These values kept falling and even dropped to around 1.0 after the last construction stage. In addition, it was found that the *FoS* values decreased after every raising phase and then increased again due to the subsequent consolidation phase. This can be explained by the increase in effective stresses at the same rate as the excess pore pressure dissipates. The shear strength increases gradually during the consolidation phase. In this sense, the lowest stability of the dam is expected immediately after a new dike construction (a raising phase). This observation is similar to that of many previous studies on the stability of an upstream tailings dam [7,12,21,33]. The possible failure mechanism of the dam is shown in Figure 5. The slip surface most likely associated with the lowest *FoS* occurred according to Figure 5. In this case, the *FoS* did not meet the adopted code requirements of the Swedish dam safety guidelines. Therefore, a slope strengthening was introduced in this study to maintain the factor of safety of 1.5 at every stage of construction.

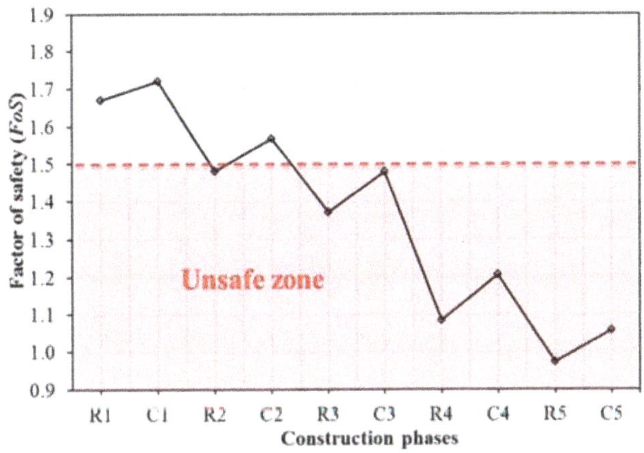

Figure 4. Factor of safety (filling rate of 5.0 m/year) during the pond filling comprising both raising phases (R) and consolidation phases (C).

Figure 5. Failure mechanism of the upstream tailings dam (filling rate of 5.0 m/year).

Strengthening an upstream tailings dam with rockfill berms at the downstream slope is a common technique that has been introduced and used in many previous works [7,15,21,25,33]. In this study, the width of the rockfill was optimized. The rockfill width was increased in the models (with the rockfill widths of 5 m, 10 m, 15 m, and 20 m) until no *FoS* became smaller than approximately 1.5 at every stage. The factors of safety for all the cases can be observed in Figure 6. These observed factors of safety were higher than those from the upstream tailings dam without rockfill berms. The weight of the rockfill berms placed next to the dikes could provide a resisting moment which increases the slope stability. As expected in all the cases, the *FoS* values decreased after every raising phase and then increased again due to the subsequent consolidation phase, regardless of rockfill widths. This trend is found to be similar to that in Figure 4. In conclusion, the rockfill width of 20 m should be used to keep the dam stable enough ($FoS \geq 1.5$) for the pond filling rate of 5.0 m/year.

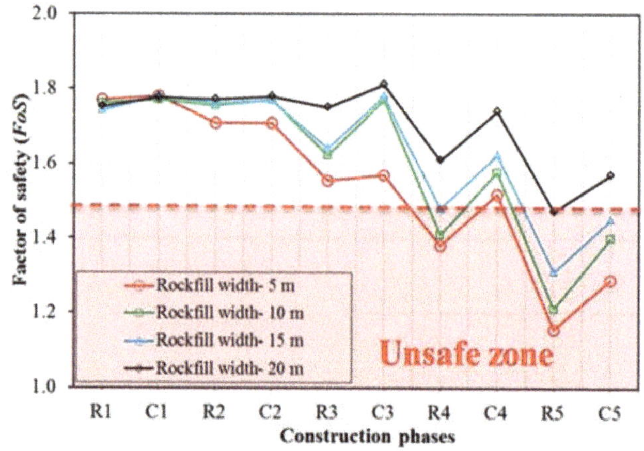

Figure 6. Rockfill width optimization based on slope stability analysis (filling rate of 5.0 m/year).

3.2. Excess Pore Water Pressure Analyses

Figure 7 shows the excess pore water pressure at the last construction stage (filling rate of 2.5 m/year and rockfill width of 20 m). There was a relatively high excess pore pressure in some zones beneath the dikes after the raising phase R5 (Figure 7a). However, this pressure decreased significantly after the consolidation phase C5 (Figure 7b). The remaining high excess pore pressure after the consolidation phase C5 occurred in the lower part of the impoundment but further away from the dam. This is due to the presence of the impermeable base in the finite element model. The trend of the excess pore water pressure distribution observed in this study is similar to that reported by Ormann et al. [7] and Saad and Mitri [14], i.e., larger remaining excess pore pressure after a consolidation phase occurred in the lower part of the impoundment but farther away from the dam.

3.3. Effects of the Pond Filling Rates on the Stability and Excess Pore Water Pressure

Figure 8 shows the effect of the pond filling rates (*PFR*) on the stability of the upstream tailings dam with the rockfill width of 20 m during the pond filling, comprising raising phases (R) and consolidation phases (C). As shown, the *FoS* decreased very slightly when the *PFR* increased from 2.5 m/year to 5.0 m/year, regardless of construction phases. At the pond filling rate of 5.0 m/year, the factor of safety of the dam at every stage of construction fulfilled the adopted code requirement of the Swedish dam safety guidelines, except for the raising phase R5 (*FoS* at R5 was 1.472). However, when the *PFR* increased from 5.0 m/year to 10 m/year, the *FoS* suddenly dropped to 1.40 at the raising phase R5. Thus, the *FoS* did not satisfy the abovementioned requirement, which might be a sign of potential instability issues for the dam.

Figure 9 shows excess pore water pressure after the last consolidation phase C5 with respect to the pond filling rates. For the ease of interpretation, the shading views of excess pore water pressure in Figure 9 are fixed at the same color intervals, except for the model with the pond filling rate of 10 m/year. As shown, an effect of the pond filling rate on excess pore water pressure was discovered: an increase in pond filling rate can lead to an increase in excess pore water pressure in the model. In particular, the maximum excess pore water pressure in the model with the pond filling rate of 2.5 m/year was 29 kPa. This value rose up to 43 kPa, 68 kPa, and 138 kPa in the models with the pond filling rates of 3.3 m/year, 5.0 m/year, and 10 m/year, respectively. In cases of high pond filling rates, there was less time for the excess pore pressure to dissipate during consolidation.

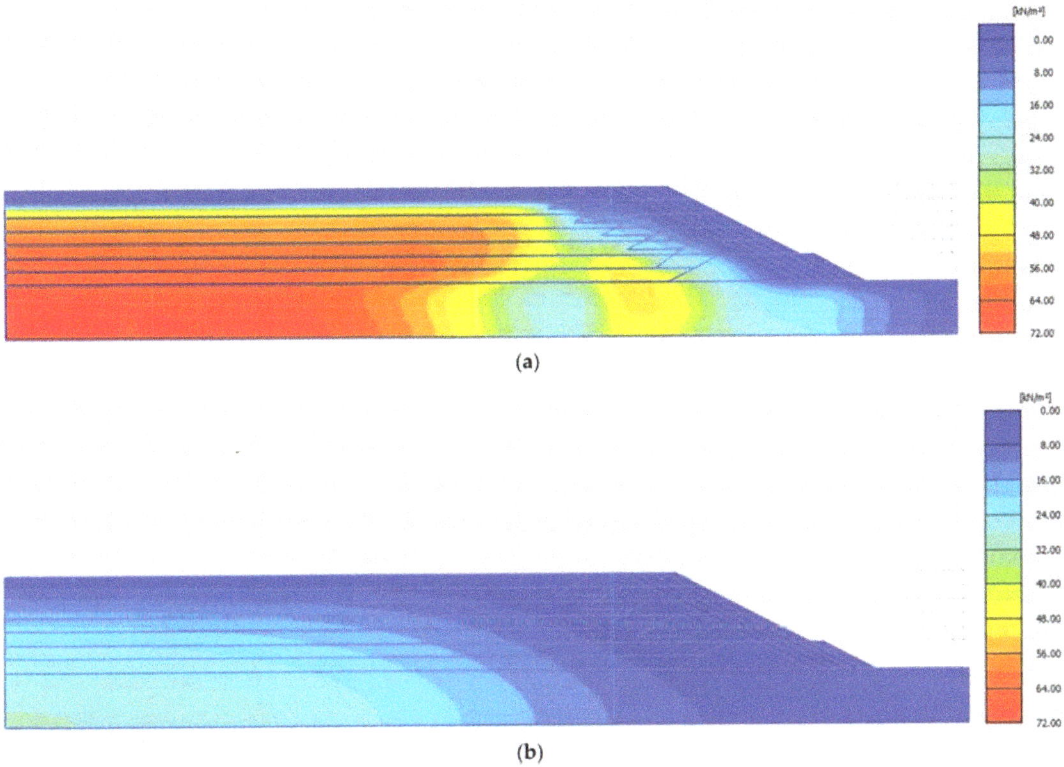

Figure 7. Excess pore water pressure at the last construction stage with the pond filling rate of 2.5 m/year and rockfill width of 20 m: (**a**) after the raising phase R5 and (**b**) after the consolidation phase C5.

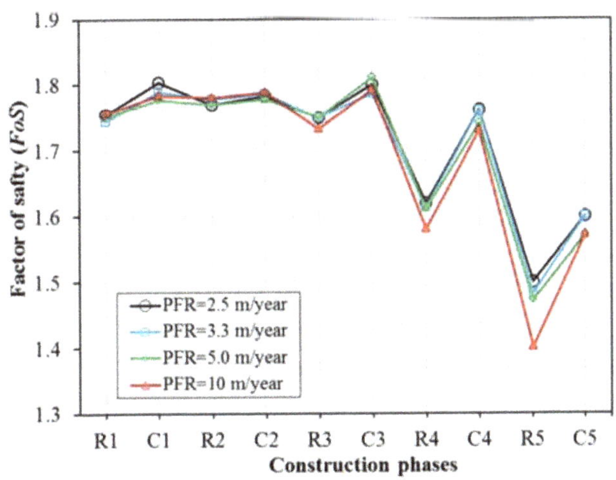

Figure 8. Effect of the pond filling rates (*PFR*) on the stability of the upstream tailings dam (rockfill width of 20 m).

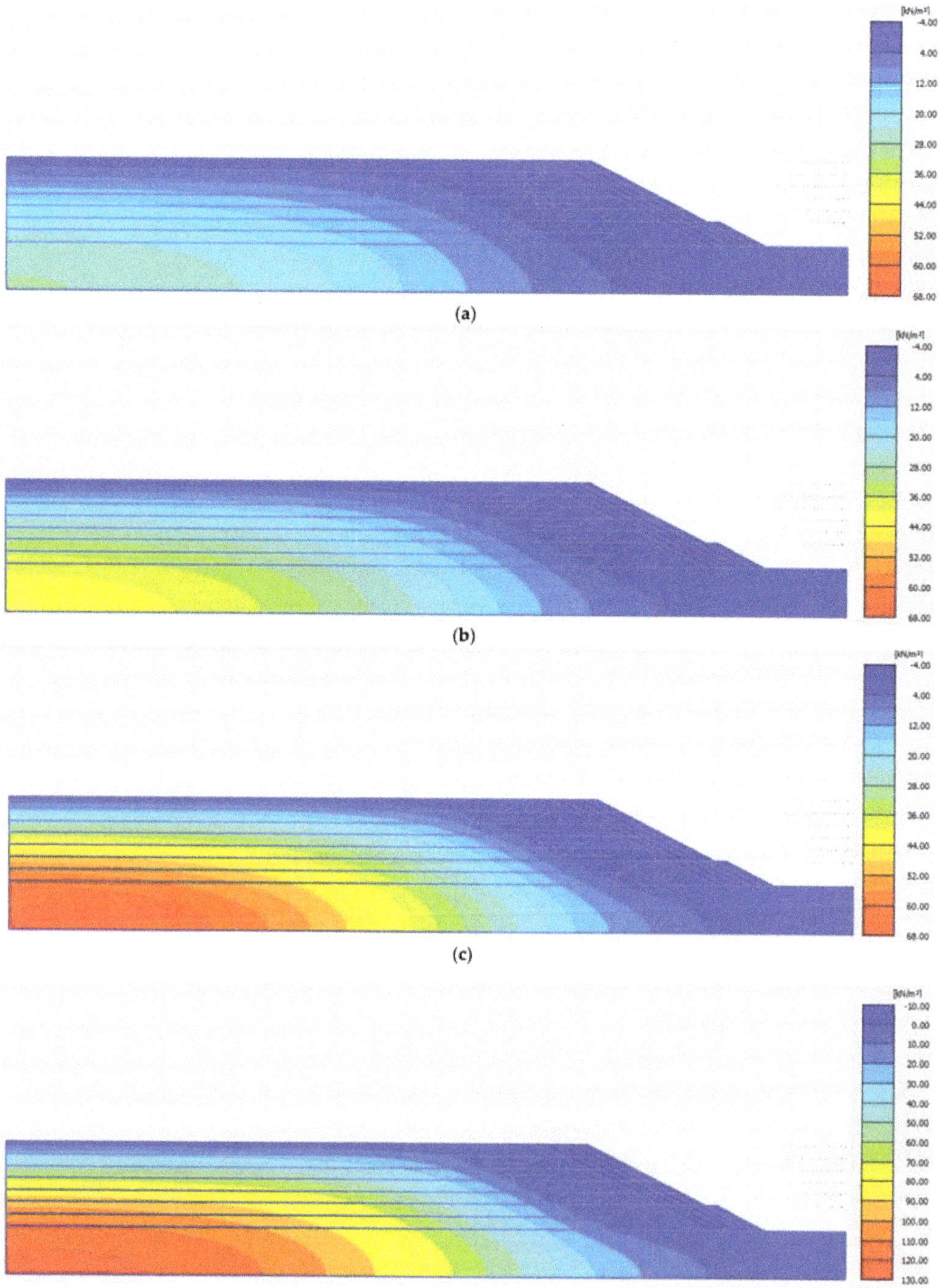

Figure 9. Excess pore water pressure after the last consolidation phase with respect to the pond filling rates: (**a**) 2.5 m/year, (**b**) 3.3 m/year, (**c**) 5.0 m/year, and (**d**) 10 m/year.

In addition, excess pore water pressure which developed along cross sections A–A′ and B–B′ (defined in Figure 1b) are plotted in Figure 10. Excess pore water pressure increased with an increase in the *PFR*, regardless of the cross sections analyzed. Take the elevation of 5 m from the initial ground level as an example: at the section A–A′, excess pore water pressure in the model with the pond filling rate of 2.5 m/year had the lowest value of 4.65 kPa, followed by 7.50 kPa (*PFR* = 3.3 m/year), 12.25 kPa (*PFR* = 5.0 m/year), and 27.25 kPa (*PFR* = 10 m/year). Similarly, at the cross section B–B′ at 5 m height, excess pore water pressure in the models with the pond filling rates of 2.5 m/year, 3.3 m/year, 5.0 m/year, and 10 m/year were 7.5 kPa, 12.1 kPa, 19.6 kPa, and 43.1 kPa, respectively. Interestingly, the same trend was observed at both cross sections, even though excess pore water pressure at the cross section B–B′ was higher than the corresponding excess pore water pressure at the cross section A–A′. It should be noted that cross section A–A′ had a draining distance shorter than cross section B–B′. It was also found from Figure 10 that excess pore water pressure decreased with increased heights from the initial ground level, regardless of the *PFR* and sections. This is because of the presence of the impermeable base in the finite element model.

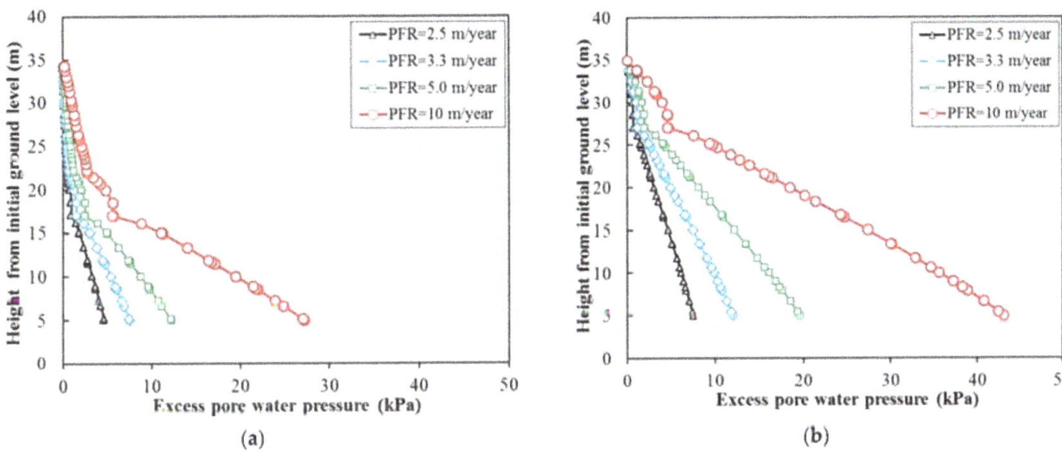

Figure 10. Excess pore water pressure after the last consolidation phase C5 with respect to the pond filling rates (*PFR*) at sections: (**a**) A–A′ and (**b**) B–B′ (defined in Figure 1b).

Finally, the build-up of excess pore pressure (*EPP*) was also considered in this study. Figure 11 shows the basic concept of excess pore water pressure build-up during the stage of construction at point B and *PFR* = 10 m/year. As shown, the excess pore water pressure increased after every raising phase and then decreased after every consolidation phase which followed. Again, this finding supports the abovementioned stability analyses, that is, the *FoS* values decrease after raising phases and then increase again after consolidation phases. More importantly, it is worth observing that *EPP* remained after every pond filling step (i.e., after consolidation phases C1, C2, C3, C4, and C5). The remaining *EPP* were built up until the last phase of consolidation C5. Based on this concept, the *EPP* build-up during the stage of construction with respect to pond filling rates (*PFR*) is presented in Figure 12. Typical points A and B (in Figure 1b) were selected for the analyses. As shown, the build-up of excess pore water pressure was dependent on the *PFR*. Take point B in Figure 12b as an example: when the *PFR* was small (i.e., *PFR* = 2.5 m/year), the remaining *EPP* were EPP_{C1} = 3.9 kPa, EPP_{C2} = 4.0 kPa, EPP_{C3} = 5.1 kPa, EPP_{C4} = 6.3 kPa, and EPP_{C5} = 7.6 kPa after the consolidation phases C1, C2, C3, C4, and C5, respectively. The remaining *EPP*, in this case, was built up, reaching only a small value of 7.6 kPa (EPP_{C5}) after the last consolidation phase C5. However, higher values of EPP_{C5} were found when the *PFR* was higher (i.e.,

12.0 kPa, 19.7 kPa, and 43.2 kPa for the cases with *PFR* = 3.3 m/year, *PFR* = 5.0 m/year, and *PFR* = 10.0 m/year). The same effect of the pond filling rates on the *EPP* build-up during the stage of construction can be observed in the results of point A (Figure 12a).

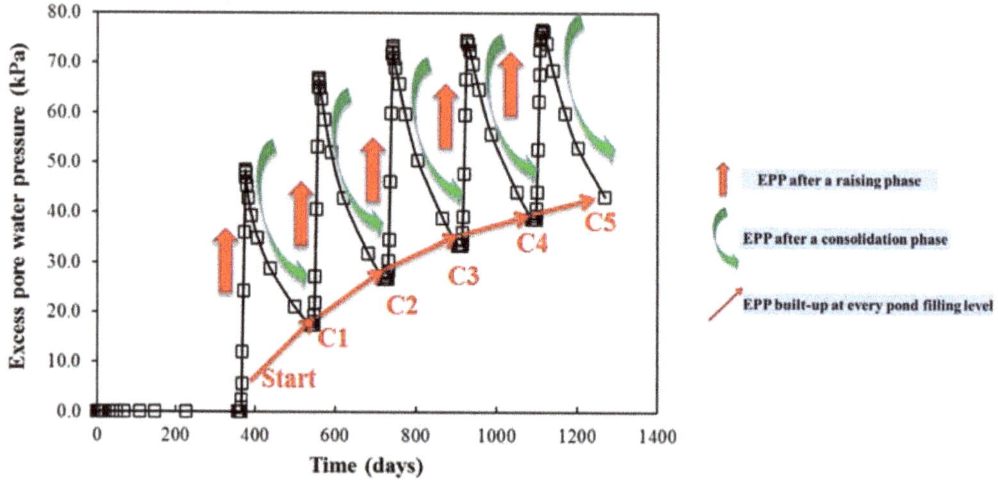

Figure 11. Concept of excess pore water pressure build-up during the stage of construction (point B and *PFR* = 10 m/year).

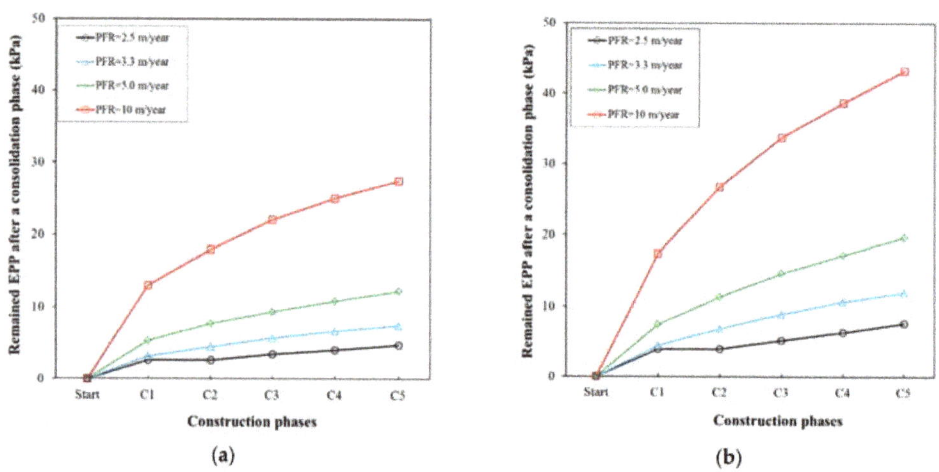

Figure 12. Effect of the pond filling rates (*PFR*) on the build-up of *EPP* during the stage of construction at: (**a**) point A and (**b**) point B (defined in Figure 1b).

4. Discussion

It is concluded by Ormann et al. [7] that the stability of an upstream tailings dam could be improved by utilizing rockfill berms as supports on the downstream side. The stability analyses obtained in this study agree with this conclusion; refer, for example, to Figures 4 and 6. In addition, it was found from this study that, in order to keep the stability of an upstream tailings dam, the exceptionally high pond filling rate is not recommended. However, the recommendation is only applicable in the case of 20 m rockfill width and the tailings type used in this study. In practice, larger rockfill width could be added next to dikes or on the downstream side to enhance the slope stability and hold the production speed. In this case, key questions remain to be answered: (i) which design of rockfill berms is needed for different production speeds to maintain the stability of a tailings dam, and (ii) how can the volume of rockfill berms be minimized (i.e., for minimum cost) while still maintaining the stability and production speed?

In practice, an upstream tailings dam is typically constructed by placing dikes on top of previously deposited soft saturated tailings, which could lead to the excess pore water pressure gradually increasing during pond filling and dam raising. Therefore, an adequate pond filling rate is required to allow excess pore water pressure to dissipate gradually during the consolidation process. The numerical results from this study showed that excess pore water pressure increased in each raising phase and decreased in the subsequent consolidation phase. However, some of the excess pore water pressure remained after every consolidation phase (i.e., the build-up of excess pore water pressure) (Figure 11). The build-up excess pore water pressure varied depending on the pond filling rates. The dam with a high pond filling rate had a shorter time for the excess pore pressure to dissipate during every consolidation phase than the dam with low ones. Therefore, the highest value was found in the dam with the pond filling rate of 10 m/year in this study (Figure 12).

Even though the numerical results obtained are encouraging, some additional work will need to be performed in the future to simulate other conditions involved in the responses of an upstream tailings dam under cyclic and seismic loading.

5. Conclusions

In this study, a numerical analysis of an upstream tailings dam subjected to pond filling rates was conducted. A finite element software was used to simulate the time-dependent pond filling process and the staged dam construction under various pond filling rates. An advanced constitutive model was adopted in the finite element models to capture the nonlinearity in the elastic deformations of the tailings materials. Based on the results of this study, the following conclusions can be drawn:

- Excess pore water pressure increased after every raising phase and then decreased after every subsequent consolidation phase in the numerical models. However, excess pore water pressure remained after every consolidation phase, leading to the build-up of excess pore water pressure and, eventually, a potentially critical situation in the stability of the dam (i.e., especially for the dam without strengthening).
- The stability of the dam decreased during the raising phase but increased during the consolidation phase. This is due to the changes of excess pore water pressure, i.e., increased values with the raising phase and decreased values with the consolidation phase. In addition, it is also vital to mention that the strengthening of an upstream tailings dam with rockfill berms at the downstream slope could improve the stability of the dam. The basic concept of excess pore water pressure build-up during the stage of construction was presented based on the remaining excess pore water pressure after every consolidation phase. The remaining excess pore water pressure varied depending on the pond filling rates, being larger in high filling rates and smaller in low filling rates. A very high pond filling rate is, therefore, not recommended for an upstream tailings dam as the dissipation of excess pore water pressure usually takes a long time.
- This study provided an in-depth analysis of the build-up of excess pore water pressure and slope stability during the stage of construction (pond filling and dam raising) of an upstream tailings dam. It is believed that the finite element method could be used to study how an upstream tailings dam should be built in order to be stable for different pond filling rates.

Author Contributions: Conceptualization, T.M.D.; J.L.; H.M.; methodology, T.M.D.; J.L.; software, T.M.D.; H.M.; validation, T.M.D.; J.L.; H.M.; Q.J.; formal analysis, T.M.D.; J.L.; H.M.; investigation, T.M.D.; J.L.; H.M.; writing—original draft preparation, T.M.D.; writing—review and editing, T.M.D.; J.L.; H.M.; Q.J.; supervision, J.L.; H.M.; Q.J.; project administration, J.L.; funding acquisition, J.L. All authors have read and agreed to the published version of the manuscript.

Funding: This research was funded by Svenskt Vattenkraftcentrum, SVC ("The Swedish Hydropower Center").

Data Availability Statement: Not applicable.

Acknowledgments: The research presented in this paper was carried out as a part of the Swedish Hydropower Center (Svenskt Vattenkraftcentrum, SVC). SVC has been established by the Swedish Energy Agency, Energiforsk, and Svenska Kraftnät, together with the Luleå University of Technology, KTH Royal Institute of Technology, Chalmers University of Technology, Uppsala University, and Lund University. The participating companies and industry associations are: Andritz Hydro, Boliden, Fortum Sweden, Holmen Energi, Jämtkraft, Karlstads Energi, LKAB, Mälarenergi, Norconsult, Rainpower, Skellefteå Kraft, Sollefteåforsens, Statkraft Sverige, Sweco Sverige, Tekniska verken i Linköping, Uniper, Vattenfall R&D, Vattenfall Vattenkraft, Voith Hydro, WSP Sverige, Zink-gruvan, and ÅF Industry. This research was funded by SVC and the Luleå University of Technology.

Conflicts of Interest: The authors declare no conflict of interest.

Abbreviations

PFR	Pond filling rate
EPP	Excess pore water pressure
R	Raising phase
C	Consolidation phase
X_{min}	Boundary at the left vertical edge
Y_{min}	Boundary at the bottom
FoS	Factor of safety
$M\text{-}C$	Mohr–Coulomb model
γ_{unsat}	Soil unit weight above phreatic level
γ_{sat}	Soil unit weight below phreatic level
E	Young's modulus
c'	Effective soil cohesion
φ'	Effective friction angle
k_x	Horizontal hydraulic conductivity
k_y	Vertical hydraulic conductivity
${K_G}^e$	Elastic shear modulus number
${K_G}^p$	Plastic shear modulus number
${K_B}^e$	Elastic bulk modulus number
m_e	Elastic bulk modulus index
n_e	Elastic shear modulus index
n_p	Plastic shear modulus index
φ_v	Constant volume friction angle
φ_p	Peak friction angle
v	Poisson's ratio
ψ	Dilatancy angle

References

1. Kossoff, D.; Dubbin, W.E.; Alfredsson, M.; Edwards, S.J.; Macklin, M.G.; Hudson-Edwards, K.A. Mine tailings dams: Characteristics, failure, environmental impacts, and remediation. *Appl. Geochem.* **2014**, *51*, 229–245. [CrossRef]
2. Davies, M.; Martin, T.; Lighthall, P. Mine tailings dams: When things go wrong. In *Tailings Dams*; Association of State Dam Safety Officials, U.S. Committee on Large Dams: Las Vegas, NV, USA, 2000; pp. 261–273.
3. Vick, S.G. *Planning, Design, and Analysis of Tailings Dams*; Wiley: New York, NY, USA, 1990.
4. Dixon-Hardy, D.W.; Engels, J.M. Guidelines and recommendations for the safe operation of tailings management facilities. *Environ. Eng. Sci.* **2007**, *24*, 625–637. [CrossRef]

5. Wills, B.A.; Finch, J.A. Tailings Disposal. In *Wills' Mineral Processing Technology*, 8th ed.; Wills, B.A., Finch, J.A., Eds.; Butterworth-Heinemann: Boston, MA, USA, 2016; pp. 439–448.
6. Wei, Z.A.; Yin, G.Z.; Wan, L.; Li, G.Z. A case study on a geotechnical investigation of drainage methods for heightening a tailings dam. *Environ. Earth Sci.* **2016**, *75*, 106. [CrossRef]
7. Ormann, L.; Zardari, M.A.; Mattsson, H.; Bjelkevik, A.; Knutsson, S. Numerical analysis of strengthening by rockfill embankments on an upstream tailings dam. *Can. Geotech. J.* **2013**, *50*, 391–399. [CrossRef]
8. Somogyi, F. Large strain consolidation of fine-grained slurries. In Proceedings of the Canadian society for civil engineering, Annual Conference, Manitoba, OT, Canada, 9–12 June 1980.
9. Gassner, F.W.; Fourie, A.B. Optimising the allowable rate of deposition on tailings dams. In Proceedings of the Fifth International Conference on Tailings and Mine Waste '98, Fort Collins, CO, USA, 26–28 January 1998; pp. 241–248.
10. Mittal, H.K.; Morgenstern, N.R. Seepage control in tailings dams. *Can. Geotech. J.* **1976**, *13*, 277–293. [CrossRef]
11. Ormann, L.; Zardari, M.A.; Mattsson, H.; Bjelkevik, A.; Knutsson, S. Numerical analysis of curved embankment of an upstream tailings dam. *Electron. J. Geotech. Eng.* **2011**, *16*, 931–944.
12. Psarropoulos, P.N.; Tsompanakis, Y. Stability of tailings dams under static and seismic loading. *Can. Geotech. J.* **2008**, *45*, 663–675. [CrossRef]
13. Zandarín, M.T.; Oldecop, L.A.; Rodríguez, R.; Zabala, F. The role of capillary water in the stability of tailing dams. *Eng. Geol.* **2009**, *105*, 108–118. [CrossRef]
14. Saad, B.; Mitri, H. Hydromechanical analysis of upstream Tailings disposal facilities. *J. Geotech. Geoenviron. Eng.* **2011**, *137*, 27–42. [CrossRef]
15. Zardari, M.A.; Mattsson, H.; Knutsson, S.; Khalid, M.S.; Ask, M.; Lund, B. Numerical analyses of earthquake induced liquefaction and deformation behaviour of an upstream tailings dam. *Adv. Mater. Sci. Eng.* **2017**, 1–12. [CrossRef]
16. Zardari, M.A.; Mattsson, H.; Knutsson, S. 3D finite element analyses of a corner at Aitik tailings dam in Sweden. In Proceedings of the International symposium on Environmental Issues and Waste Management in Energy and Mineral Production, New Delhi, India, 28–30 November 2012.
17. Zardari, M.A.; Mattsson, H.; Knutsson, S.; Ormann, L. Comparison of three dimensional and two dimensional axisymmetric finite element analyses of a corner section of a tailings dam. *Sci. Iran.* **2017**, *24*, 2320–2331.
18. Liu, C.; Shen, Z.; Gan, L.; Xu, L.; Zhang, K.; Jin, T. The seepage and stability performance assessment of a new drainage system to increase the height of a tailings dam. *Appl. Sci.* **2018**, *8*, 1840. [CrossRef]
19. Martin, T.E.; McRoberts, E.C. *Some Considerations in the Stability Analysis of Upstream Tailings Dams*; AGRA Earth & Environmental Limited: Burnaby, BC, Canada, 2002.
20. Brinkgreve, R.; Swolfs, W.; Engin, E.; Waterman, D.; Chesaru, A.; Bonnier, P.; Galavi, V. *PLAXIS 2D Reference Manual*; Delft University of Technology and PLAXIS, B.V.: Delft, The Netherlands, 2018.
21. Knutsson, R.; Viklander, P.; Knutsson, S. The use of numerical modelling in alert level set-up for instrumentation in tailings dams. In Proceedings of the ICOLD Congress 2015: International Commission on Large Dams, Stavanger, Norway, 14–20 June 2015.
22. Lee, K.; Ko, J.; Lim, H.; Lee, J.K. Stability numbers for unsupported conical excavations in multi-layered cohesive soils. *Appl. Sci.* **2020**, *10*, 8839. [CrossRef]
23. Griffiths, D.V.; Marquez, R.M. Three-dimensional slope stability analysis by elasto-plastic finite elements. *Géotechnique* **2007**, *57*, 537–546. [CrossRef]
24. Griffiths, D.V.; Lane, P.A. Slope stability analysis by finite elements. *Géotechnique* **1999**, *49*, 387–403. [CrossRef]
25. Hassellund, L.; Knutsson, R.; Mattsson, H.; Knutsson, S. Numerical simulations of stability of a gradually raised upstream tailings dam in northern Sweden. *Electron. J. Geotech. Eng.* **2016**, *21*, 4699–4720.
26. GruvRIDAS. *The Mining Industry's Guidelines for Dam Safety*; Svensk Energi AB/SveMin: Stockholm, Sweden, 2007. (In Swedish)
27. Beaty, M.H.; Byrne, P.M. *UBCSAND Constitutive Model Version 904ar*; Itasca UDM: Minneapolis, MN, USA, 2011.
28. Wiklund, V. *Fines Content and Density Effects on Tailings Behaviour: A Laboratory Study on Geotechnical Properties*; Luleå University of Technology: Luleå, Sweden, 2018.
29. Askarinejad, A.; Zhang, W.; de Boorder, M.; van der Zon, J. Centrifuge and numerical modelling of static liquefaction of fine sandy slopes. *Phys. Model. Geotech.* **2018**, *2*, 1119–1124.
30. Ferdosi, B.; James, M.; Aubertin, M. Numerical simulations of seismic and post-seismic behavior of tailings. *Can. Geotech. J.* **2015**, *53*, 85–92. [CrossRef]
31. Jonasson, F. *Geoteknisk Provtagning av Anrikningssand Damm e-F Och G-H Aitik. PM Geoteknik Uppdragsnummer 2473649*; SWECO VBB: Luleå, Sweden, 2007. (In Swedish)
32. Jonasson, F. *PM Förslag på Materialparametrar för "Övriga Material" vid Beräkning i Plaxis. Uppdragsnummer 2166133310*; SWECO VBB: Luleå, Sweden, 2008. (In Swedish)
33. Hassellund, L.; Knutsson, R.; Mattsson, H.; Knutsson, S. Numerical analysis of an upstream tailings dam. In Proceedings of the Nordic Geotechnical Meeting: Challanges in Nordic Geotechnics, Reykjavik, Iceland, 25–27 May 2016; pp. 727–736.

applied sciences

MDPI

Article

Development of Prediction Models for Shear Strength of Rockfill Material Using Machine Learning Techniques

Mahmood Ahmad [1], Paweł Kamiński [2,*], Piotr Olczak [3], Muhammad Alam [4], Muhammad Junaid Iqbal [1], Feezan Ahmad [5], Sasui Sasui [6] and Beenish Jehan Khan [7]

1 Department of Civil Engineering, University of Engineering and Technology Peshawar (Bannu Campus), Bannu 28100, Pakistan; ahmadm@uetpeshawar.edu.pk (M.A.); engrjunaidiqbal@uetpeshawar.edu.pk (M.J.I.)
2 Faculty of Mining and Geoengineering, AGH University of Science and Technology, Mickiewicza 30 Av., 30-059 Kraków, Poland
3 Mineral and Energy Economy Research Institute, Polish Academy of Sciences, 7A Wybickiego St., 31-261 Cracow, Poland; olczak@min-pan.krakow.pl
4 Department of Civil Engineering, University of Engineering and Technology, Mardan 23200, Pakistan; emalam82@gmail.com
5 State Key Laboratory of Coastal and Offshore Engineering, Dalian University of Technology, Dalian 116024, China; ahmadf@mail.dlut.edu.cn
6 Department of Architectural Engineering, Chungnam National University, Daejeon 34134, Korea; sassuikhuwaja126@gmail.com
7 Department of Civil Engineering, CECOS University of IT and Emerging Sciences, Peshawar 25000, Pakistan; beenish@cecos.edu.pk
* Correspondence: pkamin@agh.edu.pl

Citation: Ahmad, M.; Kamiński, P.; Olczak, P.; Alam, M.; Iqbal, M.J.; Ahmad, F.; Sasui, S.; Khan, B.J. Development of Prediction Models for Shear Strength of Rockfill Material Using Machine Learning Techniques. *Appl. Sci.* **2021**, *11*, 6167. https://doi.org/10.3390/app11136167

Academic Editors: Paulo José da Venda Oliveira and António Alberto Santos Correia

Received: 24 May 2021
Accepted: 1 July 2021
Published: 2 July 2021

Publisher's Note: MDPI stays neutral with regard to jurisdictional claims in published maps and institutional affiliations.

Copyright: © 2021 by the authors. Licensee MDPI, Basel, Switzerland. This article is an open access article distributed under the terms and conditions of the Creative Commons Attribution (CC BY) license (https://creativecommons.org/licenses/by/4.0/).

Abstract: Supervised machine learning and its algorithms are a developing trend in the prediction of rockfill material (RFM) mechanical properties. This study investigates supervised learning algorithms—support vector machine (SVM), random forest (RF), AdaBoost, and k-nearest neighbor (KNN) for the prediction of the RFM shear strength. A total of 165 RFM case studies with 13 key material properties for rockfill characterization have been applied to construct and validate the models. The performance of the SVM, RF, AdaBoost, and KNN models are assessed using statistical parameters, including the coefficient of determination (R^2), Nash–Sutcliffe efficiency (NSE) coefficient, root mean square error (RMSE), and ratio of the RMSE to the standard deviation of measured data (RSR). The applications for the abovementioned models for predicting the shear strength of RFM are compared and discussed. The analysis of the R^2 together with NSE, RMSE, and RSR for the RFM shear strength data set demonstrates that the SVM achieved a better prediction performance with (R^2 = 0.9655, NSE = 0.9639, RMSE = 0.1135, and RSR = 0.1899) succeeded by the RF model with (R^2 = 0.9545, NSE = 0.9542, RMSE = 0.1279, and RSR = 0.2140), the AdaBoost model with (R^2 = 0.9390, NSE = 0.9388, RMSE = 0.1478, and RSR = 0.2474), and the KNN with (R^2 = 0.6233, NSE = 0.6180, RMSE = 0.3693, and RSR = 0.6181). Furthermore, the sensitivity analysis result shows that normal stress was the key parameter affecting the shear strength of RFM.

Keywords: AdaBoost; support vector machine; k-nearest neighbor; random forest; rockfill materials; shear strength

1. Introduction

Rockfill materials (RFMs) are commonly used in civil engineering projects such as rockfill dams, slopes, and embankments as construction materials for filling. This material is either obtained from a river's alluvial deposits or by blasting available rock [1,2]. RFMs are widely being used in the construction of rockfill dams to trap the river water because of their inherent flexibility, capacity to absorb large seismic energy, and adaptability to various foundation conditions. The behavior of RFMs used in rockfill dams is important for the safe and cost-effective construction of these structures. Generally, rockfill behaves like a Mohr/Coulomb material, albeit without cohesion and with relatively high internal friction angles. Crushed rockfill, loosely layered, can behave like coarse sand. The shear strength of both types of RFM is affected by many factors such as mineral composition, surface structure, particle size, shape, relative density, individual particle strength, etc. [3–5]. Because of the variable jointing, angularity/roundness, and rock particle size distribution, the RFM can be considered the most complex material [6]. In order to know the mechanical properties of RFMs, extensive field and laboratory research is essential for understanding RFM behavior and determining shear strength parameters in order to design safe and cost-effective structures. In situ direct shear system was used to monitor the shear strength of RFM, as well as the variation in the shear strength of rockfill along with the fill lift [7]. Linero [8] carried out some large-scale shear resistance experiments to simulate the material's original grain size distribution and the expected load level. RFM with a large particle size (maximum particle size of 1200 mm) is incompatible in laboratory testing [9]. Owing to restricting the effects of large particle sizes on test apparatus, such behavior makes it much more difficult to design representative/realistic large-scale strength tests. Furthermore, determining the shear strength of RFM directly is considered a costly and difficult process. Large-scale shear tests are often time-consuming and complex, and estimating the nonlinear shear strength function without using an analytical method is difficult. As a result, several researchers have attempted to determine the mechanical properties of RFM using indirect methods based on machine learning (ML) techniques.

In recent years, several researchers used ML algorithms and achieved efficient successes in different civil engineering and other sectors such as environmental [10], geotechnical [11–18], and other fields of science [19–28]. Numerous researchers have documented the behavior of the RFM. Marsal [3], Mirachi et al. [4], Venkatachalam [5], Gupta [29], Abbas [30], and Honkanadavar and Sharma [31] carried out laboratory experiments on different rockfill materials and concluded that the behavior of stress-strain is nonlinear, inelastic and based on the level of stress. They also noted that with an increase in maximum particle size for riverbed rockfill material, the angle of internal friction increases, and a reverse pattern for quarried rockfill material is observed. Frossard et al. [32] proposed a rational approach for assessing rockfill shear strength on the basis of size effects; Honkanadavar and Gupta [9] developed power law to relate the shear strength parameter to some index properties of riverbed RFM. Describing the mechanical behavior of rockfill materials and challenges in large-scale strength tests have incited several approaches in modeling the respective behavior of such soils. In this context, the artificial neural network (ANN) approach used by Kaunda [33] needs fewer rockfill parameters and was found to be more efficient in predicting RFM shear strength. Zhou et al. [34] have recently used cubist and random forest regression algorithms and have found that both can deliver better predictive RFM shear strength results than ANN and conventional regression models. This field, however, continues to be further explored. Considering that large-scale strength tests to characterize the shear strength are challenging, ML algorithms based on support vector machine (SVM), random forest (RF), AdaBoost, and k-nearest neighbor (KNN) models are proposed. Furthermore, the ML algorithms SVM, RF, AdaBoost, and KNN have demonstrated excellent prediction efficiency in a variety of fields [35–39] because of their generalization capability. The application in civil engineering field more significantly in prediction of RFM shear strength is limited based on literature surveys.

The main intention of the present study is to explore the capability of using SVM, RF, AdaBoost, and KNN algorithms to establish a more precise and parsimonious behavioral model for predicting the RFM shear strength. A critical review of existing literature suggests that despite the successful implementation of these techniques in various domains, their implementation in the prediction of RFM shear strength is scarcely explored. One of the primary significances of this study is that the data division in the training and testing data sets has been made with due regard to statistical aspects such as maximum, minimum, mean, and standard deviation. The splitting of the data sets is made to determine the predictive capability and generalization performance of established models and later helps to better evaluate them. Additionally, sensitivity analysis is carried out to find the main parameter influencing RFM shear strength. Concisely, the present study investigated and expanded the scope of machine learning algorithms for the development of the RFM shear strength model, which will provide theoretical support for researchers to establish a basis in selecting optimal machine learning algorithms in improving the predictive performance of RFM shear strength.

The rest of this article is structured as follows: The next section introduces the description of the used database and preliminaries of the algorithms used in the proposed approach and discusses the model evaluation metrics. Development of SVM, RF, AdaBoost, and KNN models are described in Section 3. Section 4 is dedicated to the performances and comparison of proposed models. Finally, Section 5 draws conclusions and outlines promising directions for future work.

2. Materials and Methods

2.1. Data Set

In this study, 165 samples of rockfill material (RFM) shear strength case history acquired by Kaunda [33] presented in Table A1 in the Appendix A were used to develop and evaluate the effectiveness of the proposed models. The RFM shear strength case history data are summarized in Table 1, where D_{10}, D_{30}, D_{60}, and D_{90} correspond to the 10%, 30%, 60%, and 90% passing sieve sizes, C_c and C_u refer to coefficients of uniformity and curvature (C_c), respectively, *FM* and *GM* describe fineness modulus and gradation modulus, respectively, *R* represents ISRM hardness rating, UCS_{min} and UCS_{max} (MPa) indicate the minimum and maximum uniaxial compression strengths (MPa), γ is the dry unit weight (kN/m^3), σ_n is normal stress (MPa), and τ is the shear strength of RFM (MPa) as the output variable. In this study, the output parameter selected to determine shear strength was the shear stress value at the failure of test samples and was the single output variable. The database was divided into two different sets, consisting of 80 percent (132 cases) and 20 percent (33 cases) of data, respectively, represented as training and testing sets. The testing set was accustomed to determine when training should be stopped in order to avoid overfitting. In order to achieve a consistent data splitting, different combinations of training and testing sets were experienced. The abovementioned selection was in such a way that the maximum (Max), minimum (Min), mean, and standard deviation of the parameters were consistent in the training and testing data sets (Table 2).

Table 1. Rockfill materials shear strength case history data.

Case No.	Location	D_{10}/mm	D_{30}/mm	D_{60}/mm	D_{90}/mm	Cc	Cu	GM	FM	R	UCS_{min}/MPa	UCS_{max}/MPa	γ/KNm^{-3}	σ_n/MPa	τ/MPa
1	Canada	0.02	0.94	4	18	11.05	200	4.78	4.19	1	1	5	15.4	0.022	0.013
2	Canada	0.02	0.94	4	18	11.05	200	4.78	4.19	1	1	5	15.4	0.044	0.025
3	Canada	0.02	0.94	4	18	11.05	200	4.78	4.19	1	1	5	15.4	0.088	0.049
...	...	...	...	...	...	...	...	...	...	...	...	...	...	...	...
163	Netherlands	11	15	23	32	0.89	2.09	1.48	7.48	5	100	250	16.8	0.028	0.048
164	Netherlands	11	15	23	32	0.89	2.09	1.48	7.48	5	100	250	16.8	0.055	0.082
165	Netherlands	11	15	23	32	0.89	2.09	1.48	7.48	5	100	250	16.8	0.108	0.143

Table 2. Statistical parameters of the training and testing data sets.

Parameter	Data Set	Min Value	Max Value	Mean	Standard Deviation
D_{10} (mm)	Training	0.010	33.900	4.857	9.179
	Testing	0.010	33.900	2.887	7.453
D_{30} (mm)	Training	0.560	42.400	8.465	10.577
	Testing	0.560	42.400	5.442	9.050
D_{60} (mm)	Training	1.200	80.100	19.287	15.135
	Testing	1.200	50.000	14.252	10.349
D_{90} (mm)	Training	2.600	100.000	40.386	22.018
	Testing	2.600	99.000	38.091	24.289
C_C	Training	0.100	22.270	2.199	3.075
	Testing	0.100	22.270	3.226	4.492
C_U	Training	1.360	1040.000	53.324	156.064
	Testing	1.470	1040.000	134.510	294.958
GM	Training	0.200	6.000	2.788	1.243
	Testing	0.200	6.000	3.365	1.331
FM	Training	3.000	8.800	6.250	1.261
	Testing	3.000	8.800	5.709	1.374
R	Training	1.000	6.000	4.364	0.910
	Testing	1.000	5.000	4.182	1.131
UCS_{min} (MPa)	Training	1.000	250.000	75.045	39.230
	Testing	1.000	100.000	68.273	32.444
UCS_{max} (MPa)	Training	5.000	400.000	170.682	88.010
	Testing	5.000	250.000	159.545	87.957
γ (KN/m^3)	Training	9.320	38.900	20.766	4.605
	Testing	9.320	38.900	20.932	5.854
σ_n (MPa)	Training	0.002	4.205	0.729	0.780
	Testing	0.021	3.223	0.756	0.816
τ (MPa)	Training	0.005	3.921	0.660	0.662
	Testing	0.024	2.492	0.668	0.619

2.2. Support Vector Machine

Boser, Guyon, and Vapnik were the first to formulate and introduce the support vector machine (SVM) [40]. In the case of non-separable data, to accommodate errors for certain objects i, the "ideal boundary" must be introduced:

$$\begin{cases} \text{minimize}\left(\frac{1}{2}|\delta|^2 + C\sum_{i=1}^{n}\xi_i\right) \\ \text{under the constraints } y_i(b+\delta\cdot x_i)+\xi_i \geq 1 \text{ and } \xi_i \geq 0 \end{cases} \tag{1}$$

where C is the penetrating parameter; δ and b are, respectively, the normal vector and the bias of the hyperplane; and each ξ_i refers to the distance within object i and the respective margin hyperplane [41,42].

Data are implicitly mapped to a higher-dimensional space through mercer kernels, which can be broken down into a dot product to learn nonlinearly separable functions

$K(x_i, x_j) = \varphi(x_i)\cdot\varphi(x_j)$ [42]. The kernel of the radial basis function (RBF) that is used widely is listed below:

$$K(x_i, x_j) = \exp\left(-\sigma\|x_i - x_j\|^2\right) \tag{2}$$

where σ is the kernel parameter.

2.3. Random Forest

The use of a large series of low-dimensional regression trees is the basis of the random forest (RF). The theoretical development of RF is described by Breiman [43]. RF is an example of ensemble learning, which requires the development of a large number of decision trees to be implemented. In general, there are two types of decision trees: regression trees and classification trees. Regression trees were designed in the RF model since the main goal of this analysis was to predict the shear strength of RFM. Figure 1 depicts a general architecture for RF analysis. The protocol for analysis can be divided into two stages:

Stage 1: To create a sequence of sub-data sets, the bootstrap statistical technique is used to randomly sample from the initial data set (training data). The forest is then built using regression trees based on these sub-data sets. Each tree is trained by choosing a set of variables at random (a fixed number of descriptive variables selected from the random subset). Two important parameters that can be adjusted during the training stage are the number of trees (ntree) and the number of variables (mtry).

Stage 2: Once the model has been trained, a prediction can be made. In an ensemble approach, input variables are evaluated for all regression trees first, and then the final output is calculated by measuring the average value of each individual tree's prediction.

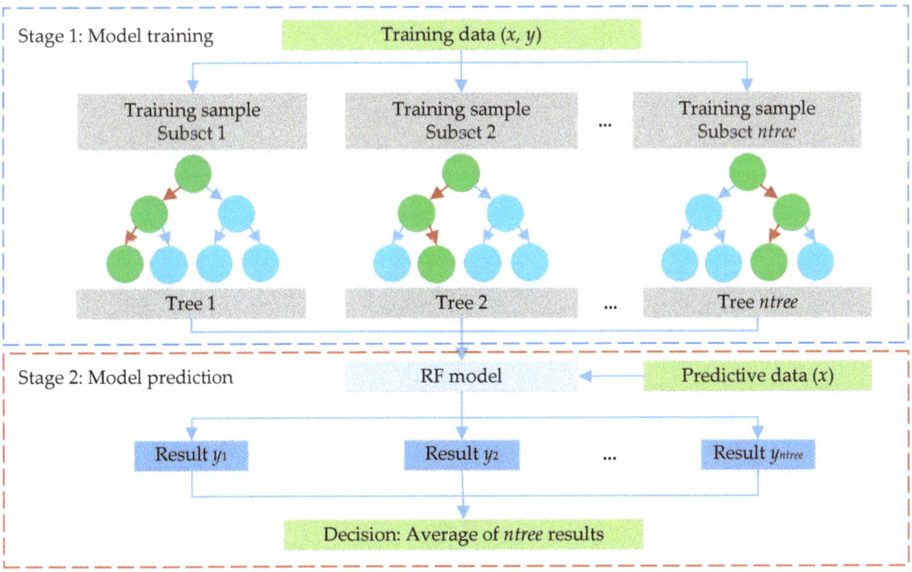

Figure 1. Schematic representation of RF analysis.

2.4. AdaBoost Algorithm

The sequential ensemble technique AdaBoost, or adaptive boosting, is based on the concept of developing many poor learners using different training sub-sets drawn at random from the original training data set. Weights are allocated during each training session, and these are used to learn each hypothesis. The weights are used to calculate the hypothesis error on the data set and are a measure of the relative importance of each instance. After each iteration, the weights are recalculated so that instances classified wrongly by the previous hypothesis obtain higher weights. This allows the algorithm to concentrate on instances that are more difficult to understand. The algorithm's most important task is to assign updated weights to instances that were wrongly labeled. In regression, the instances represent a real-value error. The AdaBoost technique can be used to mark the calculated error as an error or not an error by comparing it to a predefined threshold prediction error. Instances that have made a greater mistake on previous learners are more likely (i.e., have a higher probability) to be chosen for training the next base learner. Finally, an ensemble estimate of the individual base learner predictions is made using a weighted average or median [44].

2.5. k-Nearest Neighbor

The supervised ML algorithm k-nearest neighbor (KNN) can be used to solve both classification and regression problems. It is, however, most commonly used in classification problems [45]. In regression problems, the input data set consists of k that is nearest to the training data sets deployed in the featured set. The output is dependent if KNN is deployed to function as a regression algorithm. For KNN regression, the ensuing result is the characteristic value for the object, which is the mean figure of k's nearest neighbors. To locate the k of a data point, a parameter such as Euclidean, Mahalanobis can be used as the distance metric [46,47].

2.6. Performance Metric

The coefficient of determination (R^2), Nash–Sutcliffe efficiency (NSE) coefficient, root mean square error (RMSE), and the ratio of the RMSE to the standard deviation of measured data (RSR) were taken into account to examine the predictive capacity of the models, as shown in Equations (3)–(6) [48–50]:

$$R^2 = \left[\frac{\sum_{i=1}^{n}(O_i - \overline{O})(P_i - \overline{P})}{\sqrt{\sum_{i=1}^{n}(O_i - \overline{O})^2}\sqrt{\sum_{i=1}^{n}(P_i - \overline{P})^2}}\right] \tag{3}$$

$$NSE = \frac{\sum_{i=1}^{n}(O_i - \overline{O})^2 - \sum_{i=1}^{n}(P_i - O_i)^2}{\sum_{i=1}^{n}(O_i - \overline{O})^2} \tag{4}$$

$$RMSE = \sqrt{\frac{1}{n}\sum_{i=1}^{n}(P_i - O_i)^2} \tag{5}$$

$$RSR = \frac{\sum_{i=1}^{n}(O_i - P_i)^2}{\sqrt{\sum_{i=1}^{n}(O_i - \overline{O})^2}} \tag{6}$$

where n is the number of observations under consideration, O_i is the ith observed value, $\overline{O}$ is the mean observed value, P_i is the ith model-predicted value, and P is the mean model-predicted value.

R-squared, also called the determination coefficient, describes the change in data as the degree of fit. The normal "determination coefficient" range is (0–1). The model is considered to be efficient if the R^2 value is greater than 0.8 and is close to 1 [51]. The NSE is a normalized statistic that controls the relative extent of the residual variance relative to the variance of the data measured [52]. The NSE varies between $-\infty$ and 1. When NSE = 1, it presents a flawless match among observed and predicted values. Model predictive output with a range of $0.75 < NSE \leq 1.00$, $0.65 < NSE \leq 0.75$, $0.50 < NSE \leq 0.65$, $0.40 < NSE \leq 0.50$, or $NSE \leq 0.4$ is graded as very good, good, acceptable, or unacceptable, respectively [53,54]. The RMSE is the square root of the ratio of the square of the deviation between the observed value and the true value of the number of observations n. The RMSE has a value greater than or equal to 0, where 0 is a statistically perfect fit for the data observed [55–57]. The

RSR is interpreted as the ratio of the measured data's RMSE and standard deviation. The RSR varies between an optimal value of 0 and a large positive value. A lower RSR presents a lower RMSE, which indicates the model's greater predictive efficiency. RSR classification ranges are described as very good, good, acceptable, and unacceptable with ranges of $0.00 \leq RSR \leq 0.50$, $0.50 \leq RSR \leq 0.60$, $0.60 \leq RSR \leq 0.70$, and $RSR > 0.70$, respectively [53].

3. Model Development to Predict RFM Shear Strength

The models for RFM shear strength prediction were developed using Orange software, which is a popular open-source environment for statistical computing and data visualization. All data processing is carried out using Orange software (version 3.13). The most prevalent supervised learning classification algorithms are given by Orange. In the package documentation manuals, one can find more information about input parameters, implementation, and references.

The structure of the model was based on an input matrix identified by predictor variables, $x = \{D_{10}, D_{30}, D_{60}, D_{90}, C_c, C_u, GM, FM, R, UCS_{\min}, UCS_{max}, \gamma, \text{and } \sigma_n\}$ and output, also called target variable (y), was the RFM shear strength. In every modeling process, achieving a consistent data division and the appropriate size of the training and testing data sets is the most important task. The statistical features, such as the minimum, maximum, mean, and standard deviation of the data sets, have therefore been taken into account in the splitting process. The statistical accuracy of the training and testing data sets optimizes the performance of the models and ultimately helps to evaluate them better. On the remaining data set, the proposed models were tested. In other words, to build and test the models, 132 and 33 data sets were used, respectively. To fairly assess the predictive performance of the models, the data set used for the testing of all models was kept the same.

In order to optimize the RFM shear strength prediction, all the models (AdaBoost, RF, SVM, and KNN) were tuned based on the trial and error process. Initially, the values were chosen for model tuning parameters and gradually varied in experiments until the best fitness measurements were achieved. Figure 2 shows the schematic diagram of the proposed methodology. The optimization method aims to find the best parameters for AdaBoost, RF, SVM, and KNN in order to achieve the best prediction accuracy. Some critical hyperparameters in the AdaBoost, RF, SVM, and KNN algorithms are tuned in this study, as shown in Table 3. The definitions of these hyperparameters are also clarified in Table 3. The values for the tuning parameters of the models were first chosen and then varied in the trials until the best fitness measures mentioned in Table 3 were achieved.

Table 3. Hyperparameter optimization results.

Algorithm	Hyperparameter	Explanation	Optimal Value
AdaBoost	Number of estimators	Number of trees	2
	Learning rate	It establishes the degree to which newly acquired information can override previously acquired information	0.1
	Boosting algorithm	Updates the weight of the base estimator with probability estimates or classification results (SAMME.R/SAMME)	SAMME
	Regression loss function	Linear/square/exponential	Linear
RF	Number of trees	Number of trees in the forest	15
	Limit depth of individual trees	The depth to which the trees will be grown	03
SVM	Cost (C)	Penalty term for loss and applies for classification and regression tasks	8
	Regression loss epsilon (ε)	The distance between true and predicted values within which no penalty is applied	0.1
	Kernal type	Kernel is a function that transforms attribute space to a new feature space to fit the maximum-margin hyperplane, thus allowing the algorithm to construct the model with linear, polynomial, RBF, and Sigmoid kernels	RBF

Table 3. *Cont.*

Algorithm	Hyperparameter	Explanation	Optimal Value
KNN	Number of neighbors	Number of nearest neighbors	5
	Metric	Distance parameter—Euclidean/Manhattan/Chebyshev/Mahalanobis	Euclidean
	Weight	Uniform—all points in each neighborhood are weighted equally/distance—closer neighbors of a query point have a greater influence than the neighbors further away	Uniform

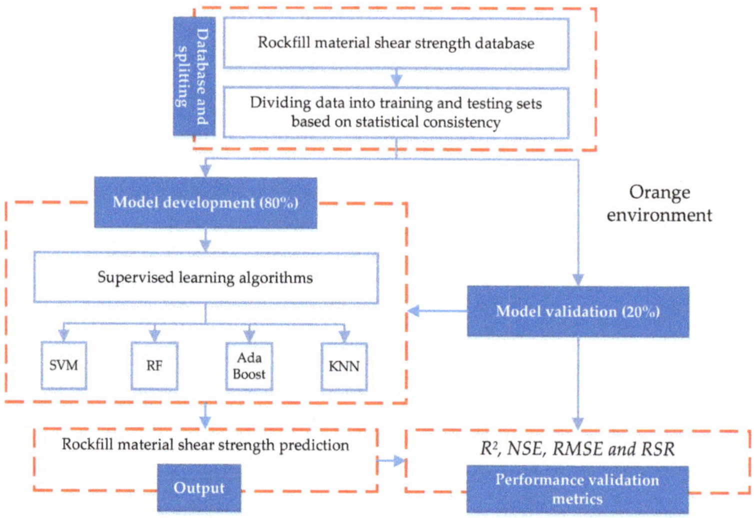

Figure 2. The flowchart of the methodology.

4. Results and Discussion

In this study, R^2, NSE coefficient, RMSE, and RMSE to standard deviation of measured data are chosen as the criterion for defining the model's output. The database is split into a training data set and a testing data set to evaluate the performance of the presented models. To make a fair comparison, all the models are developed by applying them to the same RFM shear strength training and testing data sets. Figure 3, displays the scatter plot of the actual and the predicted RFM shear strength for the training phase. The analysis of the R^2 together with NSE, RMSE, and RSR for the RFM shear strength data set demonstrates that the SVM achieved a better prediction performance with (R^2 = 0.9655, NSE = 0.9639, RMSE = 0.1135, and RSR = 0.1899) succeeded by the RF model with (R^2 = 0.9545, NSE = 0.9542, RMSE = 0.1279, and RSR = 0.2140), the AdaBoost model with (R^2 = 0.9390, NSE = 0.9388, RMSE = 0.1478, and RSR = 0.2474), and the KNN with (R^2 = 0.6233, NSE = 0.6180, RMSE = 0.3693, and RSR = 0.6181).

Figure 4, presenting the predicted RFM shear strength, is plotted with the actual RFM shear strength data. According to the test data set, all models demonstrated very good predictive potential (R^2 > 0.8) with the exception of KNN, which displayed slightly worse results (i.e., R^2 = 0.6304) for the test data set. The result of R^2 demonstrated that all SVM, RF, and AdaBoost models except KNN are appropriate, but the SVM model performed better because it had the highest R^2 value (0.9656), and after that, the RF (0.9181) and AdaBoost (0.8951) models. In comparison to the other models, the KNN model presented the worst estimates with maximum dispersion (Figure 4).

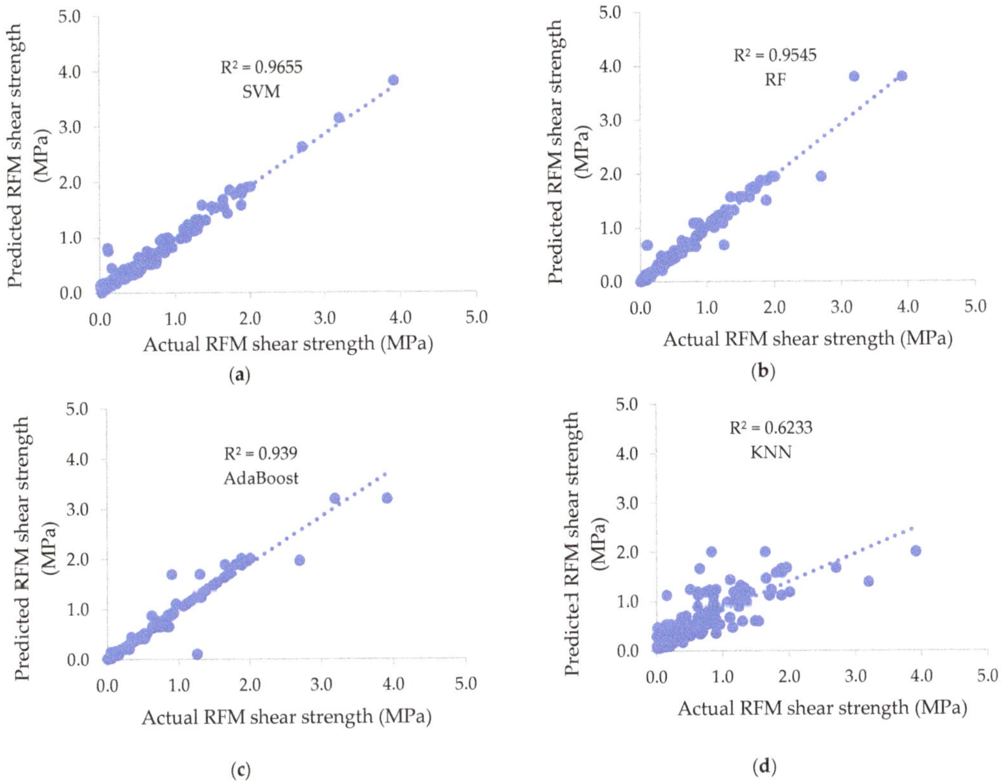

Figure 3. Scatter plots of actual vs. predicted RFM shear strength in training stage: (**a**) SVM, (**b**) RF, (**c**) AdaBoost, and (**d**) KNN.

In addition, the NSE measure was ranked from highest to lowest predictive strength, following the way: SVM (0.9654) > RF (0.9164) > AdaBoost (0.8835) > KNN (0.6076), which is similar to R^2. With regard to RMSE score, the SVM model also had the maximum predictive ability by having the lowest RMSE (0.0153), succeeded by the models RF (0.0797), AdaBoost (0.0941), and KNN (0.1727).

Finally, the reliability of all applied models was divided into four groups based on RSR values: unsatisfactory, satisfactory, good, and very good with ranges of RSR > 0.70, $0.60 \leq RSR \leq 0.70$, $0.50 \leq RSR \leq 0.60$, and $0.00 \leq RSR \leq 0.50$, respectively. The RSR value therefore demonstrates very good results throughout all our established models except the KNN model, whose performance is considered to be satisfactory. Figure 5 depicts the bar graphs comparing the R^2, NSE, RMSE, and RSR for the training and testing data sets of all the models. The R^2 defines the degree of co-linearity between our predicted and actual data. The value of RMSE is more focused on large errors than on small errors. A lower RSR indicates a lower RMSE, indicating the model's better predictive efficiency. The SVM model has high R^2 and NSE while lower RMSE and RSR values, revealing that the SVM model is preferable for predicting the RFM shear strength for the testing data. The SVM achieved a better prediction performance with (R^2 = 0.9655, RMSE = 0.0513 and mean absolute error (MAE) = 0.0184) in comparison to the cubist method (R^2 = 0.9645, RMSE = 0.0975, and MAE = 0.0644) and ANN method (R^2 = 0.9386, RMSE = 0.1320 and MAE = 0.0841) reported by Zhou et al. [34] and Kaunda [33], respectively, for the test data. Additionally, the accuracy of modeling determined by the linear regression method reported by Andjelkovic et al. [58]

between measured and calculated values of shear strength (R^2 = 0.836) was slightly lower than the proposed SVM model. In general, the generalization and reliability of the SVM algorithm perform well, and larger data sets can yield better prediction results.

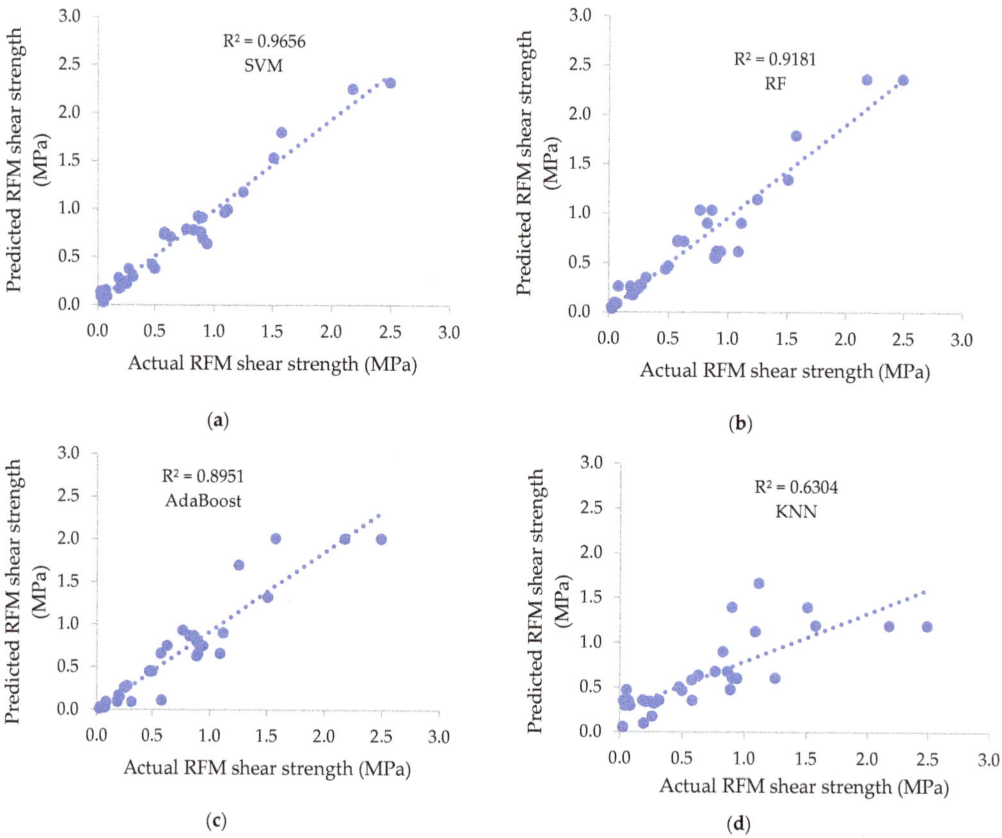

Figure 4. Scatter plots of actual vs. predicted RFM shear strength in testing stage: (**a**) SVM, (**b**) RF, (**c**) AdaBoost, and (**d**) KNN.

In the present research, a sensitivity analysis was also conducted using Yang and Zang's [59] method to evaluate the influence of input parameters on RFM shear strength. This approach has been used in several studies [60–63] and is formulated as:

$$r_{ij} = \frac{\sum_{m=1}^{n}(y_{im} \times y_{om})}{\sqrt{\sum_{m=1}^{n} y_{im}^{2} \sum_{m=1}^{n} y_{om}^{2}}} \tag{7}$$

where n is the number of data values (this study used 132 data values) and y_{im} and y_{om} are the input and output parameters. The r_{ij} value ranged from zero to one for each input parameter, and the highest r_{ij} values suggested the most efficient output parameter (which was RFM shear strength in this study). The r_{ij} values for all input parameters are presented in Figure 6. It can be seen from Figure 6 that the σ_n with r_{ij} is 0.990. Similar research of sensitivity analyses on RFM shear strength was also implemented by Kaunda [33] and Zhou et al. [34]. The findings demonstrated that normal stress is the most sensitive factor, which shows agreement with the present mentioned results.

Despite the fact that the proposed model produces desirable prediction results, certain limitations should be addressed in the future.

(1) Similar to other machine learning methods, the major disadvantages of SVM, RF, AdaBoost, and KNN models are sensitive to the fitness of the data set. Generally, if the data set is small, the generalization and reliability of the model would be influenced. However, the SVM, RF, and AdaBoost algorithms work with a limited data set, i.e., 165 cases, except for KNN. The prediction performances could be better on a larger data set. Furthermore, the developed models can always be updated to yield better results as new data becomes available.

(2) Other qualitative indicators such as the Los Angeles abrasion value and lithology may also have influences on the prediction results of the shear strength of RFM. Accordingly, it is significant to analyze the influence of these indicators on the prediction results for improving performance.

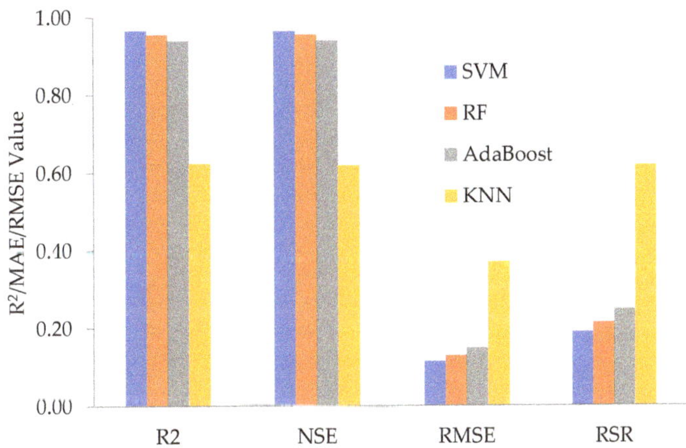

(**a**)

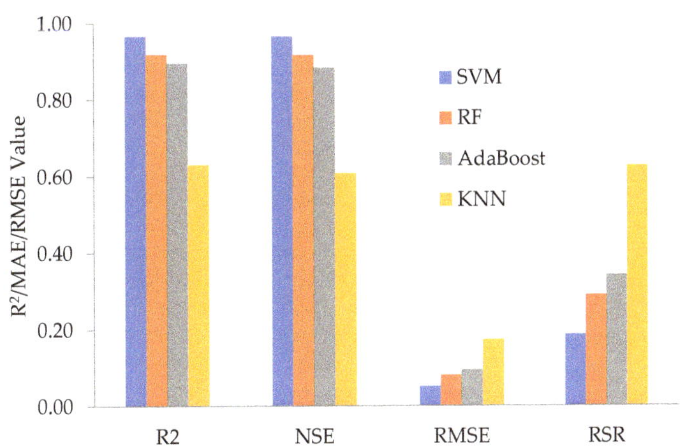

(**b**)

Figure 5. Comparison of R^2, NSE, RMSE, and RSR values from the SVM, RF, AdaBoost, and KNN models in (**a**) training phase and (**b**) testing phase.

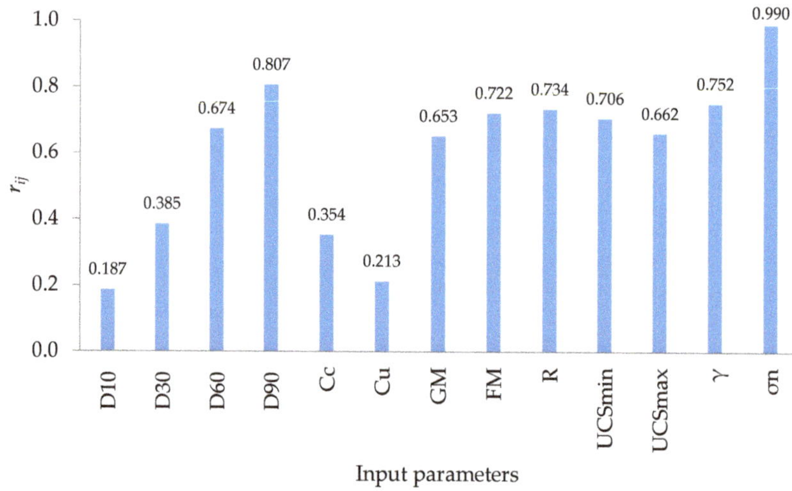

Figure 6. Sensitivity analysis results.

5. Conclusions

This study employed and examined the SVM, RF, AdaBoost, and KNN algorithms in the RFM shear strength prediction problem. To construct and validate a new model on the basis of the aforementioned algorithms, a comprehensive database containing 165 RFM case studies was collected from the available literature. Thirteen different predictive variables for rockfill characterization were selected as the input variables: D_{10} (mm), D_{30} (mm), D_{60} (mm), D_{90} (mm), C_c, C_u, GM, FM, R, UCS_{min} (MPa), γ (kN/m^3), UCS_{max} (MPa), and σ_n (MPa). The predictive performance of the proposed models is verified and compared. The conclusions can be outlined as follows:

1. In this study, the SVM model (R^2 = 0.9656, NSE = 0.9654, RMSE = 0.0153, and RSR = 0.1861) successfully achieved a high level of modeling prediction efficiency to RF (R^2 = 0.9181, NSE = 0.9164, RMSE = 0.0797, and RSR = 0.2891), AdaBoost (R^2 = 0.8951, NSE = 0.8835, RMSE = 0.0941, and RSR = 0.3414), and KNN (R^2 = 0.6304, NSE = 0.6076, RMSE = 0.1727, and RSR = 0.6264) in the test data set. As the same methodology (having the same training and test data sets) for structuring all models is taken into consideration, the SVM model resulted the best and highest performance in this aspect. This implies that this algorithm is robust in comparison with others in RFM shear strength prediction.
2. The performance (in terms of R^2) of the test data set for the SVM, RF, and AdaBoost algorithms studied falls in the range of 0.9656–0.8951 across the three models with 13 input valuables. Results conclude that it is rational and feasible to estimate the shear strength of RFM from the gradation, particle size, dry unit weight (γ), material hardness, FM, and normal stress (σ_n).
3. Sensitivity analysis results revealed that normal stress (σ_n) was the key parameter affecting the shear strength of RFM.

The findings show that the SVM model is a useful and accurate artificial intelligence technique for predicting RFM shear strength and can be used in various fields. Further, the generalization of the proposed approach for achieving improved performance results, more experimental data should be collected in future research. Finally, RFM shear strength prediction using advanced machine learning algorithms (i.e., deep learning) is left as a future research topic.

Author Contributions: Conceptualization, M.A. (Mahmood Ahmad); P.K. and P.O.; methodology, M.A. (Mahmood Ahmad), P.K.; software, M.A. (Mahmood Ahmad) and F.A.; validation, P.O., F.A., M.A. (Mahmood Ahmad), M.A. (Muhammad Alam) and F.A.; formal analysis, M.A. (Mahmood Ahmad); investigation, M.A. (Mahmood Ahmad), F.A., P.K., M.J.I. and S.S.; resources, P.K. and P.O.; data curation, M.A. (Mahmood Ahmad), M.A. (Muhammad Alam) and M.J.I.; writing—original draft preparation, M.A. (Mahmood Ahmad); writing—review and editing, M.A. (Mahmood Ahmad), B.J.K., F.A. and S.S.; supervision, P.K. and P.O.; project administration, P.K.; funding acquisition, P.K. and P.O. All authors have read and agreed to the published version of the manuscript.

Funding: This research received no external funding.

Institutional Review Board Statement: Not applicable.

Informed Consent Statement: Not applicable.

Data Availability Statement: The data used to support the findings of this study are included within the article.

Conflicts of Interest: The authors declare no conflict of interest.

Notation

ANN	Artificial neural network
AdaBoost	Adaptive boosting
KNN	k-nearest neighbor
NSE	Nash–Sutcliffe efficiency coefficient
R^2	Coefficient of determination
RF	Random forest
RFM	Rockfill material
RMSE	Root mean square error
RSR	Ratio of RMSE to the standard deviation of the measured data
ISRM	International Society of Rock Mechanics
SVM	Support vector machine
D_{10}	Sieve size at 10 percent passing
D_{30}	Sieve size at 30 percent passing
D_{60}	Sieve size at 60 percent passing
D_{90}	Sieve size at 90 percent passing
C_c	Coefficient of curvature
C_u	Coefficient of uniformity
GM	Gradation modulus
FM	Fineness modulus
R	ISRM hardness rating
UCS_{min}	Minimum uniaxial compression strength
γ	Dry unit weight
UCS_{max}	Maximum uniaxial compression strength
σ_n	Normal stress
τ	Shear strength
φ	Angle of internal friction

Appendix A

Table A1. Rockfill shear strength database.

Case No.	Location	D_{10} (mm)	D_{30} (mm)	D_{60} (mm)	D_{90} (mm)	C_c	C_u	GM	FM	R	UCS_{min} (MPa)	UCS_{max} (MPa)	γ (KN/m^3)	σ_n (MPa)	τ (MPa)
1	Canada	0.02	0.94	4	18	11.05	200	4.78	4.19	1	1	5	15.4	0.022	0.013
2	Canada	0.02	0.94	4	18	11.05	200	4.78	4.19	1	1	5	15.4	0.044	0.025
3	Canada	0.02	0.94	4	18	11.05	200	4.78	4.19	1	1	5	15.4	0.088	0.049
4	Canada	0.03	2.1	6.6	18	22.27	220	4.22	4.73	1	1	5	38.9	0.022	0.013
5	Canada	0.03	2.1	6.6	18	22.27	220	4.22	4.73	1	1	5	38.9	0.044	0.024
6	Canada	0.03	2.1	6.6	18	22.27	220	4.22	4.73	1	1	5	38.9	0.088	0.048
7	Canada	0.09	0.92	3.2	10	2.94	35.56	5	3.94	1	1	5	37	0.022	0.014
8	Canada	0.09	0.92	3.2	10	2.94	35.56	5	3.94	1	1	5	37	0.044	0.027
9	Canada	0.09	0.92	3.2	10	2.94	35.56	5	3.94	1	1	5	37	0.088	0.053
10	U.K.	1	6	19	29	1.89	19	2.61	6.36	5	100	250	19.62	0.059	0.163
11	U.K.	1	6	19	29	1.89	19	2.61	6.36	5	100	250	19.62	0.098	0.218
12	U.K.	1	6	19	29	1.89	19	2.61	6.36	5	100	250	19.62	0.198	0.367
13	U.K.	1	6	19	29	1.89	19	2.61	6.36	5	100	250	19.62	0.299	0.513
14	U.K.	0.3	3.2	16	30	2.13	53.33	3.37	5.74	5	100	250	18.0504	0.058	0.15
15	U.K.	0.3	3.2	16	30	2.13	53.33	3.37	5.74	5	100	250	18.0504	0.097	0.204
16	U.K.	0.3	3.2	16	30	2.13	53.33	3.37	5.74	5	100	250	18.0504	0.195	0.33
17	U.K.	0.3	3.2	16	30	2.13	53.33	3.37	5.74	5	100	250	18.0504	0.297	0.456
18	U.K.	1	6	19	29	1.89	19	2.65	6.36	5	100	250	19.62	0.179	0.262
19	U.K.	1	6	19	29	1.89	19	2.65	6.36	5	100	250	19.62	0.538	0.697
20	U.K.	1	6	19	29	1.89	19	2.65	6.36	5	100	250	19.62	0.887	0.112
21	U.K.	0.3	3.2	16	30	2.13	53.33	3.37	5.74	5	100	250	18.0504	0.177	0.245
22	U.K.	0.3	3.2	16	30	2.13	53.33	3.37	5.74	5	100	250	18.0504	0.529	0.666
23	U.K.	0.3	3.2	16	30	2.13	53.33	3.37	5.74	5	100	250	18.0504	0.876	0.102
24	Iran	0.1	1.2	7.5	17.3	1.92	75	4.32	7.42	4	50	100	9.3195	0.101	0.16
25	Iran	0.1	1.2	7.5	17.3	1.92	75	4.32	7.42	4	50	100	9.3195	0.301	0.34
26	Iran	0.1	1.2	7.5	17.3	1.92	75	4.32	7.42	4	50	100	9.3195	0.503	0.5
27	Iran	0.4	2.8	11	30	1.78	27.5	3.38	5.6	4	50	100	9.3195	0.172	0.207
28	Iran	0.4	2.8	11	30	1.78	27.5	3.38	5.6	4	50	100	9.3195	0.497	0.476
29	Iran	0.4	2.8	11	30	1.78	27.5	3.38	5.6	4	50	100	9.3195	0.83	0.751
30	Japan	1.3	4.6	15	32	1.09	11.54	2.68	6.28	5	100	250	17.81496	0.094	0.136
31	Japan	1.3	4.6	15	32	1.09	11.54	2.68	6.28	5	100	250	17.81496	0.177	0.242

Table A1. *Cont.*

Case No.	Location	D_{10} (mm)	D_{30} (mm)	D_{60} (mm)	D_{90} (mm)	C_c	C_u	GM	FM	R	UCS_{min} (MPa)	UCS_{max} (MPa)	γ (KN/m^3)	σ_n (MPa)	τ (MPa)
32	Japan	1.3	4.6	15	32	1.09	11.54	2.68	6.28	5	100	250	17.81496	0.351	0.415
33	Japan	1.3	4.6	15	32	1.09	11.54	2.68	6.28	5	100	250	17.81496	0.512	0.552
34	Japan	0.6	2	16	30	0.42	26.67	3.24	5.86	4	50	100	21.8763	0.093	0.165
35	Japan	0.6	2	16	30	0.42	26.67	3.24	5.86	4	50	100	21.8763	0.182	0.308
36	Japan	0.6	2	16	30	0.42	26.67	3.24	5.86	4	50	100	21.8763	0.359	0.523
37	Japan	0.6	2	16	30	0.42	26.67	3.24	5.86	4	50	100	21.8763	0.535	0.744
38	Iran	0.4	2.9	9.7	31	2.17	24.25	3.41	5.57	4	50	100	21	0.177	0.214
39	Iran	0.4	2.9	9.7	31	2.17	24.25	3.41	5.57	4	50	100	21	0.514	0.525
40	Iran	0.4	2.9	9.7	31	2.17	24.25	3.41	5.57	4	50	100	21	0.839	0.773
41	Iran	0.4	2.9	9.7	31	2.17	24.25	3.41	5.57	4	50	100	21	1.172	1.07
42	Iran	0.4	2.9	9.7	31	2.17	24.25	3.41	5.57	4	50	100	21	1.494	1.312
43	Iran	0.4	2.9	9.7	31	2.17	24.25	3.41	5.57	4	50	100	21	1.97	1.648
44	Iran	0.4	2.9	9.7	31	2.17	24.25	3.41	5.57	4	50	100	20.8	0.18	0.24
45	Iran	0.4	2.9	9.7	31	2.17	24.25	3.41	5.57	4	50	100	20.8	0.5	0.447
46	Iran	0.4	2.9	9.7	31	2.17	24.25	3.41	5.57	4	50	100	20.8	0.821	0.689
47	Iran	0.4	2.9	9.7	31	2.17	24.25	3.41	5.57	4	50	100	20.8	1.142	0.93
48	Iran	0.5	2.8	9.7	30	1.62	19.4	3.43	5.61	5	100	250	21.1	0.487	0.39
49	Iran	0.5	2.8	9.7	30	1.62	19.4	3.43	5.61	5	100	250	21.1	0.972	0.766
50	Iran	0.5	2.8	9.7	30	1.62	19.4	3.43	5.61	5	100	250	21.1	1.448	1.11
51	Iran	0.2	2.5	19.4	42.2	1.61	97	3.19	5.77	5	100	250	21	0.168	0.157
52	Iran	0.2	2.5	19.4	42.2	1.61	97	3.19	5.77	5	100	250	21	0.373	0.634
53	Iran	0.2	2.5	19.4	42.2	1.61	97	3.19	5.77	5	100	250	21	0.731	1.088
54	Iran	0.2	2.5	19.4	42.2	1.61	97	3.19	5.77	5	100	250	21	0.906	1.258
55	Iran	0.2	2.5	19.4	42.2	1.61	97	3.19	5.77	5	100	250	21	1.262	1.699
56	Iran	0.2	2.5	19.4	42.2	1.61	97	3.19	5.77	5	100	250	21	1.437	1.883
57	Iran	0.4	3.3	10.3	33.3	2.64	25.75	3.32	5.64	4	50	100	21.8	0.092	0.14
58	Iran	0.4	3.3	10.3	33.3	2.64	25.75	3.32	5.64	4	50	100	21.8	0.179	0.23
59	Iran	0.4	3.3	10.3	33.3	2.64	25.75	3.32	5.64	4	50	100	21.8	0.344	0.357
60	Iran	0.4	3.3	10.3	33.3	2.64	25.75	3.32	5.64	4	50	100	21.8	0.514	0.52
61	Iran	0.4	3.3	10.3	33.3	2.64	25.75	3.32	5.64	4	50	100	21.8	0.859	0.887
62	Iran	0.4	3.3	10.3	33.3	2.64	25.75	3.32	5.64	4	50	100	21.8	1.186	1.149
63	Iran	1.2	2.1	4.2	25.3	0.88	3.5	3.63	5.34	4	50	100	21.8	0.092	0.147
64	Iran	1.2	2.1	4.2	25.3	0.88	3.5	3.63	5.34	4	50	100	21.8	0.178	0.22

Table A1. *Cont.*

Case No.	Location	D_{10} (mm)	D_{30} (mm)	D_{60} (mm)	D_{90} (mm)	C_c	C_u	GM	FM	R	UCS_{min} (MPa)	UCS_{max} (MPa)	γ (KN/m^3)	σ_n (MPa)	τ (MPa)
65	Iran	1.2	2.1	4.2	25.3	0.88	3.5	3.63	5.34	4	50	100	21.8	0.503	0.461
66	Iran	1.2	2.1	4.2	25.3	0.88	3.5	3.63	5.34	4	50	100	21.8	1.148	0.959
67	Iran	0.4	2.9	9.7	31	2.17	24.25	3.4	5.53	4	50	100	21	0.34	0.332
68	Iran	0.4	2.9	9.7	31	2.17	24.25	3.4	5.53	4	50	100	21	0.99	0.843
69	Iran	0.4	2.9	9.7	31	2.17	24.25	3.4	5.53	4	50	100	21	1.618	1.271
70	Iran	0.4	2.9	9.7	31	2.17	24.25	3.4	5.53	4	50	100	21	2.399	1.799
71	Iran	0.4	2.9	9.7	31	2.17	24.25	3.4	5.53	4	50	100	21.5	0.342	0.342
72	Iran	0.4	2.9	9.7	31	2.17	24.25	3.4	5.53	4	50	100	21.5	0.994	0.865
73	Iran	0.4	2.9	9.7	31	2.17	24.25	3.4	5.53	4	50	100	21.5	1.63	1.321
74	Iran	0.4	2.9	9.7	31	2.17	24.25	3.4	5.53	4	50	100	21.5	2.422	1.891
75	Australia	33.9	42.4	50	60.2	1.06	1.47	0.2	8.8	5	100	250	21.7	0.163	0.23
76	Australia	33.9	42.4	50	60.2	1.06	1.47	0.2	8.8	5	100	250	21.7	0.215	0.275
77	Australia	33.9	42.4	50	60.2	1.06	1.47	0.2	8.8	5	100	250	21.7	0.412	0.424
78	Australia	30	34	40.8	50	0.94	1.36	0.52	8.5	5	100	250	21.7	0.165	0.252
79	Australia	30	34	40.8	50	0.94	1.36	0.52	8.5	5	100	250	21.7	0.215	0.28
80	Australia	30	34	40.8	50	0.94	1.36	0.52	8.5	5	100	250	21.7	0.412	0.424
81	Germany	4	11.7	36.2	98.2	0.95	9.05	1.47	7.55	4	50	100	24.2	1.039	1.114
82	Germany	4	11.7	36.2	98.2	0.95	9.05	1.47	7.55	4	50	100	24.2	2.034	1.964
83	Germany	4	11.7	36.2	98.2	0.95	9.05	1.47	7.55	4	50	100	24.2	3.004	2.705
84	Germany	3	9.1	30.4	98.2	0.91	10.13	1.67	7.28	4	50	100	24.2	0.533	0.658
85	Germany	3	9.1	30.4	98.2	0.91	10.13	1.67	7.28	4	50	100	24.2	1.039	1.114
86	Germany	3	9.1	30.4	98.2	0.91	10.13	1.67	7.28	4	50	100	24.2	2.018	1.882
87	Germany	4.2	12.8	41.2	99	0.95	9.81	1.37	7.62	4	50	100	24.2	0.512	0.512
88	Germany	4.2	12.8	41.2	99	0.95	9.81	1.37	7.62	4	50	100	24.2	1.001	0.902
89	Germany	4.2	12.8	41.2	99	0.95	9.81	1.37	7.62	4	50	100	24.2	1.987	1.728
90	USA	0.9	3	18.8	99	0.53	20.89	2.64	6.35	5	100	250	21.7	0.861	0.898
91	USA	0.9	3	18.8	99	0.53	20.89	2.64	6.35	5	100	250	21.7	1.67	1.509
92	USA	0.9	3	18.8	99	0.53	20.89	2.64	6.35	5	100	250	21.7	4.049	3.198
93	U.K.	0.44	1.5	6.99	27.5	0.73	15.89	3.82	5.16	4	50	100	18.7	0.159	0.189
94	U.K.	0.44	1.5	6.99	27.5	0.73	15.89	3.82	5.16	4	50	100	18.7	0.471	0.424
95	U.K.	0.44	1.5	6.99	27.5	0.73	15.89	3.82	5.16	4	50	100	18.7	1.13	0.905
96	Iran	0.4	2.3	12.2	44.4	1.08	30.5	3.3	5.69	4	50	100	26.2	0.815	0.66

Table A1. *Cont.*

Case No.	Location	D_{10} (mm)	D_{30} (mm)	D_{60} (mm)	D_{90} (mm)	C_c	C_u	GM	FM	R	UCS_{min} (MPa)	UCS_{max} (MPa)	γ (KN/m^3)	σ_n (MPa)	τ (MPa)
97	Iran	0.4	2.3	12.2	44.4	1.08	30.5	3.3	5.69	4	50	100	18.7	0.794	0.577
98	Iran	0.4	2.3	12.2	44.4	1.08	30.5	3.3	5.69	5	100	250	24.5	0.994	0.864
99	India	0.5	1.5	4.6	15.6	0.98	9.2	4.22	4.8	4	50	100	24.5	1.384	0.881
100	India	0.95	2.8	12.5	34.9	0.66	13.16	3.02	5.97	4	50	100	24.5	1.369	0.836
101	India	1.3	4.6	18.9	55.9	0.86	14.54	2.4	6.54	4	50	100	24.5	1.358	0.803
102	Iran (multiple)	0.5	3	10.4	31.2	1.73	20.3	3.36	5.63	4	50	100	26.2	1.056	0.625
103	Iran (multiple)	0.4	2.8	9.2	30.1	2.13	23	3.42	5.53	4	50	100	24.5	0.501	0.451
104	Iran (multiple)	0.4	2.8	9.2	30.1	2.13	23	3.42	5.53	4	50	100	24.5	0.986	0.827
105	Iran (multiple)	0.4	2.8	9.2	30.1	2.13	23	3.42	5.53	4	50	100	24.5	1.479	1.241
106	Iran (multiple)	0.5	3.3	10.2	31	2.14	20.4	3.28	5.7	5	100	250	24.5	0.485	0.379
107	Iran (multiple)	0.5	3.3	10.2	31	2.14	20.4	3.28	5.7	5	100	250	24.5	0.808	0.631
108	Iran (multiple)	0.5	3.3	10.2	31	2.14	20.4	3.28	5.7	5	100	250	24.5	1.131	0.884
109	Iran (multiple)	0.4	2.8	10.4	31.2	1.88	26	3.4	5.58	4	50	100	18.7	0.808	0.631
110	Iran (multiple)	0.4	2.8	10.4	31.2	1.88	26	3.4	5.58	4	50	100	18.7	1.131	0.884
111	USA	2.4	19.3	80.1	100	1.94	33.38	1.32	7.72	6	250	400	25.6	0.85	0.836
112	USA	2.4	19.3	80.1	100	1.94	33.38	1.32	7.72	6	250	400	25.6	1.695	1.637
113	USA	2.4	19.3	80.1	100	1.94	33.38	1.32	7.72	6	250	400	25.6	4.205	3.921
114	Iran (multiple)	0.01	1	10.4	43.9	9.62	1040	4	4.93	4	50	100	24.2	0.241	0.183
115	Iran (multiple)	0.01	1	10.4	43.9	9.62	1040	4	4.93	4	50	100	24.2	0.468	0.316
116	Iran (multiple)	0.01	1	10.4	43.9	9.62	1040	4	4.93	4	50	100	24.2	0.921	0.582
117	Iran (multiple)	0.01	1	10.4	43.9	9.62	1040	4	4.93	4	50	100	24.2	0.265	0.313

Table A1. *Cont.*

Case No.	Location	D_{10} (mm)	D_{30} (mm)	D_{60} (mm)	D_{90} (mm)	C_c	C_u	GM	FM	R	UCS_{min} (MPa)	UCS_{max} (MPa)	γ (KN/m^3)	σ_n (MPa)	τ (MPa)
118	Iran (multiple)	0.01	1	10.4	43.9	9.62	1040	4	4.93	4	50	100	24.2	0.511	0.506
119	Iran (multiple)	0.01	1	10.4	43.9	9.62	1040	4	4.93	4	50	100	24.2	1.001	0.902
120	USA	0.2	0.56	1.2	2.6	0.1	6	6	3	4	50	100	16.1	0.021	0.029
121	USA	0.2	0.56	1.2	2.6	0.1	6	6	3	4	50	100	16.1	0.042	0.051
122	USA	0.2	0.56	1.2	2.6	0.1	6	6	3	4	50	100	16.1	0.068	0.071
123	India	0.1	1.3	6.5	15	2.6	65	4.42	4.55	5	100	250	19.9	0.054	0.005
124	India	0.1	1.3	6.5	15	2.6	65	4.42	4.55	5	100	250	19.9	0.089	0.028
125	India	0.1	1.3	6.5	15	2.6	65	4.42	4.55	5	100	250	19.9	0.11	0.049
126	India	0.1	1.3	6.5	15	2.6	65	4.42	4.55	5	100	250	19.9	0.152	0.067
127	India	0.1	1.3	6.5	15	2.6	65	4.42	4.55	5	100	250	19.9	0.191	0.081
128	India	0.1	1.3	6.5	15	2.6	65	4.42	4.55	5	100	250	19.9	0.24	0.092
129	India	0.1	1	6.2	17	1.61	62	4.5	4.44	5	100	250	22.3	0.706	0.94
130	India	0.1	1	6.2	17	1.61	62	4.5	4.44	5	100	250	22.3	1.31	1.296
131	India	0.1	1	6.2	17	1.61	62	4.5	4.44	5	100	250	22.3	1.868	1.536
132	India	0.2	2.9	12.3	32	3.42	61.5	3.46	5.53	5	100	250	22.3	0.702	0.903
133	India	0.2	2.9	12.3	32	3.42	61.5	3.46	5.53	5	100	250	22.3	1.305	1.25
134	India	0.2	2.9	12.3	32	3.42	61.5	3.46	5.53	5	100	250	22.3	1.862	1.486
135	India	0.4	4.4	21.2	59.8	2.28	53	2.74	6.27	5	100	250	22.3	0.697	0.862
136	India	0.4	4.4	21.2	59.8	2.28	53	2.74	6.27	5	100	250	22.3	1.283	1.167
137	India	0.4	4.4	21.2	59.8	2.28	53	2.74	6.27	5	100	250	22.3	1.819	1.358
138	Australia	27.1	32.6	41.3	53	0.95	1.52	0.57	8.5	5	100	250	15.3	0.002	0.007
139	Australia	27.1	32.6	41.3	53	0.95	1.52	0.57	8.5	5	100	250	15.3	0.02	0.052
140	Australia	27.1	32.6	41.3	53	0.95	1.52	0.57	8.5	5	100	250	15.3	0.032	0.072
141	Australia	27.1	32.6	41.3	53	0.95	1.52	0.57	8.5	5	100	250	15.3	0.054	0.095
142	Australia	27.1	32.6	41.3	53	0.95	1.52	0.57	8.5	5	100	250	15.3	0.111	0.168
143	Australia	27.1	32.6	41.3	53	0.95	1.52	0.57	8.5	5	100	250	15.3	0.162	0.217
144	Australia	27.1	32.6	41.3	53	0.95	1.52	0.57	8.5	5	100	250	15.3	0.209	0.259
145	Australia	27.1	32.6	41.3	53	0.95	1.52	0.57	8.5	5	100	250	15.3	0.401	0.409
146	Australia	20.7	26.7	32.8	53	1.05	1.58	0.89	8.2	5	100	250	15.3	0.003	0.008
147	Australia	20.7	26.7	32.8	53	1.05	1.58	0.89	8.2	5	100	250	15.3	0.021	0.062
148	Australia	20.7	26.7	32.8	53	1.05	1.58	0.89	8.2	5	100	250	15.3	0.035	0.089

Table A1. *Cont.*

Case No.	Location	D_{10} (mm)	D_{30} (mm)	D_{60} (mm)	D_{90} (mm)	C_c	C_u	GM	FM	R	UCS_{min} (MPa)	UCS_{max} (MPa)	γ (KN/m^3)	σ_n (MPa)	τ (MPa)
149	Australia	20.7	26.7	32.8	53	1.05	1.58	0.89	8.2	5	100	250	15.3	0.058	0.116
150	Australia	20.7	26.7	32.8	53	1.05	1.58	0.89	8.2	5	100	250	15.3	0.115	0.191
151	Australia	20.7	26.7	32.8	53	1.05	1.58	0.89	8.2	5	100	250	15.3	0.155	0.206
152	Australia	20.7	26.7	32.8	53	1.05	1.58	0.89	8.2	5	100	250	15.3	0.209	0.259
153	Australia	20.7	26.7	32.8	53	1.05	1.58	0.89	8.2	5	100	250	15.3	0.394	0.401
154	Thailand	3.1	7.8	22	46.4	0.89	7.1	1.98	7.01	4	50	100	21	0.833	0.745
155	Thailand	3.1	7.8	22	46.4	0.89	7.1	1.98	7.01	4	50	100	21	1.649	1.407
156	Thailand	3.1	7.8	22	46.4	0.89	7.1	1.98	7.01	4	50	100	21	2.451	2.01
157	Thailand	3.1	7.8	22	46.4	0.89	7.1	1.98	7.01	4	50	100	21	3.223	2.492
158	Thailand	3.5	7.1	19.8	45.7	0.73	5.66	2.03	6.98	4	50	100	21	0.808	0.631
159	Thailand	3.5	7.1	19.8	45.7	0.73	5.66	2.03	6.98	4	50	100	21	1.592	1.169
160	Thailand	3.5	7.1	19.8	45.7	0.73	5.66	2.03	6.98	4	50	100	21	2.338	1.576
161	Thailand	3.5	7.1	19.8	45.7	0.73	5.66	2.03	6.98	4	50	100	21	3.14	2.178
162	Netherlands	11	15	23	32	0.89	2.09	1.48	7.48	5	100	250	16.8	0.014	0.028
163	Netherlands	11	15	23	32	0.89	2.09	1.48	7.48	5	100	250	16.8	0.028	0.048
164	Netherlands	11	15	23	32	0.89	2.09	1.48	7.48	5	100	250	16.8	0.055	0.082
165	Netherlands	11	15	23	32	0.89	2.09	1.48	7.48	5	100	250	16.8	0.108	0.143

References

1. Aghaei, A.A.; Soroush, A.; Reyhani, M. Large-scale triaxial testing and numerical modeling of rounded and angular rockfill materials. *Sci. Iran. Trans. A Civ. Eng.* **2010**, *17*, 169–183.
2. Varadarajan, A.; Sharma, K.; Venkatachalam, K.; Gupta, A. Testing and modeling two rockfill materials. *J. Geotech. Geoenviron. Eng.* **2003**, *129*, 206–218. [CrossRef]
3. Marsal, R.J. Large scale testing of rockfill materials. *J. Soil Mech. Found. Div.* **1967**, *93*, 27–43. [CrossRef]
4. Marachi, N.D. Strength and Deformation. Characteristics of Rockfill Materials. Ph.D. Thesis, University of California, Berkeley, CA, USA, 1969.
5. Venkatachalam, K. Prediction of Mechanical Behaviour of Rockfill Materials. Ph.D. Thesis, Indian Institute of Technology, Delhi, India, 1993.
6. Leps, T.M. Review of shearing strength of rockfill. *J. Soil Mech. Found. Div.* **1970**, *96*, 1159–1170. [CrossRef]
7. Liu, S.-H. Application of in situ direct shear device to shear strength measurement of rockfill materials. *Water Sci. Eng.* **2009**, *2*, 48–57.
8. Linero, S.; Palma, C.; Apablaza, R. Geotechnical characterisation of waste material in very high dumps with large scale triaxial testing. In *Proceedings of the 2007 International Symposium on Rock Slope Stability in Open Pit Mining and Civil Engineering*; Australian Centre for Geomechanics: Perth, Australia, 2007; pp. 59–75. [CrossRef]
9. Honkanadavar, N.; Gupta, S. Prediction of shear strength parameters for prototype riverbed rockfill material using index properties. In Proceedings of the Indian Geotechnical Conference, Mumbai, India, 16–18 December 2010; pp. 335–338.
10. Froemelt, A.; Duürrenmatt, D.J.; Hellweg, S. Using data mining to assess environmental impacts of household consumption behaviors. *Environ. Sci. Technol.* **2018**, *52*, 8467–8478. [CrossRef] [PubMed]
11. Ahmad, M.; Tang, X.-W.; Qiu, J.-N.; Gu, W.-J.; Ahmad, F. A hybrid approach for evaluating CPT-based seismic soil liquefaction potential using Bayesian belief networks. *J. Cent. South Univ.* **2020**, *27*, 500–516.
12. Ahmad, M.; Tang, X.-W.; Qiu, J.-N.; Ahmad, F. Evaluating Seismic Soil Liquefaction Potential Using Bayesian Belief Network and C4. 5 Decision Tree Approaches. *Appl. Sci.* **2019**, *9*, 4226. [CrossRef]
13. Ahmad, M.; Tang, X.; Qiu, J.; Ahmad, F.; Gu, W. LLDV-a Comprehensive framework for assessing the effects of liquefaction land damage potential. In Proceedings of the 2019 IEEE 14th International Conference on Intelligent Systems and Knowledge Engineering (ISKE), Dalian, China, 14–16 November 2019; pp. 527–533.
14. Ahmad, M.; Tang, X.-W.; Qiu, J.-N.; Ahmad, F.; Gu, W.-J. A step forward towards a comprehensive framework for assessing liquefaction land damage vulnerability: Exploration from historical data. *Front. Struct. Civil. Eng.* **2020**, *14*, 1476–1491. [CrossRef]
15. Ahmad, M.; Tang, X.; Ahmad, F. Evaluation of liquefaction-induced settlement using random forest and REP tree models: Taking Pohang earthquake as a case of illustration. In *Natural Hazards-Impacts, Adjustments & Resilience*; IntechOpen: London, UK, 2020.
16. Ahmad, M.; Al-Shayea, N.A.; Tang, X.-W.; Jamal, A.; Al-Ahmadi, H.M.; Ahmad, F. Predicting the pillar stability of underground mines with random trees and C4. 5 decision trees. *Appl. Sci.* **2020**, *10*, 6486. [CrossRef]
17. Pirhadi, N.; Tang, X.; Yang, Q.; Kang, F. A new equation to evaluate liquefaction triggering using the response surface method and parametric sensitivity analysis. *Sustainability* **2019**, *11*, 112. [CrossRef]
18. Pirhadi, N.; Tang, X.; Yang, Q. Energy evaluation of triggering soil liquefaction based on the response surface method. *Appl. Sci.* **2019**, *9*, 694. [CrossRef]
19. Mosavi, A.; Shirzadi, A.; Choubin, B.; Taromideh, F.; Hosseini, F.S.; Borji, M.; Shahabi, H.; Salvati, A.; Dineva, A.A. Towards an ensemble machine learning model of random subspace based functional tree classifier for snow avalanche susceptibility mapping. *IEEE Access* **2020**, *8*, 145968–145983. [CrossRef]
20. Mosavi, A.; Golshan, M.; Janizadeh, S.; Choubin, B.; Melesse, A.M.; Dineva, A.A. Ensemble models of GLM, FDA, MARS, and RF for flood and erosion susceptibility mapping: A priority assessment of sub-basins. *Geocarto Int.* **2020**, 1–20. [CrossRef]
21. Mosavi, A.; Hosseini, F.S.; Choubin, B.; Goodarzi, M.; Dineva, A.A. Groundwater salinity susceptibility mapping using classifier ensemble and Bayesian machine learning models. *IEEE Access* **2020**, *8*, 145564–145576. [CrossRef]
22. Mosavi, A.; Sajedi-Hosseini, F.; Choubin, B.; Taromideh, F.; Rahi, G.; Dineva, A.A. Susceptibility mapping of soil water erosion using machine learning models. *Water* **2020**, *12*, 1995. [CrossRef]
23. Mosavi, A.; Hosseini, F.S.; Choubin, B.; Abdolshahnejad, M.; Gharechaee, H.; Lahijanzadeh, A.; Dineva, A.A. Susceptibility Prediction of Groundwater Hardness Using Ensemble Machine Learning Models. *Water* **2020**, *12*, 2770. [CrossRef]
24. Mosavi, A.; Hosseini, F.S.; Choubin, B.; Taromideh, F.; Ghodsi, M.; Nazari, B.; Dineva, A.A. Susceptibility mapping of groundwater salinity using machine learning models. *Environ. Sci. Pollut. Res.* **2021**, *28*, 10804–10817. [CrossRef] [PubMed]
25. Mosavi, A.; Hosseini, F.S.; Choubin, B.; Goodarzi, M.; Dineva, A.A.; Sardooi, E.R. Ensemble boosting and bagging based machine learning models for groundwater potential prediction. *Water Resour. Manag.* **2021**, *35*, 23–37. [CrossRef]
26. Choubin, B.; Borji, M.; Hosseini, F.S.; Mosavi, A.; Dineva, A.A. Mass wasting susceptibility assessment of snow avalanches using machine learning models. *Sci. Rep.* **2020**, *10*, 18363. [CrossRef]
27. Lovrić, M.; Pavlović, K.; Žuvela, P.; Spataru, A.; Lučić, B.; Kern, R.; Wong, M.W. Machine learning in prediction of intrinsic aqueous solubility of drug-like compounds: Generalization, complexity, or predictive ability? *J. Chemom.* **2021**, e3349. [CrossRef]

28. Lovrić, M.; Meister, R.; Steck, T.; Fadljević, L.; Gerdenitsch, J.; Schuster, S.; Schiefermüller, L.; Lindstaedt, S.; Kern, R. Parasitic resistance as a predictor of faulty anodes in electro galvanizing: A comparison of machine learning, physical and hybrid models. *Adv. Model. Simul. Eng. Sci.* **2020**, *7*, 1–16. [CrossRef]
29. Gupta, A.K. Constitative Modelling of Rockfill Materials. Ph.D. Thesis, Indian Institute of Technology, Delhi, India, 2000.
30. Abbas, S.; Varadarajan, A.; Sharma, K. Prediction of shear strength parameter of prototype rockfill material. *IGC-2003 Roorkee* **2003**, *1*, 5–8.
31. Honkanadavar, N.; Sharma, K. Testing and modeling the behavior of riverbed and blasted quarried rockfill materials. *Int. J. Geomech.* **2014**, *14*, 04014028. [CrossRef]
32. Frossard, E.; Hu, W.; Dano, C.; Hicher, P.-Y. Rockfill shear strength evaluation: A rational method based on size effects. *Géotechnique* **2012**, *62*, 415–427. [CrossRef]
33. Kaunda, R. Predicting shear strengths of mine waste rock dumps and rock fill dams using artificial neural networks. *Int. J. Min. Mineral. Eng.* **2015**, *6*, 139–171. [CrossRef]
34. Zhou, J.; Li, E.; Wei, H.; Li, C.; Qiao, Q.; Armaghani, D.J. Random forests and cubist algorithms for predicting shear strengths of rockfill materials. *Appl. Sci.* **2019**, *9*, 1621. [CrossRef]
35. Alkhatib, K.; Najadat, H.; Hmeidi, I.; Shatnawi, M.K.A. Stock price prediction using k-nearest neighbor (kNN) algorithm. *Int. J. Bus. Humanit. Technol.* **2013**, *3*, 32–44.
36. Vijayan, V.; Ravikumar, A. Study of data mining algorithms for prediction and diagnosis of diabetes mellitus. *Int. J. Comput. Appl.* **2014**, *95*, 12–16.
37. Thongkam, J.; Xu, G.; Zhang, Y. AdaBoost algorithm with random forests for predicting breast cancer survivability. In Proceedings of the 2008 IEEE International Joint Conference on Neural Networks (IEEE World Congress on Computational Intelligence), Hong Kong, China, 1–8 June 2008; pp. 3062–3069.
38. Samui, P.; Sitharam, T.; Contadakis, M. Machine learning modelling for predicting soil liquefaction susceptibility. *Nat. Hazards Earth Syst. Sci.* **2011**, *11*, 1–9. [CrossRef]
39. Pal, M. Support vector machines-based modelling of seismic liquefaction potential. *Int. J. Numer. Anal. Methods Geomech.* **2006**, *30*, 983–996. [CrossRef]
40. Boser, B.E.; Guyon, I.M.; Vapnik, V.N. A training algorithm for optimal margin classifiers. In Proceedings of the 5th Annual ACM Workshop on Computational Learning Theory, Pittsburgh, PA, USA, 27–29 July 1992; pp. 144–152.
41. Zhou, J.; Li, X.; Shi, X. Long-term prediction model of rockburst in underground openings using heuristic algorithms and support vector machines. *Saf. Sci.* **2012**, *50*, 629–644. [CrossRef]
42. Zhou, J.; Li, X.; Wang, S.; Wei, W. Identification of large-scale goaf instability in underground mine using particle swarm optimization and support vector machine. *Int. J. Min. Sci. Technol.* **2013**, *23*, 701–707. [CrossRef]
43. Breiman, L. Random forests. *Mach. Learn.* **2001**, *45*, 5–32. [CrossRef]
44. Seo, D.K.; Kim, Y.H.; Eo, Y.D.; Park, W.Y.; Park, H.C. Generation of radiometric, phenological normalized image based on random forest regression for change detection. *Remote. Sens.* **2017**, *9*, 1163. [CrossRef]
45. Maillo, J.; Ramírez, S.; Triguero, I.; Herrera, F. kNN-IS: An iterative spark-based design of the k-nearest neighbors classifier for big data. *Knowl. Based Syst.* **2017**, *117*, 3–15. [CrossRef]
46. Chomboon, K.; Chujai, P.; Teerarassamee, P.; Kerdprasop, K.; Kerdprasop, N. An empirical study of distance metrics for k-nearest neighbor algorithm. In Proceedings of the 3rd International Conference on Industrial Application Engineering, Kitakyushu, Japan, 28–30 March 2015; pp. 280–285.
47. Prasath, V.; Alfeilat, H.A.A.; Hassanat, A.; Lasassmeh, O.; Tarawneh, A.S.; Alhasanat, M.B.; Salman, H.S.E. Distance and similarity measures effect on the performance of K-nearest neighbor classifier—A review. *arXiv* **2017**, arXiv:1708.04321.
48. Farooq, F.; Nasir Amin, M.; Khan, K.; Rehan Sadiq, M.; Faisal Javed, M.; Aslam, F.; Alyousef, R. A comparative study of random forest and genetic engineering programming for the prediction of compressive strength of High Strength Concrete (HSC). *Appl. Sci.* **2020**, *10*, 7330. [CrossRef]
49. Golmohammadi, G.; Prasher, S.; Madani, A.; Rudra, R. Evaluating three hydrological distributed watershed models: MIKE-SHE, APEX, SWAT. *Hydrology* **2014**, *1*, 20–39. [CrossRef]
50. Nhu, V.-H.; Shahabi, H.; Nohani, E.; Shirzadi, A.; Al-Ansari, N.; Bahrami, S.; Miraki, S.; Geertsema, M.; Nguyen, H. Daily water level prediction of Zrebar Lake (Iran): A comparison between M5P, random forest, random tree and reduced error pruning trees algorithms. *ISPRS Int. J. Geo. Inf.* **2020**, *9*, 479. [CrossRef]
51. Gandomi, A.H.; Babanajad, S.K.; Alavi, A.H.; Farnam, Y. Novel approach to strength modeling of concrete under triaxial compression. *J. Mater. Civ. Eng.* **2012**, *24*, 1132–1143. [CrossRef]
52. Nush, J.; Sutcliffe, J.V. River flow forecasting through conceptual models part I—A discussion of principles. *J. Hydrol.* **1970**, *10*, 282–290. [CrossRef]
53. Khosravi, K.; Mao, L.; Kisi, O.; Yaseen, Z.M.; Shahid, S. Quantifying hourly suspended sediment load using data mining models: Case study of a glacierized Andean catchment in Chile. *J. Hydrol.* **2018**, *567*, 165–179. [CrossRef]
54. Moriasi, D.N.; Arnold, J.G.; Van Liew, M.W.; Bingner, R.L.; Harmel, R.D.; Veith, T.L. Model evaluation guidelines for systematic quantification of accuracy in watershed simulations. *Trans. ASABE* **2007**, *50*, 885–900. [CrossRef]
55. Koopialipoor, M.; Fallah, A.; Armaghani, D.J.; Azizi, A.; Mohamad, E.T. Three hybrid intelligent models in estimating flyrock distance resulting from blasting. *Eng. Comput.* **2019**, *35*, 243–256. [CrossRef]

56. Asteris, P.G.; Tsaris, A.K.; Cavaleri, L.; Repapis, C.C.; Papalou, A.; Di Trapani, F.; Karypidis, D.F. Prediction of the fundamental period of infilled RC frame structures using artificial neural networks. *Comput. Intell. Neurosci.* **2016**. [CrossRef]
57. Koopialipoor, M.; Armaghani, D.J.; Hedayat, A.; Marto, A.; Gordan, B. Applying various hybrid intelligent systems to evaluate and predict slope stability under static and dynamic conditions. *Soft Comput.* **2019**, *23*, 5913–5929. [CrossRef]
58. Andjelkovic, V.; Pavlovic, N.; Lazarevic, Z.; Radovanovic, S. Modelling of shear strength of rockfills used for the construction of rockfill dams. *Soils Found.* **2018**, *58*, 881–893. [CrossRef]
59. Yang, Y.; Zhang, Q. A hierarchical analysis for rock engineering using artificial neural networks. *Rock Mech. Rock Eng.* **1997**, *30*, 207–222. [CrossRef]
60. Faradonbeh, R.S.; Armaghani, D.J.; Abd Majid, M.; Tahir, M.M.; Murlidhar, B.R.; Monjezi, M.; Wong, H. Prediction of ground vibration due to quarry blasting based on gene expression programming: A new model for peak particle velocity prediction. *Int. J. Environ. Sci. Technol.* **2016**, *13*, 1453–1464. [CrossRef]
61. Chen, W.; Hasanipanah, M.; Rad, H.N.; Armaghani, D.J.; Tahir, M. A new design of evolutionary hybrid optimization of SVR model in predicting the blast-induced ground vibration. *Eng. Comput.* **2021**, *37*, 1455–1471. [CrossRef]
62. Rad, H.N.; Bakhshayeshi, I.; Jusoh, W.A.W.; Tahir, M.; Foong, L.K. Prediction of flyrock in mine blasting: A new computational intelligence approach. *Nat. Resour. Res.* **2020**, *29*, 609–623.
63. Ahmad, M.; Hu, J.-L.; Ahmad, F.; Tang, X.-W.; Amjad, M.; Iqbal, M.J.; Asim, M.; Farooq, A. Supervised learning methods for modeling concrete compressive strength prediction at high temperature. *Materials* **2021**, *14*, 1983. [CrossRef] [PubMed]

applied sciences

MDPI

Article

Mechanical Properties of Cement-Treated Soil Mixed with Cellulose Nanofibre

Hidenori Takahashi [1,*], Shinya Omori [2], Hideyuki Asada [2], Hirofumi Fukawa [3], Yusuke Gotoh [1] and Yoshiyuki Morikawa [1]

[1] Geotechnical Engineering Department, Port and Airport Research Institute, National Institute of Maritime, Port and Aviation Technology, 3-1-1, Nagase, Yokosuka 2390826, Japan; gotou-y88s3@p.mpat.go.jp (Y.G.); morikawa@p.mpat.go.jp (Y.M.)

[2] Research and Development Center, Toa Corporation, 1-3, Anzencho, Tsurumi-ku, Yokohama 2300035, Japan; sh_omori@toa-const.co.jp (S.O.); h_asada@toa-const.co.jp (H.A.)

[3] Civil Engineering Central HQ., Design Department, Toa Corporation, 3-7-1, Nishishinjuku, Shinjuku-ku, Tokyo 1631031, Japan; h_fukawa@toa-const.co.jp

* Correspondence: takahashi-h@p.mpat.go.jp

Abstract: Cellulose nanofibre (CNF), a material composed of ultrafine fibres of wood cellulose fibrillated to nano-order level, is expected to be widely used because of its excellent properties. However, in the field of geotechnical engineering, almost no progress has been made in the development of techniques for using CNFs. The authors have focused on the use of CNF as an additive in cement treatment for soft ground, where cement is added to solidify the ground, because CNF can reduce the problems associated with cement-treated soil. This paper presents the results of a study on the method of mixing CNF, the strength and its variation obtained by adding CNF, and the change in permeability. CNF had the effect of mixing the cement evenly and reducing the variation in the strength of the treated soil. The CNF mixture increased the strength at the initial age but reduced the strength development in the long term. The addition of CNF also increased the flexural strength, although it hardly changed the permeability.

Keywords: cement-treated soil; cellulose nanofibre; strength; permeability

Citation: Takahashi, H.; Omori, S.; Asada, H.; Fukawa, H.; Gotoh, Y.; Morikawa, Y. Mechanical Properties of Cement-Treated Soil Mixed with Cellulose Nanofibre. *Appl. Sci.* **2021**, *11*, 6425. https://doi.org/10.3390/app11146425

Academic Editors: Paulo José da Venda Oliveira and António Alberto Santos Correia

Received: 17 June 2021
Accepted: 10 July 2021
Published: 12 July 2021

Publisher's Note: MDPI stays neutral with regard to jurisdictional claims in published maps and institutional affiliations.

Copyright: © 2021 by the authors. Licensee MDPI, Basel, Switzerland. This article is an open access article distributed under the terms and conditions of the Creative Commons Attribution (CC BY) license (https://creativecommons.org/licenses/by/4.0/).

1. Introduction

In recent years, it has become possible to produce cellulose nanofibre (CNF), which is a material composed of ultrafine fibres of wood cellulose fibrillated to nano-order level [1–3]. CNF is expected to be widely used because its strength is five times and its mass is one-fifth that of steel, and its thermal expansion is one-fiftieth that of glass. The characteristics of CNFs, such as high specific surface area, edibility, light weight, high strength, low thermal expansion, biodegradability, and biocompatibility, have been utilised to develop various applications. For example, transparent resins such as acrylics and epoxy resins are reinforced without much loss of their transparency (transparency reinforcement). The characteristics of light-weight and high-strength fibres are used to produce moulded products that are as strong as steel but one-fifth their mass by injecting phenolic resin into nanofibre sheets and then laminating and curing them. In addition, the thixotropy (the property that the fluidity increases when an external force is applied) of CNF solutions has already been used to commercialise ball-point pen inks, which prevent dripping and ease writing. In the field of concrete, the addition of CNFs to improve the strength of concrete has been studied [4–6]. This is attributed to its ability to accelerate the cement hydration reaction as well as to the high strength of CNF itself. The use of CNF as a stabiliser to prevent the separation of concrete compositions and bleeding before solidification, and as an additive to increase the flowability required in the pressure injection method using thixotropic properties, are being investigated and have been partly implemented.

In the field of geotechnical engineering, very little progress has been made in the development of techniques for using CNFs. The authors have focused on the use of CNF as an additive in cement treatment for soft ground, where cement is added to solidify the ground, for example, by the deep mixing method [7] to improve the ground and to construct geotechnical structures. This is because cement treatment is a relatively inexpensive and highly reliable ground improvement method. Although it is widely used, there are still some problems such as brittleness, low tensile strength, strength variability, and low short-term strength. If more cement is added to solve these problems, the environmental impact will increase. If the long-term strength of the treated soil is too high, the treated soil could make rehabilitation of the geotechnical structure difficult. The possibility of reducing these problems using the CNF effect can be considered as follows: (1) The high strength of fibres increases the strength of the treated soil and reduces the risk of brittle or tensile failure. (2) As in concrete, CNFs accelerate the cement hydration reaction. (3) Owing to its water absorption and thickening properties, shear strength can be expected to develop immediately after stirring. (4) Thixotropy improves uniformity when stirring and mixing cement, and it ensures fluidity when pumping the treated soil in the slurry state before solidification. (5) Dispersibility increases uniformity during the stirring and mixing process. Previous studies have shown that fibrous materials of a few millimetres in length can be used for mixing with the treated soil [8–15]. The mixture of long fibres can have the effect described in (1) above, although not many other effects can be expected. When using existing ground improvement machines, such as deep mixing machines, it is difficult to stir long fibres into the treated soil, which presents a number of challenges for on-site use. Carbon nanotubes (CNTs) are similar to CNF, and research has been conducted on mixing them into the treated soil [16,17]. In these studies, the addition of these materials was found to increase the initial strength of the treated soil, even at the nano level. However, CNTs are poorly dispersible and require the addition of dispersants. By contrast, CNF is hydrophilic and highly dispersible in its own right, making it easy to mix directly with water or soil. This suggests that the use of CNFs in cement treatment is useful.

As mentioned above, the effect of mixing CNF can be predicted, but it is not simply a matter of mixing CNF. The method of mixing CNF and cement needs to be studied, because adding CNF in gel form requires more cement to solidify the soil, as water is also added. It is also not clear whether CNF of increasing plasticity can be stirred evenly. In contrast to concrete, the properties of soils are diverse and need to be carefully investigated to ensure that CNFs are effective. There are different types of CNFs as well as soil, and it is necessary to know which CNF should be combined with which soils. In the first place, it has not been demonstrated that mixing CNFs can solve the problems of cement-treated soil. Therefore, the authors' group has been studying the method of adding CNF in the cement treatment method, understanding the effect of mixing CNF, and investigating its use in an actual ground improvement site. This paper presents the results of a study on the method of mixing CNF, the strength and its variation obtained by adding CNF, and the change in permeability. Strength and permeability are important parameters because cement treatments are used to increase the strength of soft ground and impervious materials. If the study finds the mixing method and the improved properties in terms of strength and permeability, CNF will clearly be of great use in the field. Incidentally, the purpose of this study is not to clarify the mechanism of the effect of CNF but to investigate how to mix CNF and whether the effect of CNF can be observed in the solidification process.

2. Materials and the Mixture Method

2.1. Characteristics of CNF Used

There are many types of CNFs in terms of raw materials and manufacturing processes, and the ones used in this study were two typical commercially available CNFs (TEMPO-CNF and ACC-CNF). Two types of ACC-CNFs with different raw materials were used, which are referred to as ACC-CNF-S and ACC-CNF-B. The tests used these three types of CNFs and two states of TEMPO-CNF: a gel containing 2% CNF and a powder containing

90% CNF. The CNFs used are listed in Table 1. CNF is generally supplied in gel form mixed with water, although when mixed with soil to produce cement-treated soil, the soil includes a large quantity of water. The addition of water, which is included in the cement slurry, and CNF gel to the soil causes the soil to soften, and more cement is required to solidify the soft ground. Powdered CNF was also used in this study to investigate whether it might be possible to mix it with soil.

Table 1. Cellulose nanofibre used for tests.

	TEMPO-CNF		ACC-CNF	
Identifier	**TEMPO-CNF**	**TEMPO-CNF**	**ACC-CNF-S**	**ACC-CNF-B**
Raw material	Wood pulp	Wood pulp	Softwood pulp	Bamboo pulp
State	Gel	Powder	Gel	Gel
Content rate	2.0%	90%	1.8% and 2.0%	2.0%
Defibrating method	Chemical fibrillation	Chemical fibrillation	Physical fibrillation	Physical fibrillation
Manufacturing method	TEMPO-mediated oxidation	TEMPO-mediated oxidation	Aqueous counter collision	Aqueous counter collision
Appearance	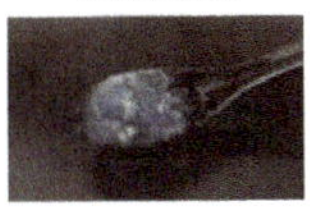	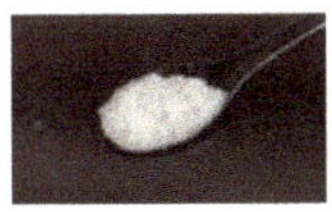	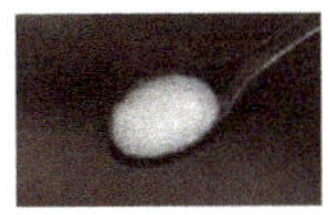	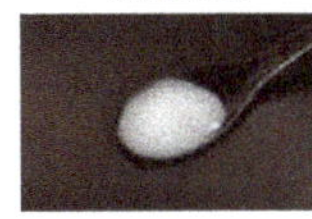

The production methods for the CNFs are different, and their degrees of fibrillation are also different. TEMPO-CNF is produced from wood pulp by a chemical defibration technique called TEMPO-mediated oxidation [18]. It is characterised by a very fine and uniform fibre width, which is as thin as approximately 3 nm. On the other hand, ACC-CNF is produced from bleached pulps of softwood or bamboo by a physical defibration technique called the aqueous counter collision (ACC) method [19]. CNF derived from soft woods is referred to as ACC-CNF-S, and that derived from bamboo is called ACC-CNF-B. These CNFs are characterised by their low chemical environmental impact as they are fibrillated using only water and raw materials, their amphiphilicity, and the concise control of the fibre width due to the gradual refinement of the fibre by repeated processing. The fibre width ranges from a few nanometres to several millimetres, depending on the number of treatments.

2.2. Characteristics of Soil and Cement

Two types of soil were used as base materials: Kasaoka clay, which is powdered clay manufactured from a mudstone in Okayama Prefecture, Japan, and Kawasaki clay, which is marine clay collected from the seabed of the Port of Kawasaki. The properties of each sample are listed in Table 2. Kasaoka clay is classified as high-liquid clay with a liquid limit of 62.1%. Kasaoka clay has few impurities and is uniform. Kawasaki clay is classified as low-liquid clay with a liquid limit of 41.0%. Kawasaki clay is a natural marine clay collected by dredging and contains impurities (e.g., small amounts of shells and gravel). A small amount of water was added as a pre-treatment to liquefy it, and it was passed through a 2 mm sieve to remove impurities before the cement treatment. Artificial seawater was added until the liquid limit ratio became 1.55.

Table 2. Physical properties of soils.

		Kasaoka Clay	Kawasaki Clay
Soil particle density, r_s (g/cm^3)		2.700	2.671
Initial water content, w_i (%)		6.8	61.8
Grain size	Gravel fraction (2–75 mm) (%)	0.0	0.0
	Sand fraction (75 µm–2 mm) (%)	1.1	13.1
	Silt fraction (5–75 µm) (%)	40.1	46.9
	Clay fraction (<5 µm) (%)	58.8	40.0
Fine fraction content, F_c (%)		98.9	86.9
Liquid limit, w_L (%)		62.1	41.0
Plastic limit, w_P (%)		20.6	23.1
Plasticity index, I_P		41.5	17.9

Three types of cement were used for the solidification process: ordinary Portland cement, high-early-strength Portland cement, and blast-furnace cement type B. Portland cement was used to determine the basic properties of the treated soil mixed with CNF. In Section 4, the results of the study on the strength variability of the treated soil are discussed by making an assumption about the treated soil in the field. Here, blast-furnace cement type B was used, assuming the soil in the field was treated. In Japan, this type of cement is frequently used for ground improvement in port and in harbour engineering. Blast-furnace cement type B is a relatively inexpensive cement that is prepared by adding blast-furnace slag powder to cement. It also has high durability against seawater and chemical substances, and low leaching of hexavalent chromium.

2.3. Mixture Procedure of Soil, Cement, and CNF

2.3.1. Method 1

An attempt was made to add and mix CNF uniformly into the soil–cement mixture in the laboratory. This was performed to reduce the variation between the specimens and to confirm the effect of CNF in the treated soil mixed with CNF. Figure 1 shows the procedure used to prepare the specimens. Three types of CNFs in gel form (TEMPO-CNF, ACC-CNF-S, and ACC-CNF-B) were prepared and watered. Considering the possibility that the gelatinous material might not be mixed uniformly into the soil, water was added until it was no longer gelatinous to ensure uniform mixing. Cement slurry was also prepared by adding water to ordinary Portland cement. As shown in the mixture procedure (Figure 1), the CNF solution and cement slurry were mixed first and then added to and mixed with dry and powdered Kasaoka clay. The reason for mixing the powdered clay with the CNF solution and cement slurry is that if they were mixed with the wet soil, too much water would be present in the treated soil, and a large quantity of cement would be needed to solidify the soil. As this is not practical, the mixing of CNF and cement into wet clay was attempted using Method 2.

The soil mixer used for mixing was of 5 L capacity in accordance with JIS R5201 [20], with a mixing time of 5 min. Figure 2 shows the soil treated immediately after mixing. As shown in the figure, it can be observed that the flowability of the mixture without CNF is high, whereas that of the mixture with CNF is low. This was due to the water absorption and thickening properties of the CNFs. However, cement was considered to have been well mixed because of its high fluidity at a high strain rate near the agitator blades, which was induced by the thixotropic nature of CNF. The mixed soil was packed in unconfined compression test moulds (50 mm diameter × 100 mm height cylindrical) and cured in a constant-temperature curing room at 20 ± 2 °C and a humidity of more than 95% for a specified period.

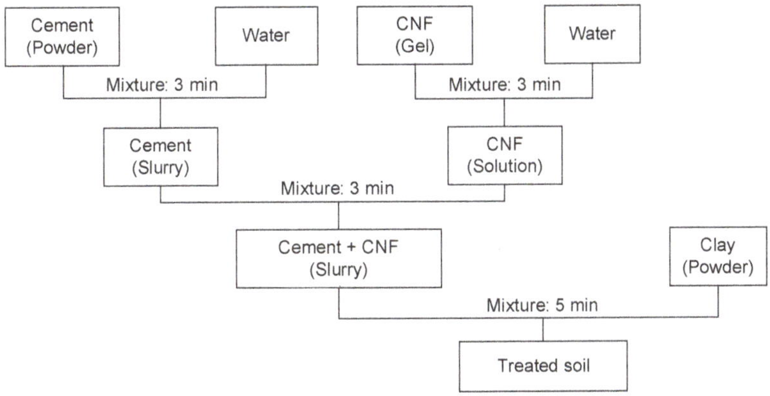

Figure 1. Mixture procedure for preparing specimens (Method 1).

Figure 2. Appearance immediately after mixture by Method 1.

Table 3 shows the mixing proportions of cement, CNF, water, and soil for the treated soil prepared by Method 1. When CNF was added, the ratio of the dry weight of CNF to the dry weight of soil was defined as the CNF addition rate, α, which was set to α = 1.0%. The reason for setting α = 1.0% is that in a previous study, the effect of CNF was not clearly visible unless it was mixed with α = 1.0%. Cement was added in the ratio of approximately 100 kg dry weight per unit volume of soil mixed with all other materials except cement. The dry weight ratio of cement to soil was 13.8%. This was slightly less than the amount added in the deep mixing method, which is one of the most common ground improvement methods used on site. Cement was not added as a powder but as a cement slurry. The water–cement ratio was set to w/c = 7.0. Considering the water contained in all soil mixtures, the water content ratio was approximately 85%. The treated soil mixed using Method 1 was subjected to unconfined compression tests, and the test results are discussed in Section 3.

Table 3. Mix proportion of cement, CNF, water, and soil in Method 1: (**a**) Treated soil without CNF; (**b**) Treated soil with TEMPO-CNF (Gel); (**c**) Treated soil with ACC-CNF-S; (**d**) Treated soil with ACC-CNF-B.

(a)	Amounts of Component									CNF Addition Rate (%)	Cement Addition (kg/m³)
	Soil (Kasaoka Clay)			Water	Cement (Ordinary Portland)	CNF -No Addition-			Total		
	Total	Soil Particle	Water			Total	Solid Content	Water			
g	769	725	45	655	100	0	0	0	1525	0.0	100.0
g/cm³	-	2.70	1.00	1.00	3.16	-	1.5	1.0	1.52		
cm³	313	268	45	655	32	0	0	0	1000		

(b)	Amounts of Component									CNF Addition Rate (%)	Cement Addition (kg/m³)
	Soil (Kasaoka Clay)			Water	Cement (Ordinary Portland)	CNF (TEMPO-CNF)			Total		
	Total	Soil Particle	Water			Total	Solid Content	Water			
g	769	725	45	300	100	363	7.25	355	1532	1.0	99.5
g/cm³	-	2.70	1.00	1.00	3.16	-	1.5	1.0	1.52		
cm³	313	268	45	300	32	360	5	355	1005		

(c)	Amounts of Component									CNF Addition Rate (%)	Cement Addition (kg/m³)
	Soil (Kasaoka Clay)			Water	Cement (Ordinary Portland)	CNF (ACC-CNF-S)			Total		
	Total	Soil Particle	Water			Total	Solid Content	Water			
g	769	725	45	260	100	403	7.25	396	1532	1.0	99.5
g/cm³	-	2.70	1.00	1.00	3.16	-	1.5	1.0	1.52		
cm³	313	268	45	260	32	400	5	396	1005		

(d)	Amounts of Component									CNF Addition Rate (%)	Cement Addition (kg/m³)
	Soil (Kasaoka Clay)			Water	Cement (Ordinary Portland)	CNF (ACC-CNF-B)			Total		
	Total	Soil Particle	Water			Total	Solid Content	Water			
g	769	725	45	300	100	363	7.25	355	1532	1.0	99.5
g/cm³	-	2.70	1.00	1.00	3.16	-	1.5	1.0	1.52		
cm³	313	268	45	300	32	360	5	355	1005		

2.3.2. Method 2

As mentioned above, CNF in gel form was used in Method 1 with the aim of ensuring uniform mixing. However, water in the gel is added to the soil to increase the amount of CNF, and consequently, more cement is required to solidify the soil. To overcome this problem, dry powdered CNF was added to wet clay in Method 2 instead of the gel form of CNF. This is a more practical method for addition in the field. It is also practical to use high concentrations of CNF, even if it is difficult to produce powdered CNF. Figure 3 shows the mixture procedure for specimen preparation using Method 2. Powdered CNF (TEMPO-CNF) and cement (blast-furnace cement type B or high-early-strength cement) were mixed well in advance. Separately, Kawasaki clay was water-adjusted to obtain a liquid limit ratio of 1.55 or 1.95; then, they were mixed and stirred. For the specimens used in the permeability tests, ACC-CNF-S mixed with high-early-strength cement was also prepared, because the powdered ACC-CNF-S was not available. The gel form was used in this type of specimen.

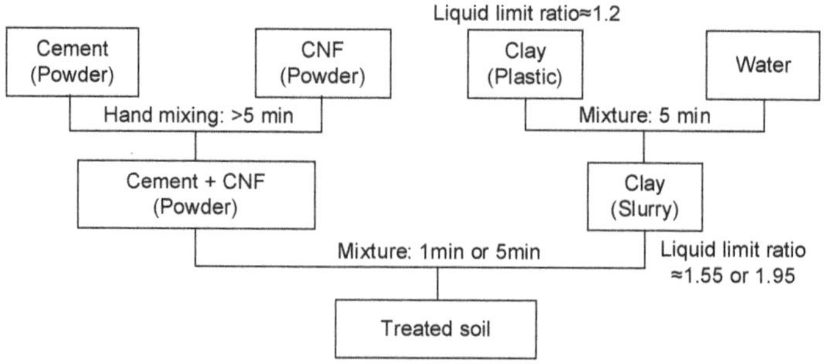

Figure 3. Mixture procedure for preparing specimens (Method 2).

The soil mixer used for mixing was the same as that used in Method 1, and the mixing time was set to 1 or 5 min. A 5 min agitation makes the cement-treated soil uniform both with and without CNF. For this reason, the mixing time was shortened to 1 min for some specimens, which deliberately created a condition of insufficient mixing. It was easy to predict that without CNF mixing, stirring would be inadequate, and this test series was used to investigate whether the addition of CNF would promote uniform mixing of the cement. The specimens used in the permeability tests required uniformly treated soil. This was why a sufficient mixing time of 5 min was provided. Figure 4 shows the immediate aftermath of a short mixing time of 1 min for the specimen. In the figure, it was easily possible to visually distinguish between cemented and uncemented areas in the soil without the CNF mixture. On the other hand, in the soil mixed with CNF, there was no unevenness of mixing as far as could be visually observed, indicating the effect of the addition of CNF. This was attributed to the effects of the thixotropy, which facilitated mixing of the soil, and the homogeneous dispersion, which mixed cement well, although the water absorption and thickening properties of CNF reduced the flowability. The mixed soil was packed in unconfined compression test moulds (50 mm diameter × 100 mm height cylindrical) and cured in a constant-temperature curing room at 20 ± 2 °C and a humidity of over 95% for a specified period.

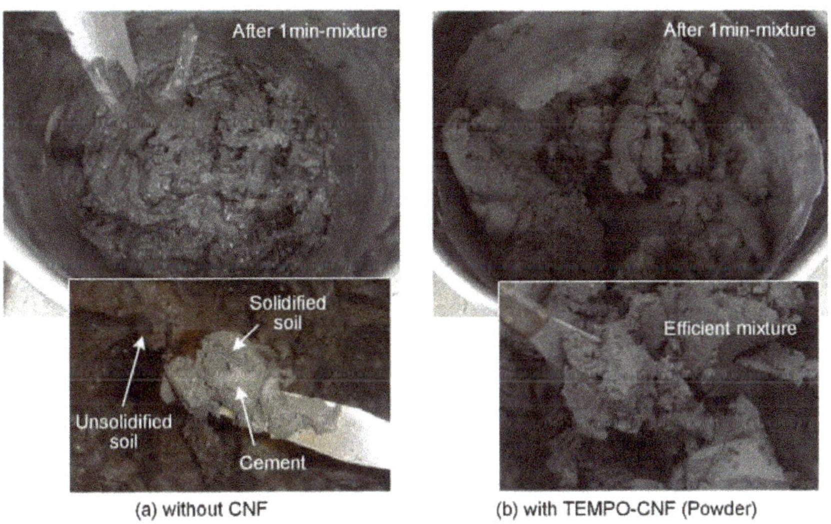

Figure 4. Appearance immediately after mixture by Method 2.

Tables 4 and 5 show the mixing proportions of cement, CNF, water, and soil for the treated soil prepared by Method 2. The tables show the proportions of the specimens for the tests to examine the strength variability and permeability tests, respectively. The ratio of CNF addition was set to α = 1.0%, which showed the effect of CNF, except for the soil treated with ACC-CNF-S, for which the ratio was set to α = 0.5%. This was because in this CNF, the powdered form was not available, and the gel form was used. The larger the amount added, the more water was added to the treated soil. Cement was added in the ratio of approximately 50 or 80 kg dry weight per unit volume of soil mixed with all other materials except cement. The dry weight ratio of cement to soil was 4.9% or 10.0%. These values are lower than the amounts added in the deep mixing method, especially the 4.9% cement addition, which was so small that the specimen could be easily broken by hand. By reducing the amount of cement in this way, the effect of CNF on cement mixing was more clearly visible. Considering the water contained in all the soil mixtures, the water content ratio was approximately 60% or 80%. Thirty specimens were prepared for each proportion with and without CNF, because the variability was expected to be large in the specimens prepared with 1 min of mixing. The treated soil prepared with these proportions was subjected to unconfined compression tests and permeability tests, and the test results are discussed in Sections 4 and 5.

Table 4. Mix proportion of cement, CNF, water, and soil in Method 2 for evaluating variability: (**a**) Treated soil without CNF; (**b**) Treated soil with TEMPO-CNF (Powder).

(a)	Soil (Kawasaki Clay)			Water	Amounts of Component Cement (Blast Furnace t. B)	CNF -No Addition-			Total	CNF Addition Rate (%)	Cement Addition (kg/m³)
	Total	Soil Particle	Water			Total	Solid Content	Water			
g	1687	1030	657	0	50	0	0	0	1737		
g/cm³	-	2.671	1.023	1.00	3.04	-	1.5	1.0	1.662	0.0	47.9
cm³	1028	386	642	0	16	0	0	0	1044		
(b)	**Soil (Kawasaki Clay)**			**Water**	**Amounts of Component Cement (Blast Furnace t. B)**	**CNF (TEMPO-CNF)**			**Total**	**CNF Addition Rate (%)**	**Cement Addition (kg/m³)**
	Total	**Soil Particle**	**Water**			**Total**	**Solid Content**	**Water**			
g	1686	1030	656	0	50	11.43	10.30	1.13	1747		
g/cm³	-	2.671	1.023	1.00	3.04	-	1.5	1.0	1.662	1.0	47.6
cm³	1027	386	641	0	16	8.00	6.87	1.13	1051		

Table 5. Mix proportion of cement, CNF, water, and soil in Method 2 for evaluating permeability: (**a**) Treated soil without CNF; (**b**) Treated soil with TEMPO-CNF (Powder); (**c**) Treated soil with ACC-CNF-S (Gel).

(a)	Soil (Kawasaki Clay)			Water	Amounts of Component Cement (Early Strength)	CNF -No Addition-			Total	CNF Addition Rate (%)	Cement Addition (kg/m³)
	Total	Soil Particle	Water			Total	Solid Content	Water			
g	2000	1237	763	325	124	0	0	0	2449		
g/cm³	-	2.671	1.023	1.00	3.14	-	1.5	1.0	1.556	0.0	78.6
cm³	1209	463	746	325	39	0	0	0	1574		
(b)	**Soil (Kawasaki Clay)**			**Water**	**Amounts of Component Cement (Early Strength)**	**CNF (TEMPO-CNF)**			**Total**	**CNF Addition Rate (%)**	**Cement Addition (kg/m³)**
	Total	**Soil Particle**	**Water**			**Total**	**Solid Content**	**Water**			
g	2000	1237	763	324	124	13.74	12.37	1.37	2461		
g/cm³	-	2.671	1.023	1.00	3.14	-	1.5	1.0	1.556	1.0	78.2
cm³	1209	463	746	324	39	9.62	8.25	1.37	1582		
(c)	**Soil (Kawasaki Clay)**			**Water**	**Amounts of Component Cement (Early Strength)**	**CNF (ACC-CNF-S)**			**Total**	**CNF Addition Rate (%)**	**Cement Addition (kg/m³)**
	Total	**Soil Particle**	**Water**			**Total**	**Solid Content**	**Water**			
g	2000	1237	763	0	124	331.9	6.64	325.3	2456		
g/cm³	-	2.671	1.023	1.00	3.14	-	1.5	1.0	1.556	0.5	78.4
cm³	1209	463	746	0	39	329.7	4.43	325.3	1578		

3. Effect of CNF on Strength Increase

3.1. Unconfined Compressive Strength

Unconfined compression tests were conducted on the specimens prepared by Method 1 to examine the strength of the treated soil. The method was in accordance with JIS A 1216 and JGS 0511 [21,22]. In the tests, a cylindrical specimen of 50 mm diameter × 100 mm height was compressed in the one-dimensional direction at a strain rate of 1%/min to investigate the axial compressive stress and strain. The strength of the soil was determined by taking the peak value of the relationship as the unconfined compressive strength. The tests were carried out 7, 28, and 112 days after preparing the specimens. The strength of treated soil is known to increase linearly with the logarithmic axis of time, and the specimens were cured for four times as many days each. The stress–strain relationship for each test case is shown in Figure 5. Figure 6 shows the time change of the unconfined compressive strength, together with the values normalised to the strength of the treated soil without CNF.

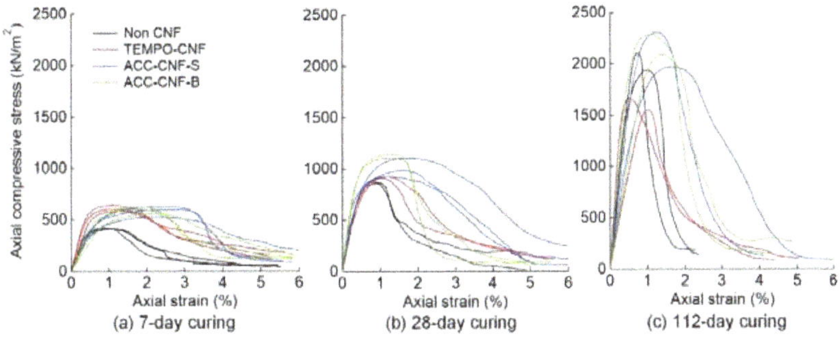

Figure 5. Relation between stress and strain in unconfined compression tests (Method 1).

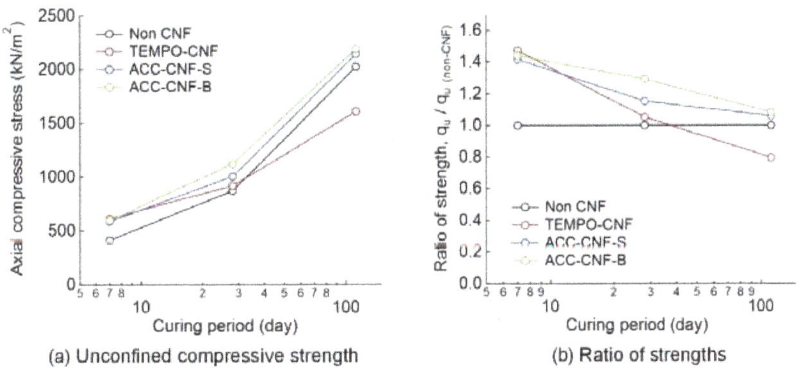

Figure 6. Time change of unconfined compressive strength (Method 1).

The strength at the age of 7 days was 1.4–1.5 times higher in the soil treated with CNF than in the soil without CNF treatment, indicating the effect of CNF. The strength of the soil at 28 days of age was 1.1–1.3 times higher than that of the soil without CNF treatment, whereas the ratio of these strengths was smaller than that of the soil at the age of 7 days. At the age of 112 days, the trend was more pronounced, and the strength of the treated soil with TEMPO-CNF was lower than that of the treated soil without CNF. The strength of the soil treated with ACC-CNF was similar to that of the treated soil without CNF. This tendency can be clearly observed in Figure 6b, where the strength ratios of the treated soil without CNF are shown. It was found that the CNF mixture increased the strength at the initial age of the treated soil but reduced the strength development in the long term. This trend is more evident for the TEMPO-CNF. Increasing the initial strength and reducing the long-term strength is advantageous in cement treatment. As mentioned in Section 1, one of the challenges of cement treatment is the low short-term strength, and if the long-term strength is too high, the treated soil can hinder the rehabilitation of geotechnical structures. CNFs have the potential to solve these problems. It is not clear how the strength development was suppressed with age, although it was possible that in TEMPO-CNF, the carboxylic groups of CNF reacted with the calcium ions of cement and aggregated, suppressing the cement hydration reaction and reducing the increasing rate of long-term strength. On the other hand, the failure strains (axial strains at the peak value) of the soil treated with TEMPO-CNF were not very large, whereas those of the soil treated with ACC-CNF were large. The addition of ACC-CNF increased the failure strain by a factor of 2, indicating that some types of CNFs could mitigate the brittleness, which is a major problem in cement-treated soil.

The soil treated with CNF was photographed using an optical microscope. Figure 7 shows micrographs of the sheared surfaces (peeled and sliding surfaces) produced by the unconfined compression tests. In the photographs of the soil treated with CNF, no fibrous material could be observed in the TEMPO-CNF specimen, whereas it could be observed in the ACC-CNF specimen. The fibrous material was considered to be a CNF. TEMPO-CNF has a high degree of friability, which cannot be captured by optical microscopy, but a bunch of CNFs was seen in the void. Fibrous material could be observed at any given position in the ACC-CNF specimen, suggesting that the CNF was uniformly agitated. It was assumed that the uniformly mixed CNFs contributed to the strength increase. In addition, a number of small voids appeared in the soil treated with CNF, especially in the case of the TEMPO-CNF mixture. This was attributed to the reduced fluidity of the CNF mixture, which allowed air to enter the soil. A few small voids did not contribute significantly to the strength loss, and therefore would not affect the strength development much.

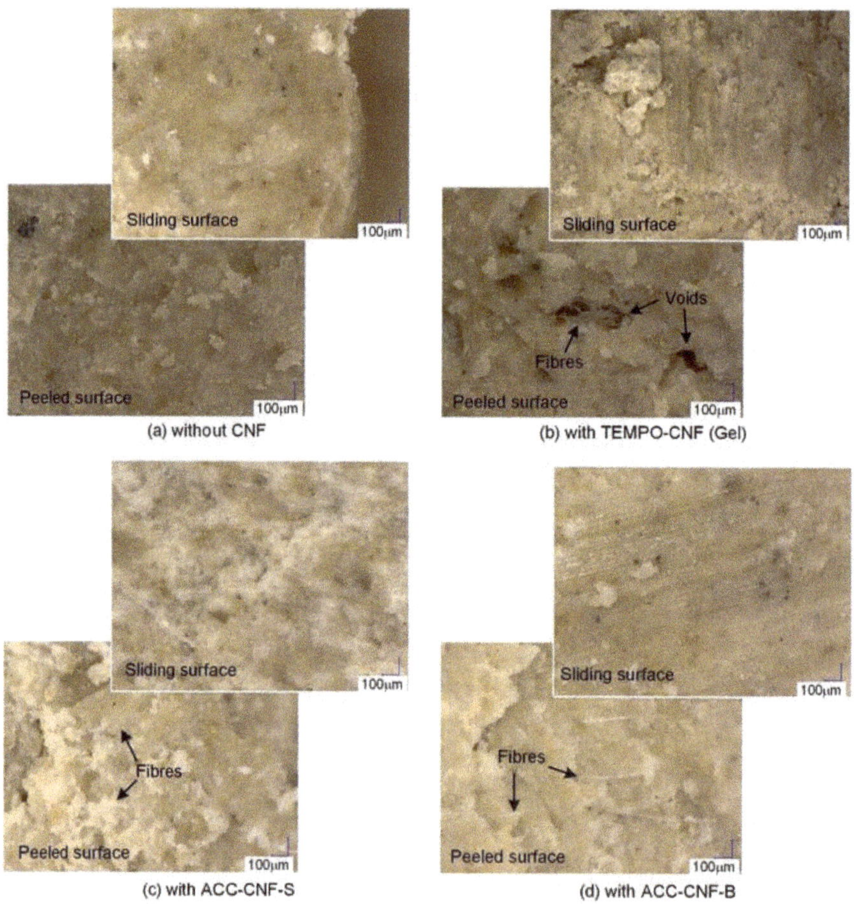

Figure 7. Appearance of sheared surface (Method 1).

3.2. Flexural Strength

One of the problems of cement-treated soil is its low tensile and flexural strengths. As it was anticipated that the CNF mixture would improve these properties, some tests were conducted to investigate the flexural strength. Since the flexural failure of treated soil depends on its tensile strength, its flexural strength can indirectly indicate the tensile strength of the treated soil. The method was implemented in accordance with JIS R 5201 [20]. In this test, a prismatic specimen (40 mm high × 40 mm deep × 160 mm long) was placed on support rolls placed at a distance of 100 mm, and the roll was loaded in one direction at the centre of the specimen at a loading rate of 50 N/s. Then, the relationship between bending stress and deflection was investigated, and the peak value was taken as the flexural strength. Figures 8 and 9 show the relationship between the bending stress and deflection for each test case and the micrographs of the tensile fracture surfaces, respectively.

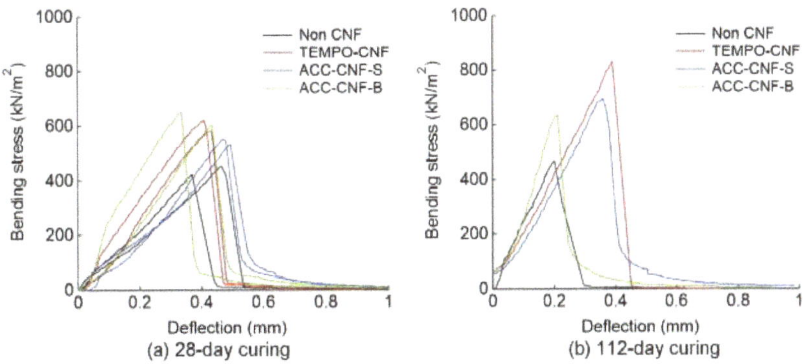

Figure 8. Relation between bending stress and deflection in bending strength tests (Method 1).

Figure 9. Appearance of tensile fracture surface (Method 1).

The flexural strength of the specimens without CNF was approximately 400–470 kN/m^2, which was 25–50% of the unconfined compressive strength, regardless of the age of the specimens. Previous studies [23–25] showed that the flexural strength was approximately 20–50% of the unconfined compressive strength (on average around 30%), and our results were consistent with these findings. The flexural strength of the specimens with TEMPO-CNF was 590 kN/m^2 (at 28 days of age) and 810 kN/m^2 (at 112 days of age), indicating that the flexural strength increased with the addition of CNF. The ratio of flexural strength to unconfined compressive strength was as large as 50–65%, which was not the case for normal cement-treated soil. The increase in flexural strength can be attributed to the increase in the tensile strength. The flexural strength of the specimens with ACC-CNF was 560 kN/m^2 (at 28 days of age) and 660 kN/m^2 (at 112 days of age), and the ratio to the unconfined compressive strength was 33–56%. The addition of ACC-CNF also increased the flexural strength, although not as much as that of TEMPO-CNF. The low flexural and tensile strength of the treated soil is frequently a problem in the field, and the addition of CNF was found to be a possible solution to this problem. The micrographs showed fibrous material in the ACC-CNF specimen, as well as in the sheared surfaces from the unconfined compression test, while the TEMPO-CNF specimen showed voids and material that could be CNF in the void.

4. Effect of CNF on Strength Variability

4.1. Dispersion of Unconfined Compressive Strength

CNF has dispersive properties that make substances in liquids homogeneous, and this property has been considered for use in a variety of applications. CNF is also expected to improve the dispersibility of cement when it is mixed with the soil, and its thixotropy property increases the uniformity of cement. The uniform mixing of cement with soil could reduce the variation in strength. This section presents the test results regarding the effect of CNF mixing on the strength variability. It is important to consider variability, as a reduction in the strength variability can reduce the design strength. Figure 10 shows the relationship between the frequency distribution of intensity occurrences and the design strength. During design, the percent defective is determined at the beginning, and the mean strength (design strength) is determined so that it is satisfied. As shown in the figure, the smaller the variance in strength, the lower the mean strength for the same percent defect. The advantages of reducing the strength and dispersion are significant. This reduces the risk that large dispersals will result in treated soil of extremely high strength as well as high cost of cement and high environmental impact of cement production. As mentioned before, large-strength treated soil can hinder the rehabilitation of geotechnical structures.

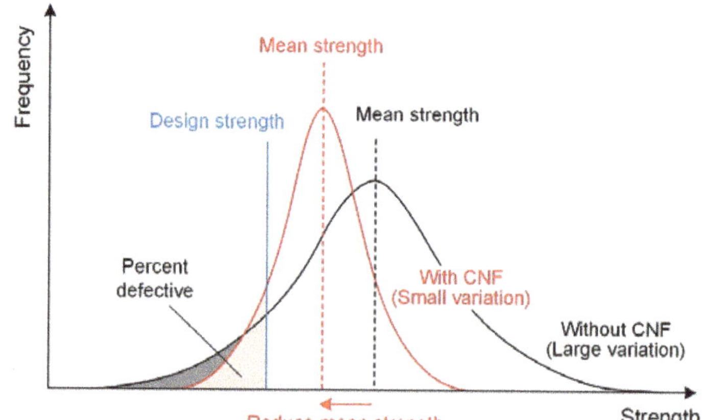

Figure 10. Conceptual diagram of frequency distribution of strength and design strength.

Unconfined compression tests were conducted at 28 days of age on a total of 60 specimens prepared by Method 2 with a short mixing time of 1 min. The axial compressive stress–strain relationships obtained from the tests are shown in Figure 11a,b. Figure 11c shows the results of the unconfined compression test of the treated soil used for the permeability test (preparation method: Method 2, mixing time: 5 min), which is discussed later in Section 5. Most of the peak strengths (unconfined compressive strengths) of the specimens without CNF were below

50 kN/m^2, which was quite low. By contrast, the peak strengths of the specimens with CNF were relatively higher, with unconfined compressive strengths ranging from 100 to 350 kN/m^2. Comparing the mean values, the unconfined compressive strength was 43 kN/m^2 (without CNF) and 219 kN/m^2 (with CNF); the strength ratio was significantly larger: 5.1 times. As the strength ratio of the well-mixed specimens with and without CNF at 28 days of age was 1.1–1.3 times, as mentioned in Section 3, the large strength ratio of 5.1 indicates that the addition of CNF caused the cement to mix evenly with the soil. The failure strains of the specimens with CNF were relatively low owing to the higher strength of the specimens. The average failure strains were 3.9% (without CNFs) and 1.4% (with CNFs).

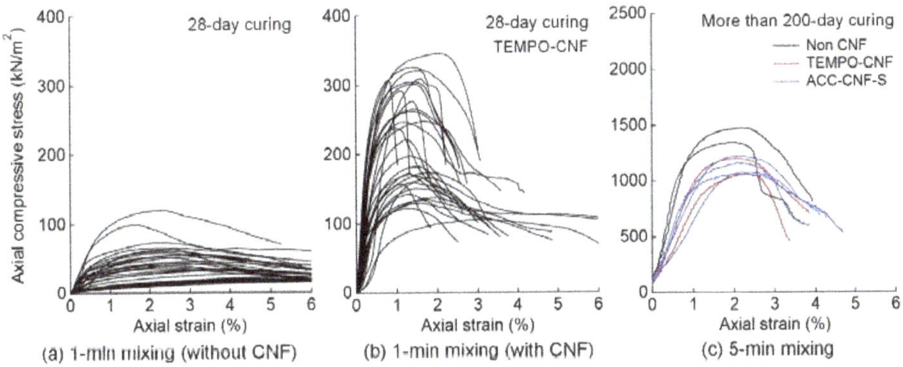

Figure 11. Relation between stress and strain in unconfined compression tests (Method 2).

Figure 12 shows photographs of the sheared surfaces generated by the unconfined compression tests. The sheared surface of the specimen without CNF showed that the solidified portions were scattered within the poorly solidified soil. The unconfined compressive strength was extremely low, which was probably because the solidified portions were not continuous. On the other hand, the sheared surface of the specimen with CNF appeared to be homogeneous, suggesting that the cement spread over the whole specimen, and that this condition led to the development of a relatively high unconfined compressive strength.

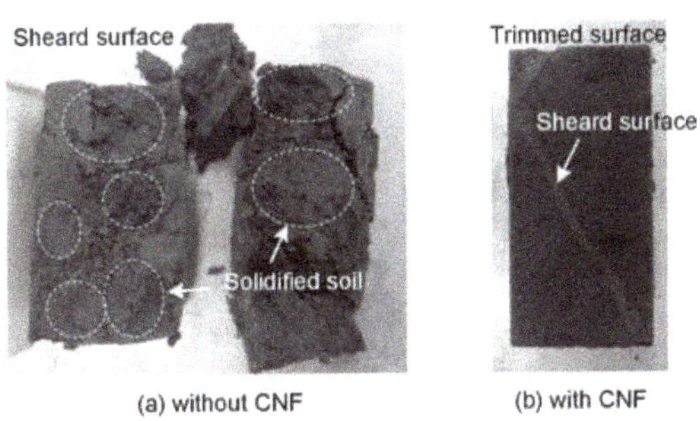

Figure 12. Appearance of sheared and trimmed surfaces (Method 2).

4.2. Evaluation of Variability

Here, variations in the unconfined compressive strength and failure strain were discussed. Figures 13 and 14 show the frequency distributions around the mean values of the unconfined compressive strength and failure strain of the specimens, respectively. Although the strengths of the specimens without CNF were generally low, some specimens developed significant strength, resulting in a strength coefficient of variation of 0.56. On the other hand, the variation coefficient of the strength of the specimens with CNF was

0.33, and the frequency distribution was in the range of 2σ, as shown in the figure. Thus, it was confirmed that the addition of CNF reduced the variation in strength in terms of the coefficient of variation of strength. The coefficients of variation of failure strain for the specimens without and with CNF were 0.51 and 0.26, respectively. For failure strain too, the addition of CNFs reduced the variability. The density and water content of the specimens were also measured, and no differences were observed between the specimens with and without CNF. This suggests that there was no difference in the properties of the base soil or the amount of cement added between the specimens with and without CNF, indicating that CNF had an effect on the strength of the specimens.

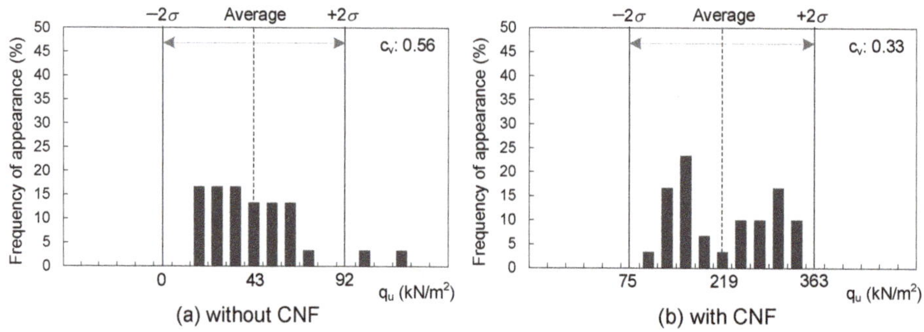

Figure 13. Frequency of appearance of unconfined compressive strength.

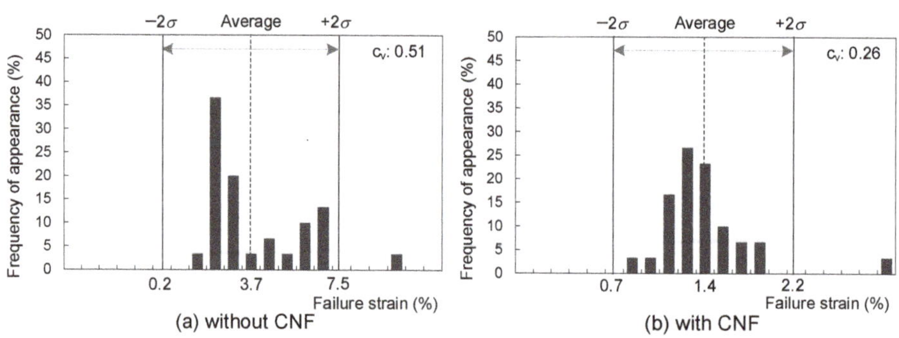

Figure 14. Frequency of appearance of failure strain.

In this study, the amount of cement added was reduced so that the effect of CNF on cement mixing could be more clearly expressed. Considering the strength of the soil treated by the deep mixing method, the amount of cement was small. It will be important in the future to compare the variability of the specimens with increasing cement content. However, there are some methods that produce treated soil with lower strength than the deep mixing method. Examples include the pre-mixing method [26] and the pneumatic flow mixing method [27]. In these methods, the soil is mixed with cement in a mixer or in a pneumatic flow in a pipe and poured in situ before solidification occurs. Strengths of 100–400 kN/m^2 at 28 days of age are generally expected with cement additions of approximately 50–100 kg/m^3. The amount of cement added and the expected strength were similar to those in this study. If the mixing time is short for some reason, the soil could be extremely weak, as described above. Mixing CNFs could prevent such problems.

5. Effect of CNF on Permeability

5.1. Test Method

In addition to the strength of the treated soil, its permeability was also investigated. Previous studies showed that the permeability of clayey soils decreases when they are mixed with cement [28–30]. The treated soil is occasionally used as an impervious material, and permeability is one of the most important aspects of the treated soil. In addition, a high permeability indicates that the calcium ions in the treated soil can easily diffuse to the outside, and therefore, permeability is an important parameter for evaluating the deterioration of the treated soil. It is known that treated soils start degrading from the exposed surface [31–36], which is one of the points that must be understood when using cement treatment in the field. This section discusses how the permeability changes with the addition of CNFs.

The specimens prepared using Method 2 were cured for more than 200 days. The unconfined compression test results of the specimens are shown in Figure 11c. As part of the specimen was used for the permeability test, the unconfined compression tests in this study did not use specimens of the size specified in JIS A 1216 and JGS 0511 [21,22]. It should be noted that because the compression tests were conducted on small prismatic specimens (15 mm wide × 15 mm deep × 30 mm high), the test results could contain errors. All other methods were implemented in accordance with the standards. The unconfined compression test results showed that the strength of the specimens with CNF was lower than that without CNF. This was in agreement with the trend of the results described in Section 3, where it was found that the addition of CNF reduced the long-term strength.

The remainder of the treated soil in the unconfined compression test was subjected to a permeability test. As there is no standard method for permeability testing of treated soil, the tests were conducted using the method utilised in the past study [37]. Figure 15 shows a conceptual diagram of the test apparatus. In this test method, a cylindrical specimen with a diameter of 50 mm and height of 30 mm was permeated by percolating water above the specimen, applying confining pressure from the side through a membrane. The lateral confining pressure prevented the upper pressurised water from leaking at the side of the specimen. The water pressure above the specimen was 400 kN/m^2, and the average hydraulic gradient inside the specimen increased to 1361. This was because the permeability of the treated clayey soil was low, and the hydraulic gradient had to be increased to increase the drainage rate. The drainage water that passed through the specimen was collected to determine the permeability per unit time. The pH, calcium ion concentration, and sodium ion concentration were measured periodically in the collected effluent. The permeability period depended on the specimen: 114 days for 'no CNF', 114 days for TEMPO-CNF, and 49 and 97 days for ACC-CNF-S, respectively. The reason for the long-term measurements was to check whether the addition of CNF would change the condition of the effluent, as it is known that continued permeability degrades the treated soil.

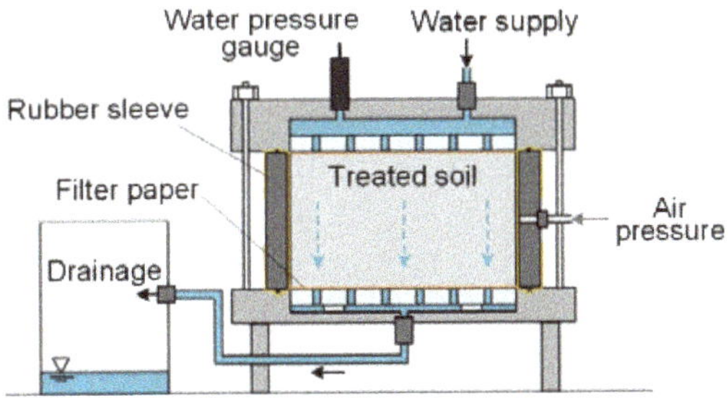

Figure 15. Schematic view of test apparatus.

5.2. Test Results

Figure 16 shows the drainage rate per unit time and the chemically measured values for each day. Note that these values indicate the effluent collected at the time of measurement and not the accumulated effluent. The drainage rate per unit time, which indicates the permeability, showed that the permeability of the specimens with TEMPO-CNF was approximately 35% lower and that of the specimens with ACC-CNF was approximately 25% higher, depending on the type of CNF. It was not simply that the CNF itself clogged the pores and reduced the permeability. Figure 17 shows scanning electron microscopy (SEM) images of the permeable and non-permeable specimens. As shown in the images, the specimen with ACC-CNF-S contained fibrous material that could be considered as CNF, and the fibrous material formed a network. On the other hand, CNF did not appear in the specimen with TEMPO-CNF, because TEMPO-CNF has a high degree of fibrillation and cannot be observed at a magnification of 500 times. In any case, the amount of CNF added was found to be small and insufficient to close the pore space. The pore conditions of the soil particles were similar in all cases, and the reason for the change in permeability could not be determined from the SEM photographs. The reason for the change in permeability is not understood, although CNF may have influenced the cement hydration reaction, changing the amount of hydrate formed, or the hydrophilicity of the CNF itself may also have affected the permeability. Incidentally, a difference of 25–35% is not generally considered to be significant, as the permeability frequently varies by orders of magnitude depending on the pore conditions and the amount of cement added. It is necessary to conduct more tests on more specimens to improve the accuracy of the results. In other words, it could be said that the addition of CNF did not change the permeability significantly, at least on a per-digit basis.

The pH, calcium ion concentration, and sodium ion concentration of the drainage water were also measured and are shown in Figure 16. The drainage was the result of the pore water in the treated soil being pushed out, which indicates the state of the pore water. The pH values were similar, indicating that the addition of CNF did not change the alkaline atmosphere of pore water, and that the environment in all specimens was conducive to cement solidification. The concentration of calcium ions was higher in the TEMPO-CNF specimens. The low permeability of the specimens may have resulted in a higher supply of calcium from the cement hydrate to pore water, or the calcium bound to the carboxylic groups may have been more easily leached than the cement hydrate. Sodium ions were detected early, but not thereafter. The fact that the base soil was marine clay and that the water content ratio was adjusted with artificial seawater suggested that the treated soil contained sodium. The concentration of sodium ions in the specimens with TEMPO-CNF was higher than that in the specimens without TEMPO-CNF, suggesting that the sodium bound to the carboxylic groups may have leached out. The effect of CNF was not apparent in terms of the chemical properties of the effluent.

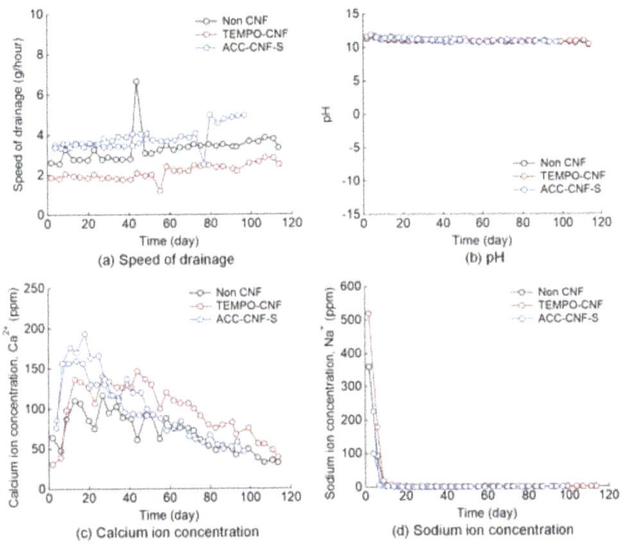

Figure 16. Chemical properties of drained water.

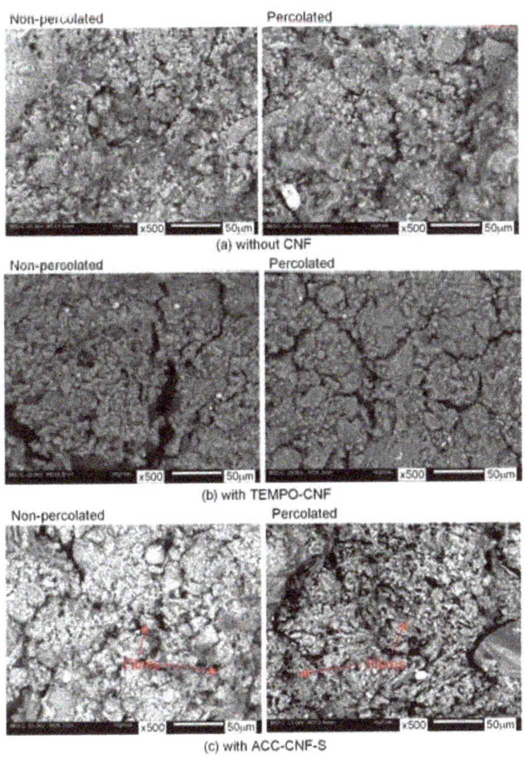

Figure 17. SEM microphotograph of specimens.

6. Conclusions

This study investigated methods of mixing CNF, the increase in strength and the reduction of variability by adding CNF, and the change in permeability. The results can be summarised as follows:

1. The method of mixing either the soil or the additive (cement, CNF) in a dry state was attempted to reduce the amount of water in the treated soil. Both methods could be used, but for on-site mixing, it was considered more practical to mix cement and CNF in dry powder or highly concentrated gel form.
2. During the mixing of cement and soil, the water absorption and thickening properties of the CNF reduced the flowability, but as a result, it was possible to mix the cement evenly, which was probably because of the other effects at work. In addition, when the mixing time was extremely short, unevenness was observed between the solidified portions and those that did not solidify in the specimen without CNF, whereas the treated soil was produced uniformly in the specimen with CNF.
3. In regard to the strength development characteristics over time, it was found that the mixture of CNF increased the strength at the initial age, but it reduced the strength development in the long term. Increasing the initial strength and reducing the long-term strength can be advantageous for cement treatment. Observation of the sheared surface showed that a fibrous material, which was considered to be CNF, was widely mixed in the specimens to which CNF with a relatively low degree of friability was added.
4. The effect of shortening the mixing time on the variation in the strength of CNF was investigated. It was observed that the cement tended to mix more evenly in the treated soil with CNF, and the average strength ratio at 28 days of age was 5.1 (strength with CNF/strength without CNF). The coefficients of variation of the strength and failure strain were also smaller when CNF was added. Reducing the variation in the strength of the treated soil has the advantage that the amount of cement added can be reduced, and unwanted high-strength treated soil is not produced.
5. The addition of CNFs increased the flexural strength. This could be attributed to the increased tensile strength. Fibrous material appeared on the tensile fracture surface of the specimen to which CNF with low fibrillation was added. The low flexural and tensile strength is one of the drawbacks of cement-treated soil, and this problem could be solved.
6. The addition of CNF caused a slight change in permeability, although the number of tests in this study was not sufficient to determine a clear effect. It would suggest that the addition of CNF does not significantly change the permeability. The pore water conditions were estimated from the effluent, and it was found that the alkaline atmosphere was the same regardless of the presence or absence of CNF.

Thus, the addition of CNF to cement treatment had beneficial effects, depending on how it was used. This result is a quantum leap forward in encouraging the use of CNF in cement-treated soil. However, at present, the production volume of CNFs is limited, and their unit price is high. When CNFs are used in large quantities for civil engineering, the production volume will increase, and the price will decrease significantly. In practice, the price must be reduced before CNFs can be utilised. In addition, it is important to conduct trial mixing tests before the actual addition of CNF in the field, as the degree of effectiveness and properties may vary depending on the type of soil and the type of CNF. Moreover, if the chemical mechanism, although complex, can be clarified, it will be possible to add CNF more reliably to achieve the required effect. More detailed investigations of properties other than strength and permeability will be conducted in the future.

7. Patents

This paper includes the contents of the following patents applied in Japan: Manufacturing method of ground improvement soil and ground improvement method (JP2021011728A, JP2021011729A).

Author Contributions: Project administration and writing—original draft preparation, H.T.; data curation, S.O.; validation, H.A.; investigation, H.F.; investigation, Y.G.; validation, Y.M.; writing—review and editing, All authors. All authors have read and agreed to the published version of the manuscript.

Funding: This research received no external funding.

Institutional Review Board Statement: Not applicable.

Informed Consent Statement: Not applicable.

Acknowledgments: The authors would like to thank DKS Co. for the supply of TEMPO-CNFs and Marubeni Corporation and Chuetsu Pulp & Paper Co. for the supply of ACC-CNFs. In addition, the authors wish to thank Yosuke Tanaka (Toa Corporation) for his supports during the tests.

Conflicts of Interest: The authors declare no conflict of interest.

References

1. Saito, T.; Nishiyama, Y.; Putaux, J.L.; Vignon, M.; Isogai, A. Homogeneous suspensions of individualized microfibrils from TEMPO-catalyzed oxidation of native cellulose. *Biomacromolecules* **2006**, *7*, 1687–1691. [CrossRef]
2. Abe, K.; Iwamoto, S.; Yano, H. Obtaining cellulose nanofibers with a uniform width of 15 nm from wood. *Biomacromolecules* **2007**, *8*, 3276–3278. [CrossRef]
3. Iwamoto, S.; Abe, K.; Yano, H. The effect of hemicelluloses on wood pulp nanofibrillation and nanofiber network characteristics. *Biomacromolecules* **2008**, *9*, 1022–1026. [CrossRef]
4. Jiao, L.; Su, M.; Chen, L.; Wang, Y.; Zhu, H.; Dai, H. Natural cellulose nanofibers as sustainable enhancers in construction cement. *PLoS ONE* **2016**, *11*, e0168422. [CrossRef] [PubMed]
5. Fu, T.; Moon, R.J.; Zavattieri, P.; Youngblood, J.; Weiss, W.J. Cellulose nanomaterials as additives for cementitious materials. *Cellul. Reinf. Nanofibre Compos.* **2017**, 455–482. [CrossRef]
6. Mejdoub, R.; Hammi, H.; Suñol, J.J.; Khitouni, M.; M'nif, A.; Boufi, S. Nanofibrillated cellulose as nanoreinforcement in Portland cement: Thermal, mechanical and microstructural properties. *J. Compos. Mater.* **2017**, *51*, 2491–2503. [CrossRef]
7. Kitazume, M.; Terashi, M. *The Deep Mixing Method*; CRC Press/Balkema: London, UK, 2013; pp. 1–436. [CrossRef]
8. Consoli, N.C.; Prietto, P.D.M.; Ulbrich, L.A. Influence of fiber and cement addition on behavior of sandy soil. *J. Geotech. Geoenvironmental Eng.* **1998**, *124*, 1211–1214. [CrossRef]
9. Consoli, N.C.; Montardo, J.P.; Donato, M.; Prietto, P.D.M. Effect of material properties on the behaviour of sand-cement-fibre composites. *Ground Improv.* **2004**, *8*, 77–90. [CrossRef]
10. Consoli, N.C.; de Moraes, R.R.; Festugato, L. Split tensile strength of monofilament polypropylene fiber-reinforced cemented sandy soils. *Geosynth. Int.* **2011**, *18*, 57–62. [CrossRef]
11. Estabraph, A.R.; Namdar, P.; Javadi, A.A. Behavior of cement-stabilized clay reinforced with nylon fiber. *Geosynth. Int.* **2012**, *19*, 85–92. [CrossRef]
12. Correia, A.A.S.; Oliveira, P.J.V.; Custodio, D.G. Effect of polypropylene fibres on the compressive and tensile strength of a soft soil, artificially stabilised with binders. *Geotext. Geomembr.* **2015**, *43*, 97–106. [CrossRef]
13. Correia, A.A.S.; Oliveira, P.J.V.; Teles, J.M.N.P.C.; Pedro, A.M.G. Strength of a stabilised soil reinforced with steel fibres. *Geotech. Eng.* **2017**, *170*, 312–321. [CrossRef]
14. Tang, Q.; Shi, P.; Zhang, Y.; Liu, W.; Chen, L. Strength and deformation properties of fiber and cement reinforced heavy metal-contaminated synthetic soils. *Adv. Mater. Sci. Eng.* **2019**, 1–9. [CrossRef]
15. Wang, W.; Zhang, C.; Guo, J.; Li, N.; Li, Y.; Zhou, H.; Liu, Y. Investigation on the triaxial mechanical characteristics of cement-treated subgrade soil admixed with polypropylene fiber. *Appl. Sci.* **2019**, *9*, 4557. [CrossRef]
16. Correia, A.A.S.; Casaleiro, P.D.F.; Rasteiro, M.G.B.V. Applying multiwall carbon nanotubes for soil stabilization. *Procedia Eng.* **2015**, *102*, 1766–1775. [CrossRef]
17. Correia, A.A.S.; Rasteiro, M.G. Nanotechnology applied to chemical soil stabilization. *Procedia Eng.* **2016**, *143*, 1252–1259. [CrossRef]
18. Saito, T.; Kimura, S.; Nishiyama, Y.; Isogai, A. Cellulose nanofibers prepared by TEMPO-mediated oxidation of native cellulose. *Biomacromolecules* **2007**, *8*, 2485–2491. [CrossRef]
19. Kondo, T.; Kose, R.; Naito, H.; Kasai, W. Aqueous counter collision using paired water jets as a novel means of preparing bio-nanofibers. *Carbohydr. Polym.* **2014**, *112*, 284–290. [CrossRef]
20. Japanese Standards Association. *JIS R 5201—Physical Testing Methods for Cement*; Japanese Standards Association: Tokyo, Japan, 2019; pp. 1–107.
21. Japanese Standards Association. *JIS A 1216—Method for Unconfined Compression Test of Soils*; Japanese Standards Association: Tokyo, Japan, 2020; pp. 1–16. (In Japanese)
22. Japanese Geotechnical Society. JGS 0511—Method for unconfined compression test of soils. In *Japanese Geotechnical Society—Laboratory Testing Standards of Geomaterials*; Japanese Geotechnical Society: Tokyo, Japan, 2020; pp. 1–6.

23. Terashi, M.; Tanaka, H.; Mitsumoto, T.; Niidome, Y.; Honma, S. *Fundamental Properties of Lime and Cement Treated Soil*, 2nd ed.; Port and Airport Research Institute: Kanagawa, Japan, 1980; Volume 19, pp. 33–62. (In Japanese)
24. Kanda, T.; Tanaka, T.; Nohara, H.; Kubota, J. Development of high-performance fiber-reinforced soil cement. *Annu. Rep. Kajima Tech. Res. Inst.* **1999**, *47*, 79–85. (In Japanese)
25. Namikawa, T.; Koseki, J. Evaluation of tensile strength of cement-treated sand based on several types of laboratory tests. *Soils Found.* **2007**, *47*, 657–674. [CrossRef]
26. Coastal Development Institute Technology. *The Premixing Method*; CRC Press: London, UK, 2003; pp. 1–152. [CrossRef]
27. Kitazume, M. *The Pneumatic Flow Mixing Method*; CRC Press/Balkema: London, UK, 2017; pp. 1–233. [CrossRef]
28. Chew, S.H.; Kamruzzaman, A.H.M.; Lee, F.H. Physicochemical and Engineering behavior of cement treated clays. *J. Geotech. Geoenvironmental Eng.* **2004**, *130*, 696–706. [CrossRef]
29. Takahashi, H.; Kitazume, M. Consolidation and permeability characteristics on cement treated clays from laboratory tests. In Proceedings of the International Symposium on Engineering Practice and Performance of Soft Deposits, Toyonaka, Japan, 2–4 June 2004; pp. 187–192.
30. Vähäaho, I.; Kangas, H.; Takahashi, H.; Kitazume, M. Effect of permeability and stiffness of treated column on consolidation phenomenon of improved ground. In Proceedings of the 16th International Conference on Soil Mechanics and Geotechnical Engineering, Osaka, Japan, 12–16 September 2005; pp. 1269–1274.
31. Terashi, M.; Tanaka, H.; Mitsumoto, T.; Honma, S.; Ohhashi, T. *Fundamental Properties of Lime and Cement Treated*, 3rd ed.; Port and Airport Research Institute: Kanagawa, Japan, 1983; Volume 22, pp. 69–96. (In Japanese)
32. Kitazume, M.; Nakamura, T.; Terashi, M.; Ohishi, K. Laboratory tests on long-term strength of cement treated soil. In Proceedings of the 3rd International Conference on Grouting and Ground Treatment, New Orleans, LA, USA, 10–12 February 2003; Volume 1, pp. 586–597.
33. Hayashi, H.; Nishikawa, J.; Ohishi, K.; Terashi, M. Field observation of long-term strength of cement treated soil. In Proceedings of the 3rd International Conference on Grouting and Ground treatment, New Orleans, LA, USA, 10–12 February 2003; Volume 1, pp. 598–609.
34. Ikegami, M.; Ichiba, T.; Ohishi, K.; Terashi, M. Long-term strength change of cement treated soil at Daikoku Pier. In Proceedings of the Soft Ground Engineering in Coastal Areas, Yokusuka, Japan, 28–29 November 2002; pp. 241–246.
35. Committee on Geocement. A study on long term stability of soil-cement columns improved by geocement. *Cem. Concr.* **2014**, *804*, 9–14. (In Japanese)
36. Takahashi, H.; Morikawa, Y.; Fujii, N.; Kitazume, M. Thirty-seven-year investigation of quicklime-treated soil produced by deep mixing method. *Ground Improv.* **2018**, *171*, 135–147. [CrossRef]
37. Takahashi, H.; Morikawa, Y.; Uemura, T. Trial tests to promote degradation of cement treated soil by percolation technique. In Proceedings of the Deep Mixing 2021, Online, 15–17 June 2021.

applied sciences

MDPI

Article

Novel Approach for Suppression of Ettringite Formation in Sulfate-Bearing Soil Using Blends of Nano-Magnesium Oxide, Ground Granulated Blast-Furnace Slag and Rice Husk Ash

Khaled Ibrahim Azarroug Ehwailat, Mohd Ashraf Mohamad Ismail * and Ali Muftah Abdussalam Ezreig

School of Civil Engineering, Universiti Sains Malaysia, Engineering Campus, Nibong Tebal 14300, Seberang Prai Selatan, Pulau Pinang, Malaysia; Khaledazrog@student.usm.my (K.I.A.E.); Aliezreig@student.usm.my (A.M.A.E.)
* Correspondence: ceashraf@usm.my

Citation: Ehwailat, K.I.A.; Ismail, M.A.M.; Ezreig, A.M.A. Novel Approach for Suppression of Ettringite Formation in Sulfate-Bearing Soil Using Blends of Nano-Magnesium Oxide, Ground Granulated Blast-Furnace Slag and Rice Husk Ash. *Appl. Sci.* **2021**, *11*, 6618. https://doi.org/10.3390/app11146618

Academic Editors: Paulo José da Venda Oliveira and António Alberto Santos Correia

Received: 8 June 2021
Accepted: 11 July 2021
Published: 19 July 2021

Publisher's Note: MDPI stays neutral with regard to jurisdictional claims in published maps and institutional affiliations.

Copyright: © 2021 by the authors. Licensee MDPI, Basel, Switzerland. This article is an open access article distributed under the terms and conditions of the Creative Commons Attribution (CC BY) license (https://creativecommons.org/licenses/by/4.0/).

Abstract: The treatment of sulfate-bearing soil with calcium-based stabilizers such as cement or lime often results in ettringite formation, consequently leading to swelling and strength deterioration. Ettringite formation has negative environmental and economic effects on various civil engineering structures. This study was conducted to investigate the use of different materials (nano–magnesium oxide (M), ground granulated blast-furnace slag (GGBS), and rice husk ash (RHA)) for gypseous soil stabilization to prevent ettringite formation. Various tests were performed, including flexural strength, unconfined compression strength, linear expansion, and microstructure analysis (SEM/EDX), on lime (L)-, (M)-, (M-RHA)-, (M-GGBS)-, and (M-GGBS-RHA)-stabilized gypseous soil samples to determine their properties. The results indicated that the swelling rates of the soil samples mixed with 20% M-RHA, M-GGBS, and M-GGBS-RHA binders were much lower (less than 0.01% of volume change) than those of the soil samples mixed with 10% and 20% lime-stabilized binders after a curing period of 90 days. Meanwhile, the strengths of the soil samples mixed with 20% of M-RHA, M-GGBS, and M-GGBS-RHA soil specimens after soaking of 90 days were obviously higher (with a range from 2.7–12.8 MPa) than those of the soil samples mixed with 20% of lime-stabilized binder. The SEM and EDX results showed no ettringite formation in the M-RHA-, M-GGBS-, and M-GGBS-RHA-stabilized soils. Overall, the test results proved the potential of M-RHA, M-GGBS, and M-GGBS-RHA as effective soil stabilizers.

Keywords: sulfate-bearing soil; ettringite; soil stabilization; gypseous soil; calcium-based stabilizer; non-calcium-based stabilizer

1. Introduction

Gypseous soils are soils containing significant amounts of calcium sulfates ($CaSO_4 \times 2H_2O$). Such soils are widely distributed in many countries, such as the US, UK, India, Africa, and Australia [1,2]. Gypseous soils are common in highway pavement construction, but they often pose many problems, such as cracking, collapsing, and pavement layer settlement, thereby making pavement and highway construction on such soil a challenge to geotechnical engineers [1,3,4]. One of the proven economical ways to improve the engineering properties of soil samples is via chemical soil stabilization [5–7]. The most common soil stabilization method is the addition of cement or lime to increase the soil's durability and strength [1,3–5,8–11]. However, adverse effects of ground heave have been reported when using cement and lime to stabilize sulfate-rich soil. Sulfate-induced heave is well known to have serious negative impacts on the general performance of runways, highways, and other structures that are constructed on lime- or cement-stabilized sulfate-rich soil layers [1,4,11]. The application of lime or cement on sulfate-rich soils frequently elicits a series of reactions that result in the formation of ettringite [$Ca_6Al_2(SO_4)_3(OH)_{12} \times 26H_2O$], which is an expansive sulfate-mineral-based compound [2,4,12]. Ettringite contains 26 water molecules and can resist expansive increases in volume when exposed to hydration, thereby leading to a significant heave effect, swelling, and decline in compressive strength [13]. Many types of equipment and tests have been utilized to determine the negative impact of sulfate on soil stabilization. The most common equipment and tests are unconfined compressive strength tests used to reveal the influence of ettringite on the strength of soil samples containing sulfate, and linear expansion to observe the magnitude of swelling induced by ettringite formation. Adeleke et al. [4] demonstrated understanding of the mechanisms of sulfate-bearing soil when stabilized with a calcium-based stabilizer. Unconfined compressive strength and linear expansion tests were conducted on gypseous soil containing a high level of calcium sulfate (10%, 15%, and 20%) treated with 7%, 8%, 9%, and 10% calcium-based stabilizer (cement). The outcomes illustrated that the strength reduced by a factor range of 6–47% at 28 days of curing, while the swelling magnitude depended mainly on the sulfate concentration, with the highest rate of swelling observed in the presence of sulfate at 10%.

Treatment of sulfate-containing soils (SO_4) with calcium-based stabilizers encourages a reaction between the calcium content of the stabilizer (lime or cement) and the sulfate/aluminum content of the soil. With the presence of water, this generates large quantities of ettringite (a crystalline mineral), causing further changes in volume and damage to the structures built on such soil (see Equation (1)) [2,14,15].

$$6Ca^{2+} + 12OH^- + 2Al^{3+} + 3SO_4^{2-} + 26H_2O \rightarrow Ca_6Al_2(SO_4)_3(OH)_{12} \times 26H_2O \quad (1)$$

Ettringite can transform into another expanding mineral: thaumasite[$Ca_6Si_2(CO_3)_2(SO_4)_2(OH)_{12} \times 24H_2O$]. The transformation occurs due to the replacement of aluminum and sulfate with silica and carbonate at temperatures < 15 °C (see Equation (2)) [2,12].

$$\begin{aligned} Ca_6[Al(OH)_6]_2 \times \ & (SO_4)_3 26H_2O + 2H_2SiO_4^{2-} + 2CO_3^{2-} + O_2 \\ & \rightarrow Ca_6[Si(OH)_6]_2 \times (SO_4)_{23} \times (CO_3)_2 \times 24H_2O + 2Al(OH)_4^- + SO_4^{2-} + 4OH^- + 2HO_2 \end{aligned} \quad (2)$$

Chemical modification remains the most effective method of soil stabilization. Among the available chemical admixtures, the most extensively used for soil stabilization is lime [16]. Aldawood et al. [1] performed unconfined compressive strength, X-ray diffraction, and scanning electron microscopy tests on gypseous soil (with different gypsum amounts: 5%, 15%, and 25%) to investigate the influence of lime applied at different dosages (0%, 3%, 5%, and 10%) to stabilize sulfate-bearing soil. The results noted that the mechanical properties of soil specimens were not only affected by the curing time, but also depended on the lime dosage. The optimum reduction in strength was detected to be 5%, and the results of micro tests confirmed the ettringite formation and the pozzolanic products. Researchers and engineers are now progressively exploring novel materials that can serve as soil stabilizers to overcome the lime-induced heave problems often encountered in sulfate-bearing soils during the construction of civil structures. Novel methods should be able to repress heave and ettringite formation in sulfate-rich soils. Materials that exhibit pozzolanic properties, such as rice husk ash (RHA) and ground granulated blast-furnace slag (GGBS), are believed to be suitable sulfate soil stabilizers because such materials generally consume lime [17], thereby reducing its availability for the formation of expansive products while improving the strength of the soil [18].

Annually, millions of tons of industrial and agricultural waste are produced globally, creating serious problems such as environmental pollution and the deterioration of the health of many inhabitants of the world [10,19]. Due to this fact, studies are needed to find alternative uses for waste materials for economic purposes. Concurrently, the sources of cement, lime, and other raw materials, especially near large urban centers, are depleting and thus becoming more expensive, in addition to existing transportation costs and environmental restrictions [10]. Considering these facts, waste materials such as RHA and GGBS can be used as alternative raw materials in the production of building materials to reduce the cost of building projects while protecting the environment [9,10,17,19–21].

The processes involved in the production of lime and Portland cement (PC) contribute significantly to environmental pollution; for instance, the production of one ton of (PC) leads to the emission of almost 0.95 tons of CO_2 and requires almost 5000 MJ of energy to complete [9,22–24]; for lime production, these values are ~0.79 tons of CO_2 and ~3200 MJ of energy [9]. This calls for the use of waste materials as a complete or partial replacement of conventional binders, as proposed by various researchers [6,10,12,20,25–29]. GGBS is a waste product from the iron industry and has been considered a suitable material for soil stabilization [19]. In terms of CO_2 emission and energy consumption, only 1300 MJ of energy is required to produce one ton of GGBS, and the process emits only 0.07 tons of CO_2 [2,26]. GGBS can be activated via lime for soil stabilization, as it is a latent hydraulic material [16,30]. This process involves three basic reactions: the lime–soil reaction, lime–GGBS reaction, and GGBS–soil reaction [30]. Calcium silicate hydrates (CSH), calcium aluminosilicate hydrates (CASH), and calcium aluminate hydrates (CAH) are the major hydration products of lime–GGBS-stabilized soil [6,24,26].

GGBS can serve as a partial or complete replacement for cement or lime to prevent or reduce ettringite formation in stabilized sulfate soils. The alumina and silica contents of GGBS can quickly react with the calcium content in soil to form a cementitious gel, thereby reducing or preventing ettringite formation [6,15]. GGBS can also reduce water availability and permeability through the formation of a denser cementitious matrix, which improves resistance to both external and internal sulfate attacks [17,21,31,32].

One agricultural waste material is the husk separated from rice. Due to its limited applications, this material is usually burnt and destroyed. The product remaining after the burning phase, called RHA, has unique characteristics, such as a high silica content [10,33]. Bazyar [10] indicated that RHA has positive impacts on the improvement of gypsum clays' mechanical properties with lime. The author also recommended 6–8% lime and 8–10% RHA as the optimum amounts for stabilization of gypseous soil. RHA has an adequate amount of silica with a high specific surface area, which is very suitable for activating the soil and lime reaction.

Reactive magnesia (M) is a more effective substance in GGBS activation compared to lime, as it facilitates a higher rate of strength development [2,6,20,25,26,28]. Reactive M is commonly obtained from the calcination of magnesite in the temperature range of 700–800 °C (below the temperature for dead-burned M (>1400 °C)). About 2400 MJ of energy is required to produce one ton of reactive M, and the process emits about 1.4 tons of CO_2 [23,24,27]. Reactive M is used at low concentrations during GGBS activation; hence, the overall CO_2 emission during the reaction of M with GGBS is considered relatively lower than that of PC. For instance, the production of one ton of M-GGBS with M at the optimum M:GGBS ratio of 1:9–1:4 only emits about 0.20–0.34 tons of CO_2 [9,26]. The study by Li et al. [2] also showed that M-GGBS binder protected gypseous soil samples from swelling and contributed to better strength after soaking compared to PC.

This work aimed to establish a new method to stabilize soils subjected to internal sulfate attacks by gypsum ($CaSO_4$) using a combination of nano–magnesium oxide (M), RHA, and GGBS. For this purpose, unconfined compressive strength (UCS), linear expansion (LP), and flexural strength (FS) tests were performed before and after exposure to water. In this study, two steps were taken to achieve the objectives:

Step 1: Running compaction characteristics for each mixture.

Step 2: Performing UCS before and after sulfate exposure, LP, and FS tests.

The curing times applied were 7, 28, and 90 days. In addition to the UCS, LP, and FS tests, SEM and EDS were performed to study the stabilized soil samples to determine their mineralogical and microstructural performance.

2. Materials and Methods

2.1. Materials

The materials utilized for this research were kaolin clay (K), hydrated lime (L), calcium sulfate (gypsum) (G), (M), (RHA), and (GGBS). Kaolin clay was supplied by KAOLIN

(Malaysia) SDN BHD under the brand name MK40 as a white, finely ground, odorless powder. Kaolin clay was used due to the following considerations: (1) it is one of the major compositions of natural clay minerals; (2) it has a uniform composition, minimal organic content, and consistent and uniform mineralogy; (3) it has a low cation exchange capacity [4,20]; (4) it has a higher alumina content than most other costly minerals and can thus release more alumina in high pH environments, thereby participating in ettringite mineral formation and increasing susceptibility of sulfate attack [12]. For these reasons, kaolin clay is a suitable control soil in the soil stabilization process. Table 1 summarizes the major properties of kaolin clay. A hydrometer test of clay was conducted according to BS EN ISO 17892–4:2016 [34] with the grading curve shown in Figure 1.

Table 1. Characteristics of the kaolin clay used in this study.

Properties	Value
Liquid limit (%)	57.78
Plastic limit (%)	38.13
Shrinking limit (%)	4.4
Plasticity index	19.65
Sand (%)	-
Silt (%)	88.35
Clay (%)	11.65
Electric conductivity (μS/cm)	320
pH	5
Specific gravity (G_s)	2.46
Water content (%)	1.01
Optimum moisture content (%)	29
Maximum dry density (Mg/m^3)	1.326

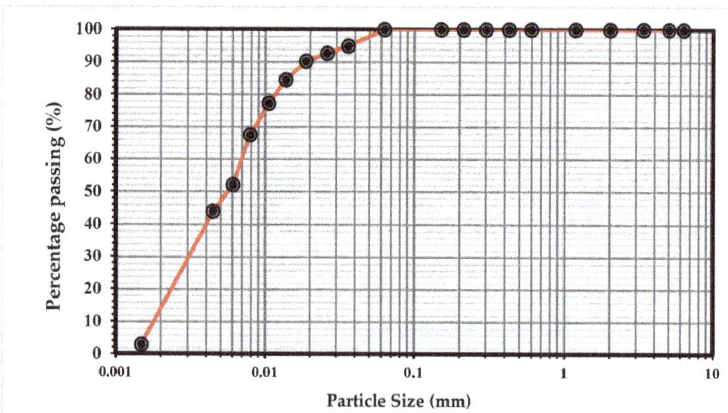

Figure 1. Particle size distribution (PSD) of the kaolin clay used in this study.

L, G, and RHA were collected from Sungai Jawi, 14200 Penang, Malaysia. However, G ($CaSO_4 \times 2H_2O$) was selected in this research due to its much lower solubility compared to other sulfate types (potassium sulfate, magnesium sulfate, and sodium sulfate), as presented in Table 2. Moreover, it is one of the sulfates that contain calcium, and it is logical that it would form ettringite if a non-calcium-based stabilizer had been used, as presented in Table 2.

Table 2. Most common types of sulfate found in soil.

Sulfate Type	Common Name	Chemical Formula	Solubility
Calcium	Selenite, gypsum	$CaSO_4 \times 2H_2O$	1.44
Potassium	Arcanite	K_2SO_4	130
Magnesium	Epsomite	$MgSO_4 \times 7H_2O$	225
Sodium	Themadite, mirabilite	$Na_2SO_4 \times 10H_2O$	>225

RHA is a byproduct of the rice milling industry that is produced from the burning of rice husk. The processing of 1000 kg of rice grain is estimated to generate about 200 kg of RHA, which, when burnt, should produce about 40 kg of RHA [35]. Currently, RHA is commercially used in the extraction of silica and as a pozzolanic material [36]. The phase of the silica content of RHA (amorphous or crystalline) is a function of the duration and temperature of the burning process. Fine particle-sized amorphous silica is more reactive, and many researchers [36,37] have shown that burning RHA at a temperature range of 500–650 °C results mostly in the crystalline form of silica, as it can be effectively transformed from its active state into a crystalline form. In this research, RHA was prepared by burning RHA at a temperature range of 350 to 550 °C for three hours to retain the amorphous silica content.

GGBS is a byproduct of the pig-iron manufacturing process and is formed through the rapid cooling of molten iron slag to retain its amorphous structure, followed by grinding to increase its specific surface area. The GGBS used in this study was collected from MDC Sungai Pentani Company, Malaysia. It was chosen to reduce the rate of ettringite formation through the provision of more Al and Si, which react with Ca^{2+} to form complex cementing gels [12,31,32]. The denser structure and lower calcium ion content of GGBS should contribute to superior sulfate resistance [6,38].

In this research, M was chosen, as it is a green and low-carbon stabilizer for clay soil. It was collected from Hang Zhou Jiu Peng New Material Co., Ltd., China as a white, fine, crystalline powder. The production of reactive M requires low temperatures, consumes less fuel, and emits less CO_2 than that of Portland cement (PC) [2]. Nanostructured binders have been receiving more attention recently because the main hydrate CSH gel of cement is also a natural nanostructured material [5,22]. Furthermore, it reduces sulfate-induced expansion and has a lower crystallinity and higher reactivity [2,15], thus reacting significantly more quickly with water [25,39].

An X-ray fluorescence test (XRF) was carried out to determine the chemical properties of all materials (K, L, M, GGBS, RHA, and G), which are summarized in Table 3.

Table 3. Chemical compositions of kaolin clay, L, M, G, GGBS, and RHA.

Oxides	Characteristic (%)					
	Kaolin Clay	L	M	G	GGBS	RHA
CaO	-	-	-	-	37	0.41
$CaOH_2$	-	92	-	-	-	-
SiO_2	58	2.5	-	-	32.7	93.1
Al_2O_3	38	0.9	-	-	15.3	0.21
Ca_2SO_4	-	0.1	-	99	-	-
SO_3	-	-	0.03	-	4.7	-
MgO	-	3.5	99.5	-	8.1	1.59
Cl	-	-	0.01	0.005	-	-
Fe	-	0.06	0.01	0.005	-	0.21
H_2O	-	0.7	0.2	-	-	-
Loss on ignition	11–14	0.24	0.25	0.99	2.2	4.48
pH	5	11.85	10.83	7.5	10.23	8.97
Specific gravity	2.46	2.23	3.58	2.34	2.96	2.11

2.2. Sample Preparation

Sulfate-bearing soil or gypseous soil was prepared artificially by mixing kaolin clay with 10% gypsum (by dry weight of soil). The concentration of sulfate was determined as the worst case according to AASHTO [40–42]. The risk of different sulfate levels is summarized in Table 4.

Table 4. Severity of sulfate levels.

Risk Level	Sulfate Concentration	
	Parts per Million	Percentage of Dry Weight
Low risk	>3000 ppm	>3%
Moderate risk	3000–5000 ppm	3–5%
Moderate to high risk	5000–8000 ppm	5–8%
High to unacceptable risk	>8000 ppm	>8%
Unacceptable risk	>10,000 ppm	>10%

Cylinder samples were 50 mm in diameter and 100 mm in height and were prepared as reported by [2,4,8] for tests of linear expansion (LP), flexural strength (FS), and unconfined compressive strength (UCS). Each mixture system was compacted with an optimum moisture content (OMC) and maximum dry density (MMD) following the BSEN 13286–2:2010 standard [43]. After compaction of samples, each sample was covered using cling film to reduce moisture loss.

The total binder contents were fixed at 10% and 20% based on the weight of the soil (see Table 5) for each system (unary, binary, and ternary). This was achieved using activator (L) calcium-based stabilizer and (M) non-calcium-based stabilizer at dosages of 10% and 20%, with (G) calcium sulfate dosed at 10% (as a worst case) into kaolin clay. The ratios of M stabilized with (RHA and GGBS) (Figure S1 in supplementary materials shows a comparison of particle size distribution between RHA and GGBS) were set as 1:3, 1:1, and 3:1 in a binary system, and 1:05:05, 1:1:2, 1:2:1, and 1:2.5:0.5 in a ternary system. Figure 2 summarizes the experimental process. In total, 462 cylinder samples were prepared: 207 for testing UCS, 138 for LP, and 117 for FS.

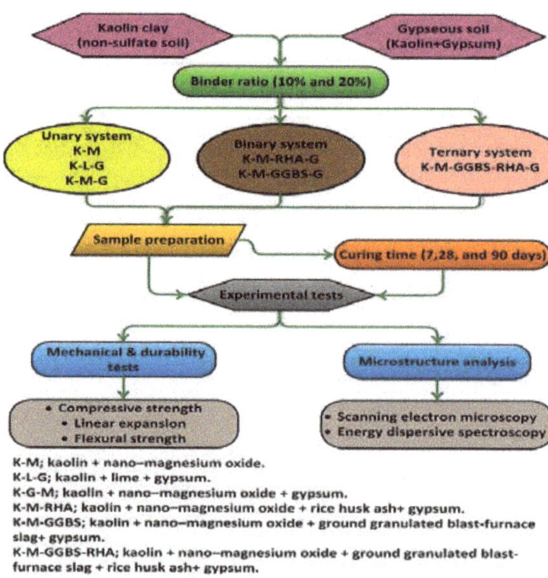

Figure 2. Schematic representation of the experimental procedure.

Table 5. Mixture designs of stabilizer agents.

Mix Code	Binder Composition	Binder Ratio (%)	Dosage (%)
Unary			
K	K	-	-
K-M	M	100	10, 20
K-L-G	L	100	10, 20
K-M-G	M	100	10, 20
Binary			
K-M-RHA-G	M:RHA	3:1, 1:1, 1:3	10, 20
K-M-GGBS-G	M:GGBS	3:1, 1:1, 1:3	10, 20
Ternary			
K-M-GGBS-RHA-G	M:GGBS:RHA	1:0.5:0.5, 1:1:2, 1:2:1, 1:2.5:0.5	10, 20

(K) Kaolin, (G) gypsum, (L) hydrated lime, (M) nano–magnesium oxide, (RHA) Rice husk ash, and (GGBS) ground granulated blast-furnace slag.

2.3. Experiments

2.3.1. Linear Expansion Test (Swelling) (LP) Test

Two cylindrical samples were prepared for each mix proportion, and were cured for 7, 28, and 90 days to determine their vertical swelling ratio (%). Swelling readings were recorded every 24 h until no significant swelling ratio was observed after partial soaking in water. The LP test was conducted following the BS EN 13286–49:2004 standard [44].

2.3.2. Unconfined Compressive Strength (UCS) Test

The UCS test was conducted according to BS EN ISO 17892–7:2018 [45] for each mix proportion; three cylindrical samples were tested for compressive strength before and after soaking. "Before soaking" implies that the samples had been cured for 7, 28, and 90 days without having been immersed in water, while "after soaking" implies after the linear expansion test, which was 52 days after the preparation of the samples for all soil stabilization systems. A constant rate of compression strain of 1 mm/min was applied until sample failure occurred.

2.3.3. Flexural Strength (FS) Test

For this test, specimens similar to those described for the UCS and LP tests were used. A three-point bending test was performed to determine the flexural strength for kaolin clay and the optimum mixing of sulfate-bearing soil mixed with the calcium-based stabilizer (lime) and non-calcium-based stabilizer. For each test, three samples were prepared using a compaction mold at the maximum dry density (MMD) and optimum moisture content (OMC) and cured for 7 and 28 days. The flexural strength test was performed according to ASTM D1635 at a constant strain rate of 0.1 mm/min. The flexural stress for the circular specimens of the outer layer of each specimen was calculated as follows:

$$FS = \frac{PL}{\pi r^3} \tag{3}$$

where FS = flexural strength, P = maximum applied load, r = radius of the sample, and L = support span.

2.3.4. Microstructure Analysis (SEM/EDX)

Scanning electron microscopy (SEM) was performed to inspect the surface morphology of the specimens, which was followed by energy dispersive spectroscopy (EDS) to reveal the presence and composition of various elements within. Small portions of soil specimens were collected from carefully hand-broken samples after UCS and then dried in the oven at 105 °C for 24 h prior to testing.

3. Results and Discussion

3.1. Unconfined Compressive Strength (UCS) Test

3.1.1. Effects of M-RHA on UCS

Figures 3 and 4 depict the effects of L and M additions on the UCS values after 7, 28, and 90 days of curing for the gypseous soil samples. The UCS values for M-treated samples increased from 885.33 to 1108.67 kPa when the M content was increased from 10% to 20% without the presence of sulfate at 90 days of curing. Samples treated with L and M (10% and 20%) showed a decrease in UCS values in the presence of sulfate (see Figures 3 and 4). Furthermore, after soaking, the UCS values deteriorated from 885.33 to 635 kPa for M10%, (reducing by 28.27%) and 1108.67 to 1035.33 kPa with 20% M content in the presence of sulfate (reducing by 6.62%). However, this deterioration, as compared to that of the lime stabilizer, was much less. The observed deterioration in the UCS results can be attributed to ettringite production, the growth of which between the particles of the sample soil would lead to the destruction of the soil structure. In addition, the residual sulfate chelates formed calcium in CSH through a decalcification process, which accounted for the lower strengths of the soaked and unsoaked samples. However, the higher strength of the M-treated samples might have been due to the hydration process and formation of pozzolanic products that might have improved the strength of the bonds between the soil particles.

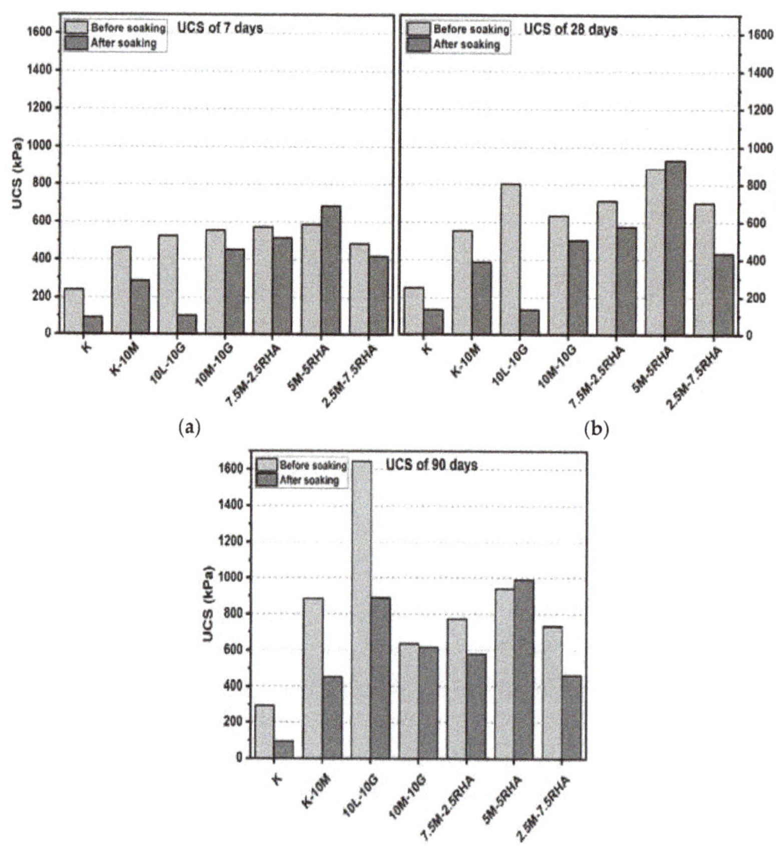

Figure 3. UCS of soils stabilized with 10% of L, M, and M-RHA before and after soaking in the presence of sulfate: (**a**) after 7 days of curing, (**b**) after 28 days of curing, (**c**) after 90 days of curing.

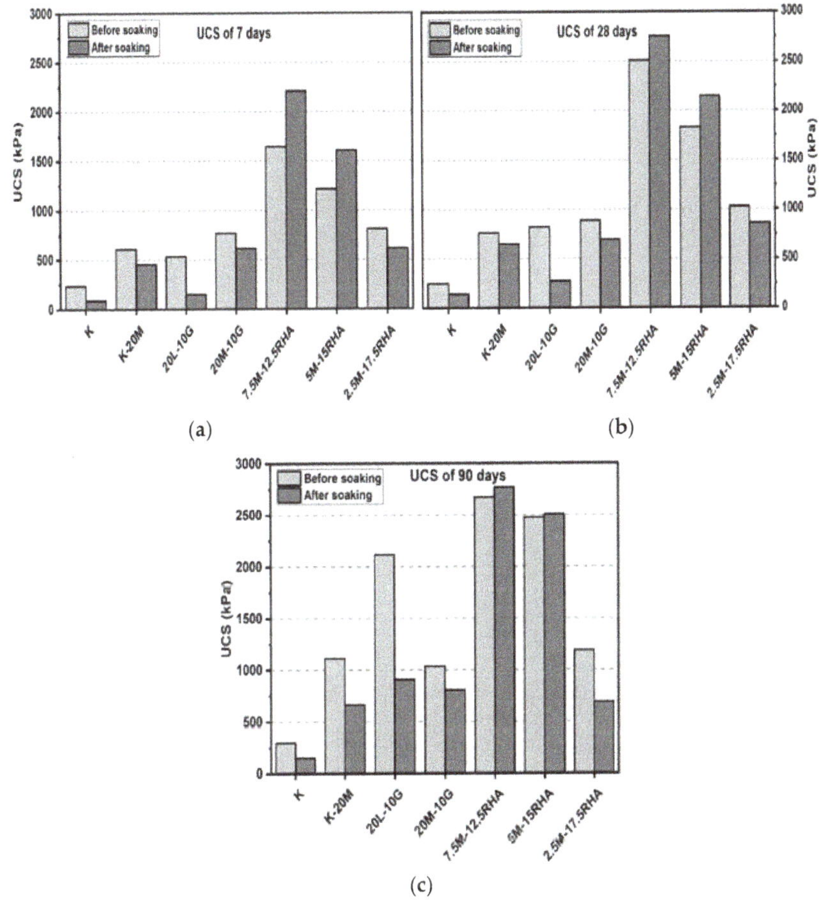

Figure 4. UCS of soils stabilized with 10% of L, M, and M-GGBS before and after soaking in the presence of sulfate: (**a**) after 7 days of curing, (**b**) after 28 days of curing, (**c**) after 90 days of curing.

Figures 3 and 4 show the effects of the RHA content on enhancing the USC values for the M-treated samples (soaked and unsoaked) after 7, 28, and 90 days of curing in the presence of sulfate. For example, Figure 4b indicates that the UCS values of the samples treated with 7.5M-12.5RHA increased from 2510 kPa to 2753 kPa when immersed in water after 28 days of curing, where UCS increased by 9.68%. This impact was also observed for 5M-15RHA (increased from 1830 to 2148 kPa), which was improved by 17.37%, and 5M-5RHA (increased from 884 to 929 kPa), which was enhanced by 5.09%. These improvements could be due to pozzolanic product (CSH) formation, which would have improved the USC value for the samples with higher M-RHA content.

3.1.2. Effects of M-GGBS on UCS

For the M-GGBS stabilizer, all specimens showed enhancement of their UCS values. Figures 5 and 6 presented the maximum UCS of the M-GGBS-stabilized soils at the M:GGBS ratio of 1:3 when a 20% binder was used to stabilize the sulfate-bearing soil: 3327.33, 7125, and 12,115 kPa at 7, 28 and 90 days of curing, respectively. This improvement may have been because of the high content of highly reactive M, and because an excessive residual can negatively impact the strength of stabilized soils. Nevertheless, the UCS values of the stabilized samples before and after soaking indicated enhancement in the mechanical performance of the samples after wetting, except ratios of 3:1 and 1:1 when used to treat gypseous soil samples with 10% M-GGBS-stabilizer (see Figure 5); for example, the samples treated with 5M-5GGBS deteriorated by 13.02%, 33.11% and 29.76% at 7, 28 and 90 days of curing, respectively. The UCS of the treated soil sample with 2.5M:17.5GGBS was 11,275 kPa after soaking, which was significantly higher than that before soaking (914.23%) after 7 days of curing. After soaking, the 5M-15GGBS ratio of 1:3 exhibited an optimum UCS value of 12,831 kPa, which was increased by 5.91% after 90 days of curing. However, the UCS values showed a decline at higher M-GGBS ratios. These findings agree with the report by Li et al. [2], where highly reactive M was found to have better activation efficiency with GGBS compared to low-reactive M; therefore, lower M-GGBS ratios are recommended to achieve higher UCS values.

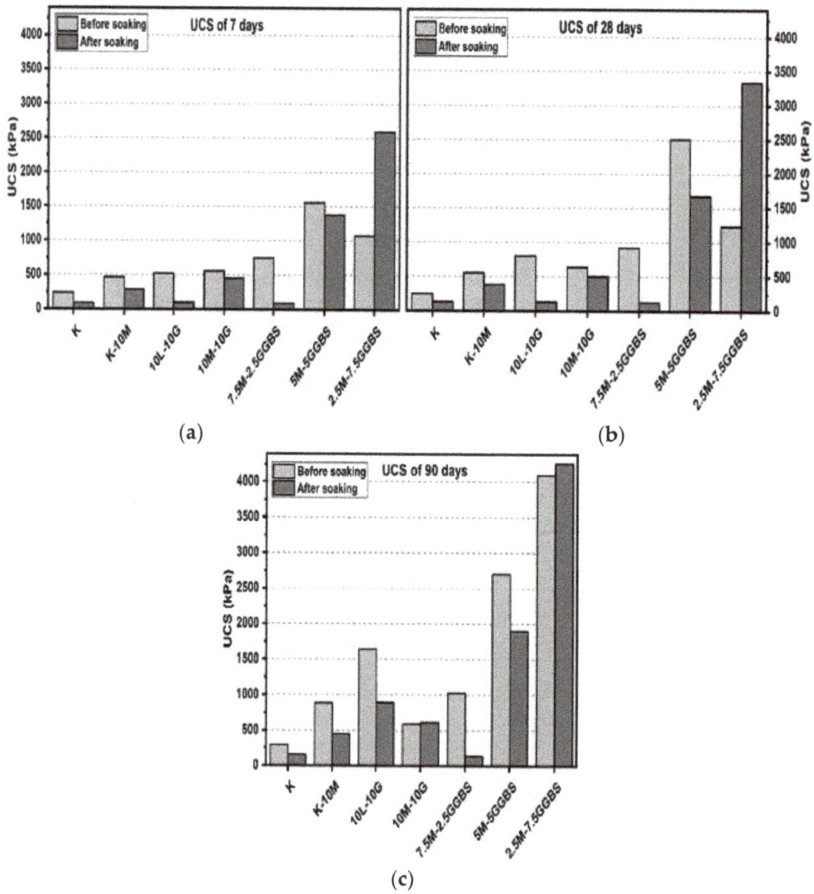

Figure 5. UCS of soils stabilized with 20% of L, M, and M-RHA before and after soaking in the presence of sulfate: (**a**) after 7 days of curing, (**b**) after 28 days of curing, (**c**) after 90 days of curing.

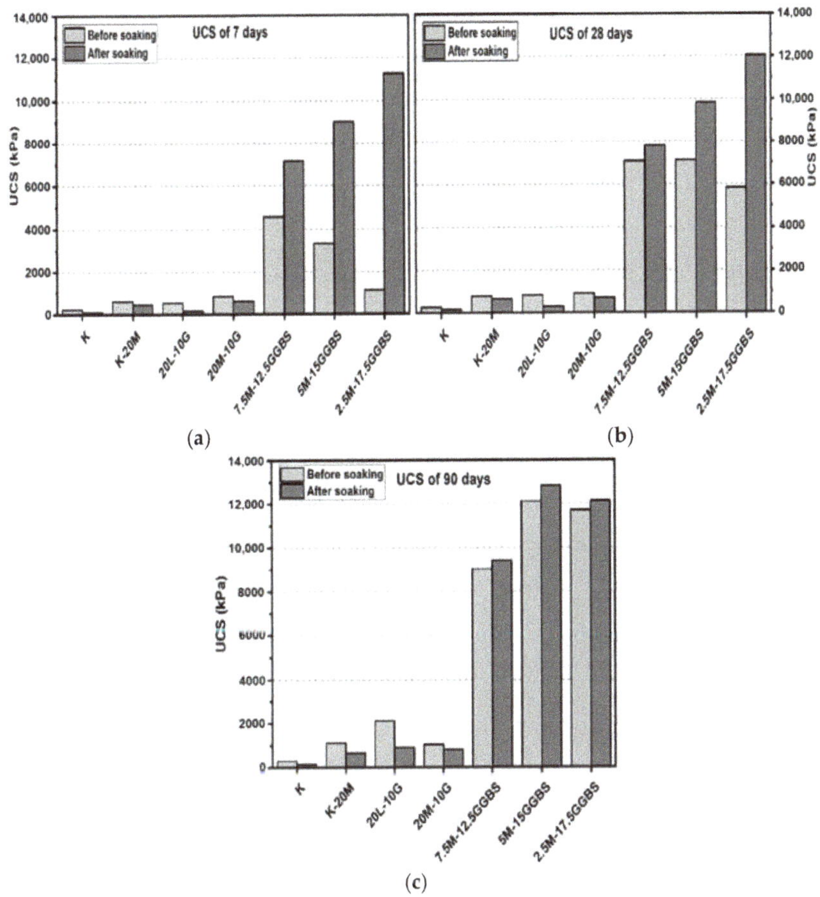

Figure 6. UCS of soils stabilized with 20% of L, M, and M-GGBS before and after soaking in the presence of sulfate: (**a**) after 7 days of curing, (**b**) after 28 days of curing, (**c**) after 90 days of curing.

3.1.3. Effects of M-GGBS-RHA on UCS

As shown in Figures 7 and 8, all ternary binder compositions improved the UCS values in the stabilized gypseous soil. The highest values of UCS obtained by the M-GGBS-RHA stabilizer ratio of 1:2.5:0.5 were 2050, 5232, and 9142 kPa after 7, 28, and 90 days of curing, respectively. However, the lowest enhancements after 7, 28, and 90 days of curing were 563.67, 758.67, and 975 kPa, respectively, when a stabilizer–binder ratio of 1:0.5:0.5 was used.

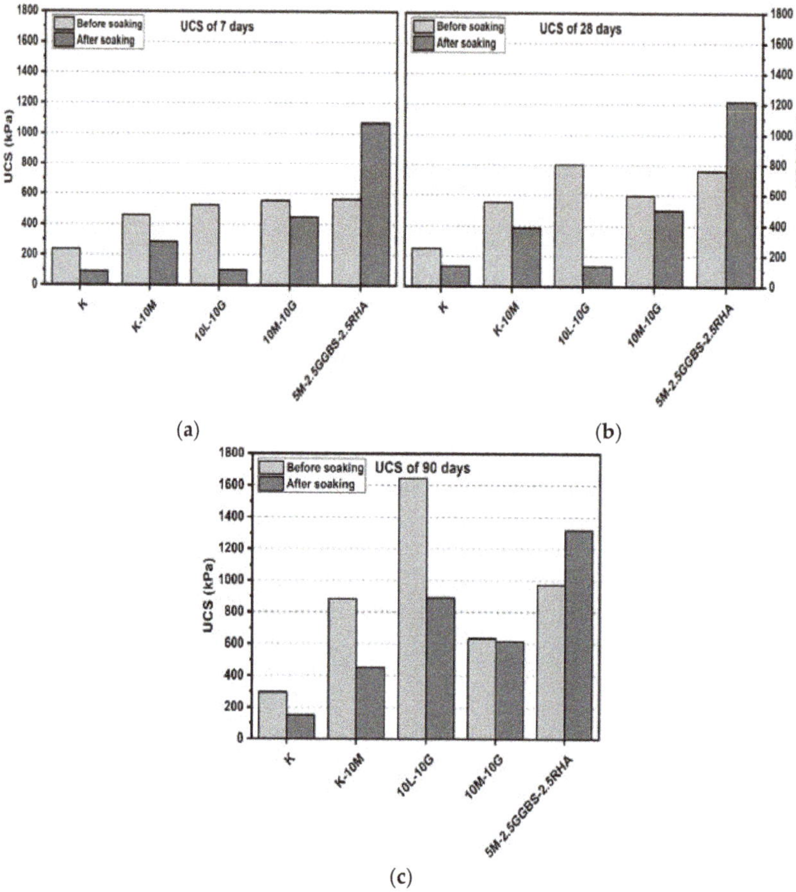

Figure 7. UCS of soils stabilized with 10% of L, M, and M-GGBS-RHA before and after soaking in the presence of sulfate: (**a**) after 7 days of curing, (**b**) after 28 days of curing, (**c**) after 90 days of curing.

After soaking, the M-GGBS-RHA ratio of 1:2.5:0.5 produced the highest UCS of 9721 kPa among all the M-GGBS-RHA-stabilized soils, an increase of 6.33%; the UCS decreased with the reduction in the GGBS ratio. This enhancement in the UCS values might have been due to the hydration process following the prolonging of the curing time from 7 days to 90 days; this should ensure proper hydration, improved strength, and resistance against sulfate attacks [8]. Furthermore, the production of more cementitious gels, such as CSH, CAH, and CSAH, contributed to consumption of the available calcium, leading to the inhibition of ettringite formation [4].

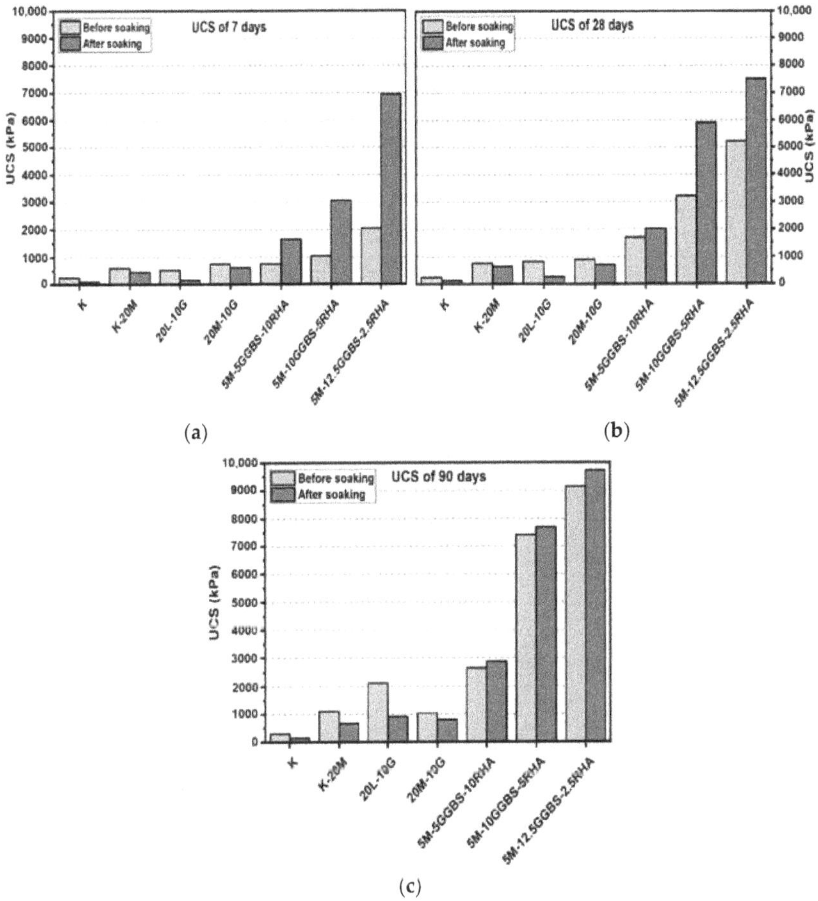

Figure 8. UCS of soils stabilized with 20% of L, M and M-GGBS-RHA before and after soaking in the presence of sulfate: (**a**) after 7 days of curing, (**b**) after 28 days of curing, (**c**) after 90 days of curing.

3.2. Linear Expansion Test (Swelling) (LP) Test

Figures 9–14 illustrate the typical swelling plots for K-M clay and K-L-G and K-M-G clay systems dosed with 10% wt of gypsum and stabilizer with 10% and 20% wt of L and M for observatory periods of 7, 28, and 90 days, respectively. Swelling was observed immediately after soaking the cylinder samples in water, which was sustained throughout monitoring until the cessation of swelling. A higher rate of swelling was observed compared to the reported expansion in lime-stabilized sulfate-bearing soils [4,10,15].

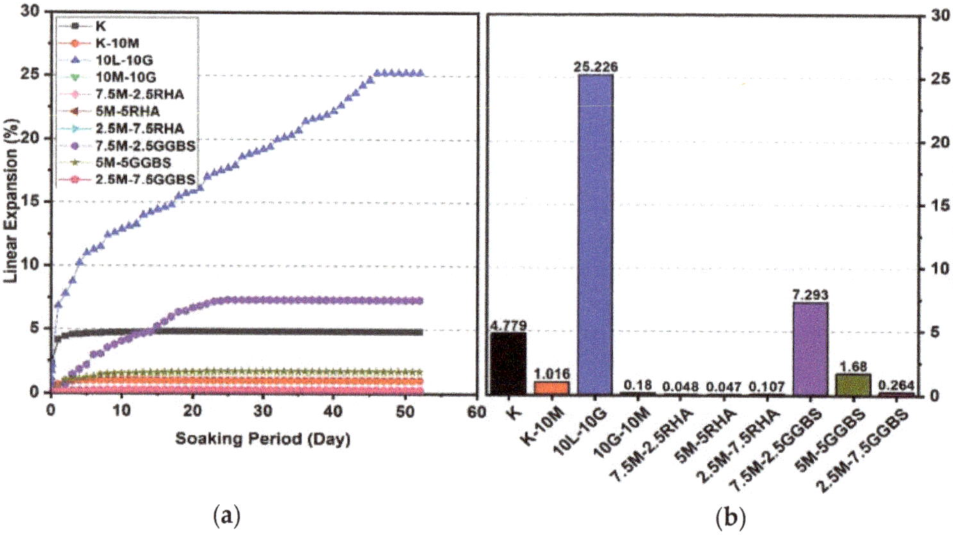

Figure 9. Vertical swelling strain of soils stabilized with 10% of L, M, M-RHA, and M-GGBS after 7 days of curing: (**a**) results presented as a line curve, (**b**) results presented as columns.

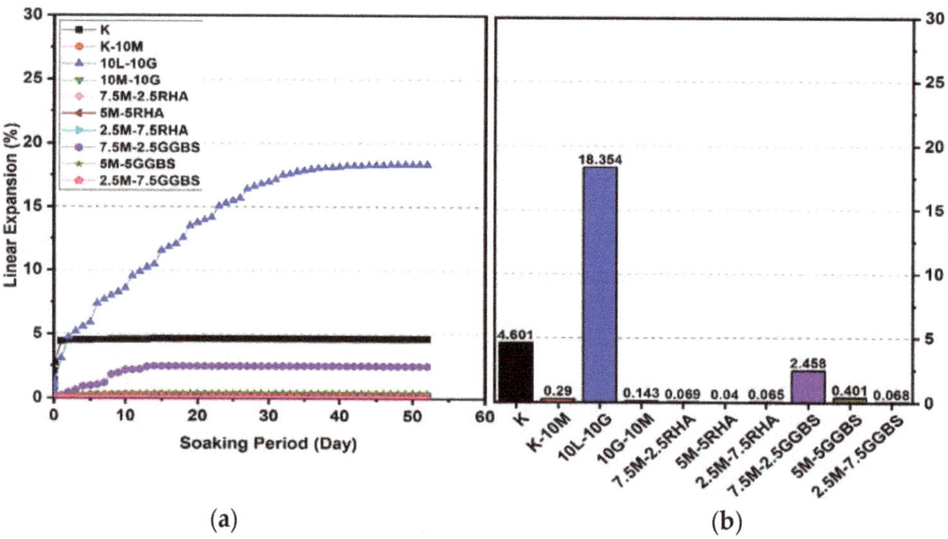

Figure 10. Vertical swelling strain of soils stabilized with 10% of L, M, M-RHA, and M-GGBS after 28 days of curing: (**a**) results presented as a line curve, (**b**) results presented as columns.

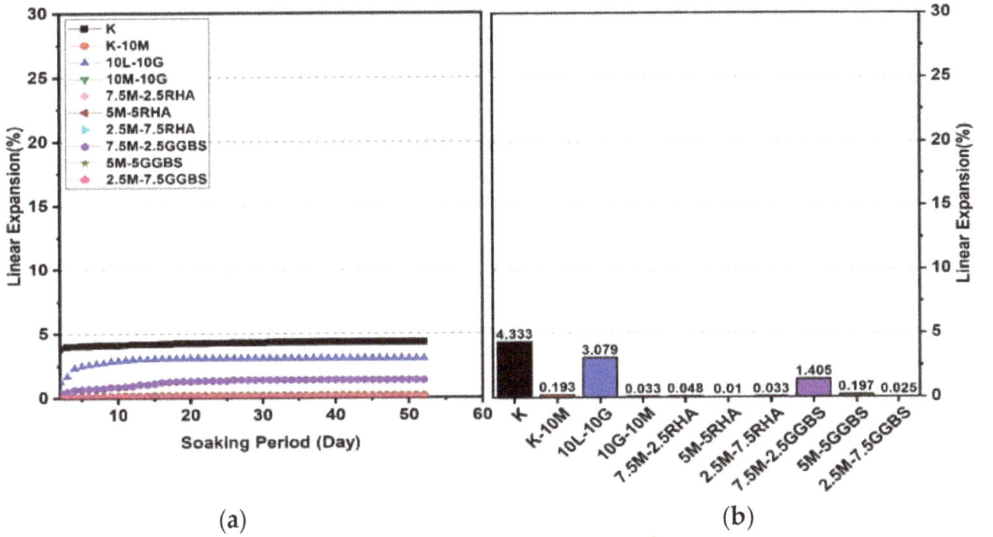

Figure 11. Vertical swelling strain of soils stabilized with 10% of L, M, M-RHA, and M-GGBS after 90 days of curing: (**a**) results presented as a line curve, (**b**) results presented as columns.

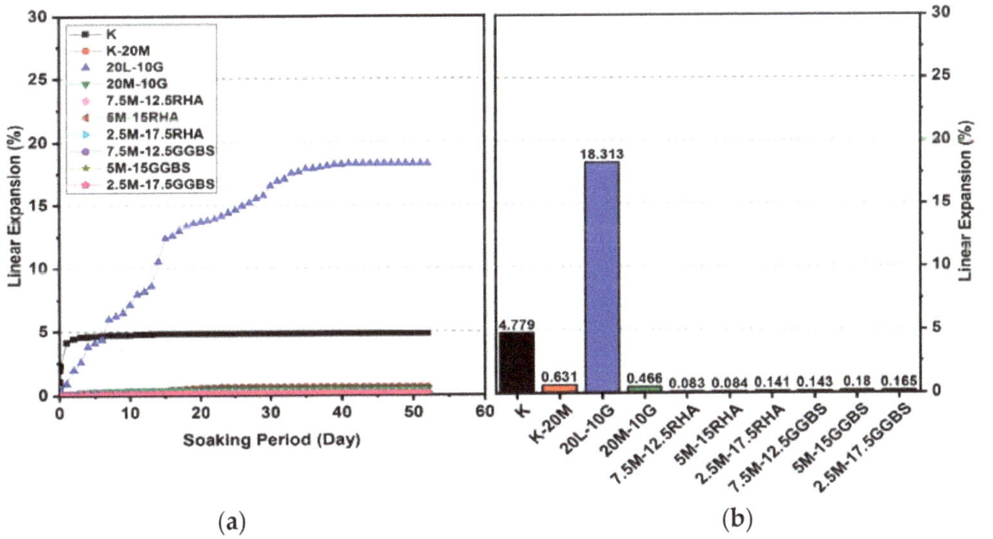

Figure 12. Vertical swelling strain of soils stabilized with 20% of L, M, M-RHA, and M-GGBS after 7 days of curing: (**a**) results presented as a line curve, (**b**) results presented as columns.

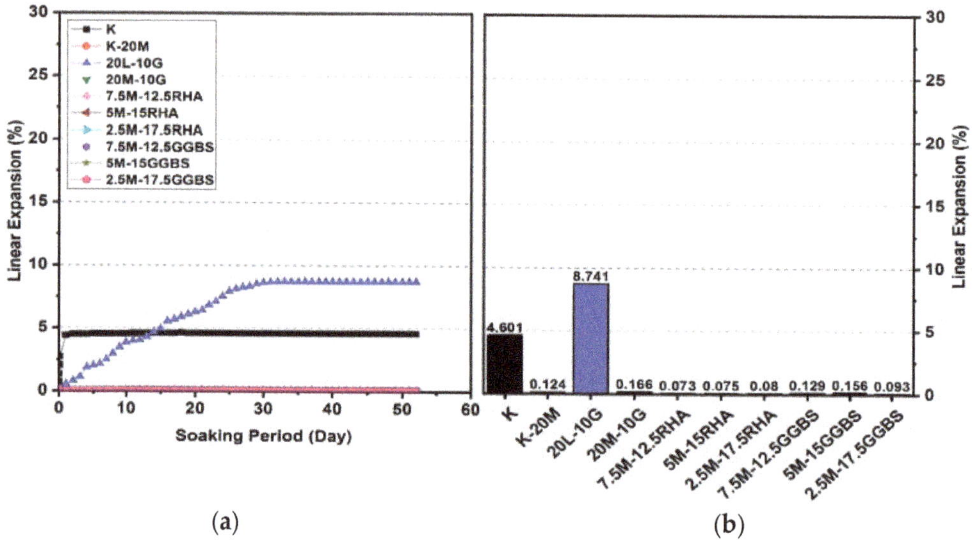

Figure 13. Vertical swelling strain of soils stabilized with 20% of L, M, M-RHA, and M-GGBS after 28 days of curing: (**a**) results presented as a line curve, (**b**) results presented as columns.

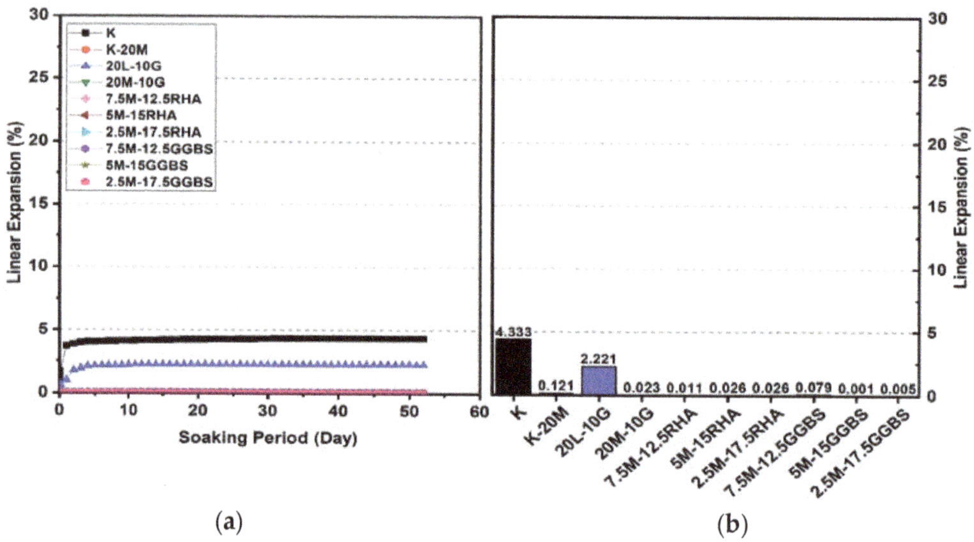

Figure 14. Vertical swelling strain of soils stabilized with 20% of L, M, M-RHA, and M-GGBS after 90 days of curing: (**a**) results presented as a line curve, (**b**) results presented as columns.

Figures 9 and 12 show the highest degree of swelling obtained with a calcium-based stabilizer (10% wt): which was 25.226% with 10% lime after 7 days curing; meanwhile, the lowest swelling was observed with 20% lime: 18.313% after seven days of curing. Nevertheless, the comparison of the swelling values of the treated soils between the calcium-based stabilizer and non-calcium-based stabilizer clearly showed that the non-calcium-based stabilizer (M) had a significant effect on restricting the swelling values. For example, when 10% and 20% M were used, the swelling values were 0.18% and 0.466%;

hence, the swelling was restricted to 99.28% and 97.45%. This result was due to the absence of calcium in non-calcium-based stabilizer (M), which led to the suppression of ettringite formation.

3.2.1. Effects of M-RHA on Swelling

Utilizing the M-RHA stabilizer resulted in a restriction in the volume change (swelling) for all gypseous soil specimens, as presented in Figures 9–14. The lowest minimum values of swelling, achieved with a 5M-5RHA ratio of 1:1, were 0.047%, 0.04%, and 0.01% after 7, 28, and 90 days of curing, respectively, with the swelling inhibited 99.81%, 99.78%, and 99.67%, respectively. Furthermore, the 20% M-RHA-stabilizer led to a low volumetric change after 90 days of curing at 0.011%, 0.026%, and 0.026%, with the swelling suppressed to 99.5%, 98.82%, and 98.82%, respectively. This reduction in swelling can be attributed to the restriction of ettringite formation and the production of more CSH compounds in the absence of calcium.

3.2.2. Effects of M-GGBS on Swelling

Figures 9–14 present the vertical volume change (swelling) of gypseous soil treated with the M-GGBS stabilizer. The maximum vertical volume change was 7.293%, obtained after applying a 7.5M-2.5GGBS ratio of 3:1 over 7 days of curing. However, all other specimens exhibited low swelling. For example, the swelling of soil samples stabilized with 5M-15GGBS and 2.5M-17.5GGBS was 0.001% and 0.005% after 90 days of curing, with swelling magnitude almost wholly suppressed to 99.95% and 99.77%. The reduction in the swelling values was because of the use of GGBS, which exhibited superior sulfate resistance along with its denser structure and lower presence of calcium ions [6,38].

3.2.3. Effects of M-GGBS-RHA on Swelling

Figures 15–17 reveal that the use of the M-GGBS-RHA stabilizer (10 and 20 wt. %) modified the behavior of the volumetric change and resulted in a significant decrease in the swelling magnitude of the gypseous soil. After 28 days of observation, the swelling magnitudes for 5M-2.5GGBS-2.5RHA, 5M-5GGBS-10RHA, 5M-10GGBS-5RHA, and 5M-12.5GGBS-2.5RHA were 0.04%, 0.089%, 0.044%, and 0.001%, respectively, with swelling values suppressed to 99.78%, 98.98%, 99.5%, and 99.98% respectively. Nevertheless, the longer curing period (90 days) produced further declines in the swelling rate, thereby achieving minimum swellings of 0.01%, 0.058%, 0.017%, and 0.001%, where the swelling was roughly inhibited to 99.67%, 97.38%, 99.23%, and 99.95% respectively.

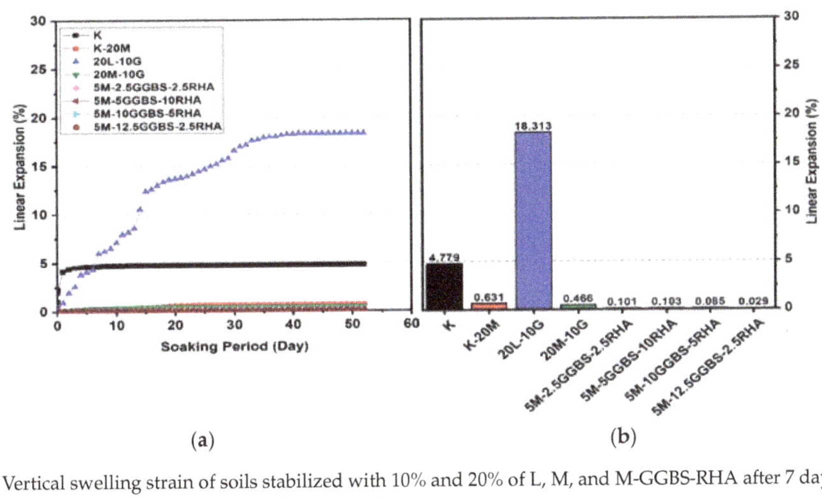

(a) (b)

Figure 15. Vertical swelling strain of soils stabilized with 10% and 20% of L, M, and M-GGBS-RHA after 7 days of curing: (**a**) results presented as a line curve, (**b**) results presented as columns.

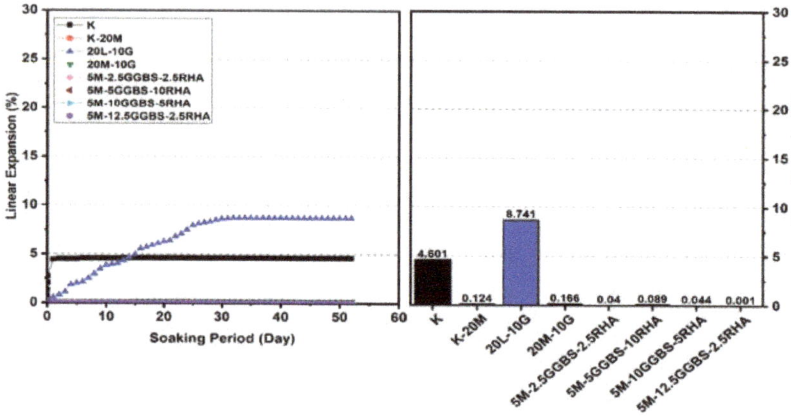

(a) (b)

Figure 16. Vertical swelling strain of soils stabilized with 10% and 20% of L, M, and M-GGBS-RHA after 28 days of curing: (**a**) results presented as a line curve, (**b**) results presented as columns.

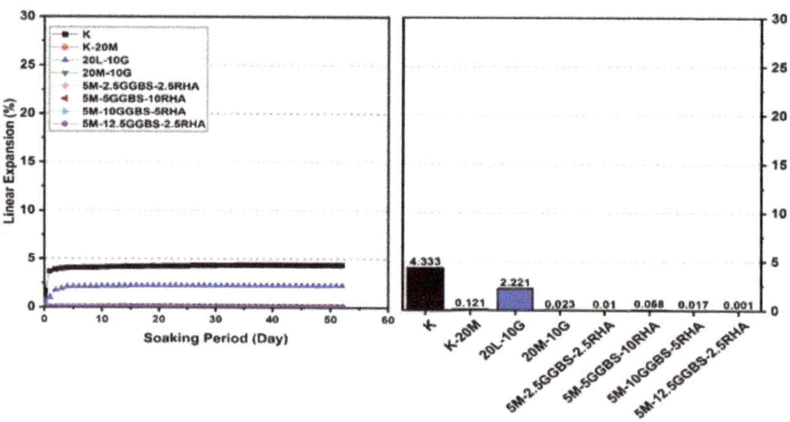

(a) (b)

Figure 17. Vertical swelling strain of soils stabilized with 10% and 20% of L, M, and M-GGBS-RHA after 90 days of curing: (**a**) results presented as a line curve, (**b**) results presented as columns.

Results from the linear expansion test indicated that swelling behavior disappeared in the high-sulfate soil, which was treated with the M-GGBS-RHA-stabilizer. The volume changes of the treated soils were less than those of the lime-treated soils, indicating that the swelling characteristics were enhanced in the treated high-sulfate soils.

These results indicate that the M-GGBS-RHA-stabilized soil resisted sulfate attacks better than the lime-stabilized soil. Moreover, the results showed that the swelling of the lime-stabilized soil was higher than that of the M-GGBS-RHA-stabilized soil with an increase in the soaking time. The phenomenon can be explained by two factors: (1) a larger amount of calcium in the lime-stabilized soil contributed to ettringite formation in the presence of sulfate and (2) the consumption of calcium by sulfate, which led to a reduction in the formation of cementitious material such as CSH, CAH, and CSAH. Adeleke et al. [4] reported that there are various levels of risk in relation to swelling, as shown in Table 6. The risk level was "very strong" in comparison with the maximum expansions of 25.226% and 18.313% obtained in the current study from the mixing of K-10L-10G and K-20L-10G, respectively.

Table 6. Risk levels for swelling of clay soil.

Swelling (%)	Risk
0	No swell
0–0.1	Negligible
0.1–0.5	Light
0.5–1.0	Medium
1.0–2.0	Strong
Over 2.0	Very strong

3.3. Flexural Strength Flexural (FS)

Flexural load curves for the kaolin clay and sulfate-bearing soil along with the calcium-based stabilizer (L) and non-calcium-based stabilizer (M) with the inclusion of RHA, GGBS, and GGBS-RHA ratios of 10% and 20% after curing for 7 and 28 days, are shown in Figures 18 and 19. The kaolin clay curve exhibited a flexural strength of 121.44 kPa after curing for 28 days. In contrast, the curves produced by the M stabilizer mixtures showed a slight improvement with and without the presence of sulfate, which is a direct measure of the flexural strength of the specimens. When compared to specimens treated with the L stabilizer, the flexural strength was less with the presence of sulfate. Furthermore, using 10% and 20% of the M-RHA, M-GGBS, and M-GGBS-RHA stabilizers resulted in a significant increase in the peak flexural load. This effect was especially relevant regarding the 20% M-GGBS content. The flexural strength of the M-GGBS mixture with 10% and 20% content reached 1270.37 and 1922.55 kPa after curing for 28 days, respectively, gaining values 303.4% and 302.9% higher than those of the L stabilizer (314.89 and 477.11 kPa) using 10% and 20% mixtures, respectively, after 28 days of curing.

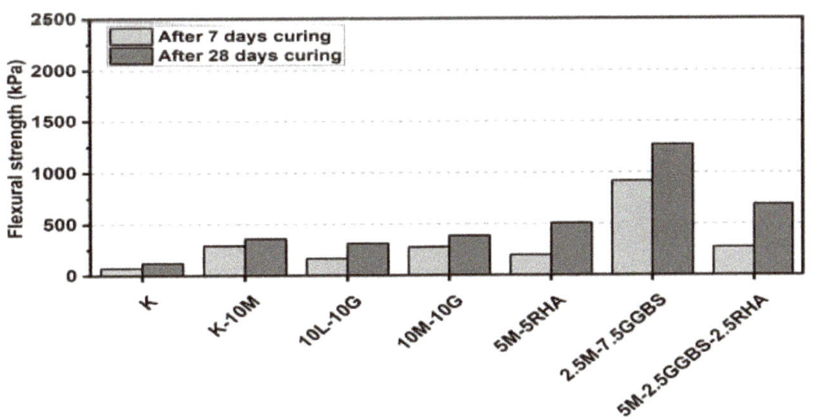

Figure 18. Flexural load curves of test gypseous soils treated with 10% mixture content after 7 and 28 days of curing.

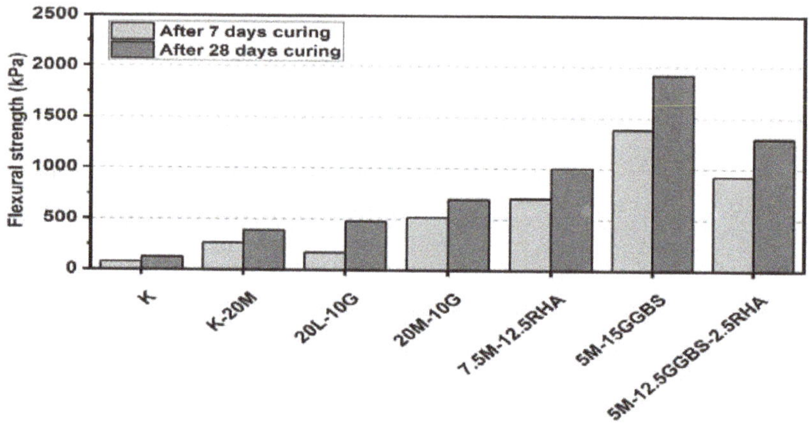

Figure 19. Flexural load curves of test gypseous soils treated with 20% mixture content after 7 and 28 days of curing.

3.4. SEM and EDX Analysis

The SEM images of the gypseous soil–lime mixture after 90 days of curing showed that a significant amount of crystalline mineral needles (ettringite) had formed, as shown in Figures 20a and 21a. This resulted in a decrease in strength, a high capacity for swelling, and the destruction of the soil specimen's structure. The improvement in the gypseous soil samples was determined by measuring the Si/Al ratio (%) utilizing EDX to explain the strength development. Generally, an increase in the Si/Al ratio refers to the forming of a high number of Si–O–Si bonds. Consequently, the mechanical strength of the soil sample structural matrix was enhanced [45]. The Si/Al ratio in the (10% and 20%) L-G specimens obtained from the EDX pattern ranged from 1.06 to 1.12, which was low due to the ettringite formation and high porosity in the microstructure of the samples. The EDX pattern of the gypseous soil–lime mixture and its elemental composition is represented in Figures 20b and 21b.

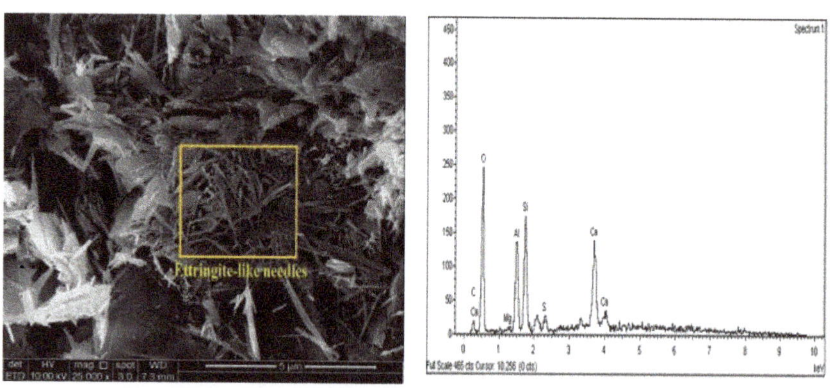

(a) (b)

Figure 20. SEM photo and EDX spectrograph of stabilized gypseous soils after 90 days of curing: (**a**) SEM photo of 10L:10G and (**b**) EDX spectrograph of 10L:10G.

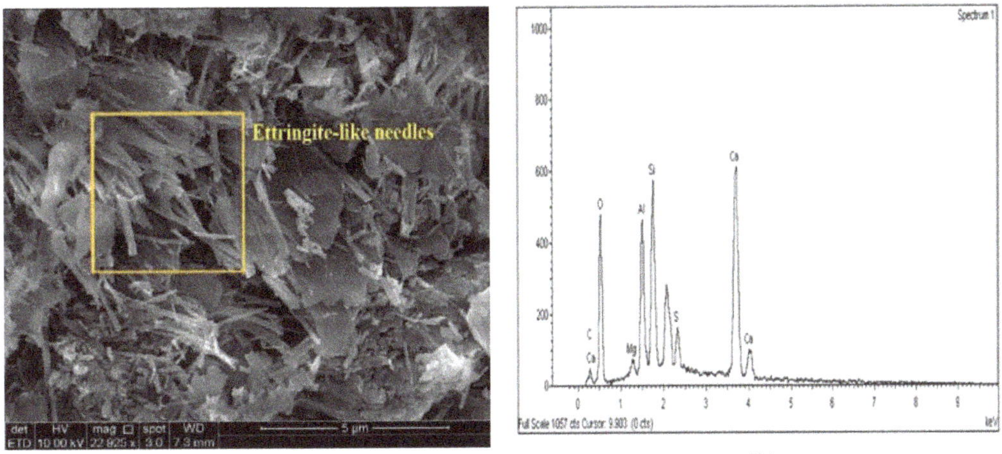

(a) (b)

Figure 21. SEM photo and EDX spectrograph of stabilized gypseous soils after 90 days of curing: (**a**) SEM photo of 20L:10G and (**b**) EDX spectrograph of 20L:10G.

Figures 22a and 23a show SEM observation images of the sulfate soil–M-G mixture, revealing spherical and irregular particles of magnesium silicate hydrate (MCH). The EDX information showed a slight increase in the Si/Al ratio (1.1 and 1.14) for M-G-stabilizer specimens with 10% and 20% when compared to the L-G specimens (1.06 and 1.12), which was probably due to the absence of ettringite and the formation of MSH product. Figures 22b and 23b, respectively, present the EDS spectrum and elemental composition of the gypseous soil–M-G mixture.

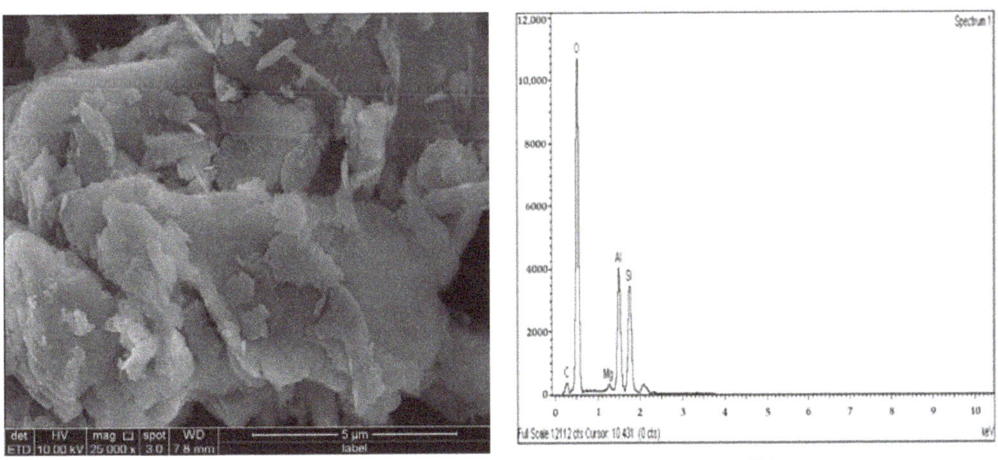

(a) (b)

Figure 22. SEM photo and EDX spectrograph of stabilized gypseous soils after 90 days of curing: (**a**) SEM photo of 10M:10G and (**b**) EDX spectrograph of 10M:10G.

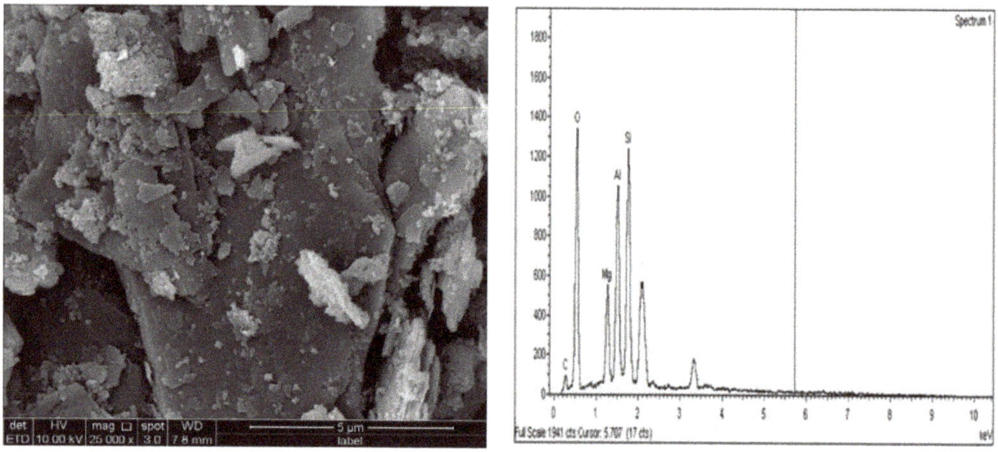

(a) (b)

Figure 23. SEM photo and EDX spectrograph of stabilized gypseous soils after 90 days of curing: (**a**) SEM photo of 20M:10G and (**b**) EDX spectrograph of 20M:10G.

Figures 24a, 25a and 26a show the SEM observation images of M-GGBS, M-RHA, and M-GGBS-RHA, respectively. In terms of particle shape and surface features, M-GGBS was quite different from M-RHA and M-GGBS-RHA. M-RHA and M-GGBS-RHA particles were spherical, as shown in Figures 25a and 26a. The particles had shiny surfaces without any dust covering them. The M-GGBS particles were rather angular, irregular in shape, and the presence of plate-shaped particles was evident (Figure 24a), which had rough, gritty surface textures. Figures 24a, 25a and 26a show the surface morphologies of the soil–M-GGBS, –M-RHA, and –M-GGBS-RHA specimens after 90 days of curing. As indicated, no ettringite formation was observed. This proves that the addition of M with GGBS and RHA as a stabilizing agent contributed to the formation of CSH gel to a significant degree. The presence of some crystalline structures, such as MSH (magnesium silicate hydrate), was also observed. The pozzolanic activity occurring between M, GGBA, and RHA as a stabilizer and clay was attributed to the formation of CSH gel. Based on the results derived from the EDX spectrum, the highest Si/Al ratios (1.23, 1.59, and 1.41) were observed in the 7.5M-12.5RHA, 5M-15GGBS, and 5M-12.5GGBS-2.5RHA samples, respectively. This improvement was due to GGBS and RHA dissolving in an alkaline activator to generate extra soluble alumina and silica, which increased the Si/Al ratio of the specimen, leading to the complete activation of the particles and producing calcium–alumino–silicate hydrate (CASH) in the matrix. Figures 24b, 25b and 26b show the EDS spectrum and elemental composition of the gypseous soil–M-GGBS, –M-RHA, and –M-GGBS-RHA mixtures, respectively.

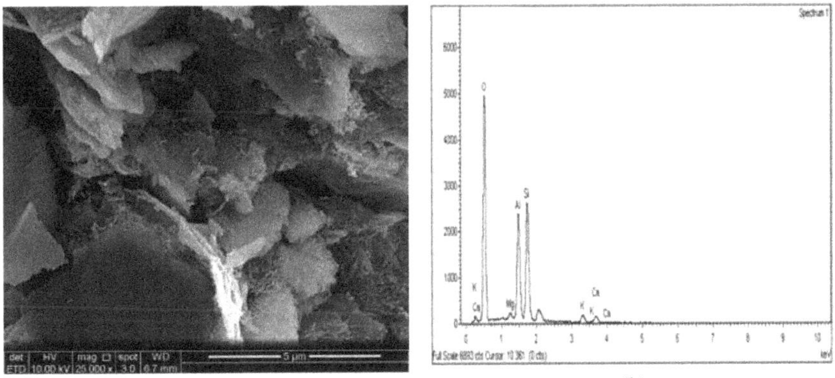

(**a**) (**b**)

Figure 24. SEM photo and EDX spectrograph of stabilized gypseous soils after 90 days of curing: (**a**) SEM photo of 5M:15GGBS and (**b**) EDX spectrograph of 5M:15GGBS.

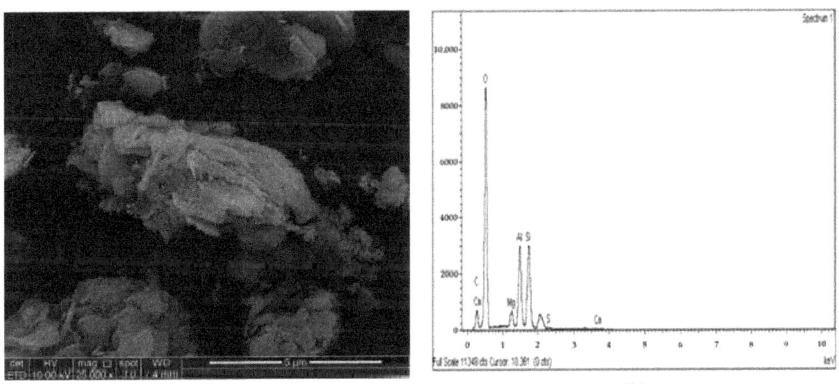

(**a**) (**b**)

Figure 25. SEM photo and EDX spectrograph of stabilized gypseous soils after 90 days of curing: (**a**) SEM photo of 7.5M:12.5RHA and (**b**) EDX spectrograph of 7.5M:12.5RHA.

(**a**) (**b**)

Figure 26. SEM photo and EDX spectrograph of stabilized gypseous soils after 90 days of curing: (**a**) SEM photo of 5M:12.5GGBS:2.5RHA and (**b**) EDX spectrograph of 5M-12.5GGBS:2.5RHA.

The increase in UCS value with prolonged curing periods could have been due to cementing gel formation from the interaction of the soil with the nanoparticles, which improved the UCS of the treated samples as a result of the improved soil particle bonding [46,47].

4. Conclusions

Sulfate-bearing soils cause major issues in pavement and various civil engineering infrastructures due to their causing significant swelling and strength damage. The results of this study portrayed the impacts of (M) as a non-calcium-based stabilizer, rice husk ash, and GGBS as abundant byproducts in treating gypseous soil with a high level of sulfate. Based on the results of the study, the following conclusions were made:

- The minimum swelling values of M-RHA-, M-GGBS-, and M-GGBS-RHA-stabilized gypseous soils with 20% doses (0.011%, 0.001%, and 0.001%) were lower than that of lime-stabilized soil.
- The optimum strength values of M-RHA-, M-GGBS, and M-GGBS-RHA-stabilized soils after soaking with 20% (2.7, 12.8, and 9.7 MPa) were notably higher than that of lime-stabilized soil (0.9 MPa).
- The curing time had a significant impact on the resistance to sulfate attack as a result of the hydration process; an increase in the curing period from 7 to 90 days resulted in adequate hydration and improved strength in the presence of sulfate (gypsum).
- The results of the SEM and EDX showed that no ettringite was formed in the M-RHA-, M-GGBS-, or M-GGBS-RHA-stabilized soils.
- The formation of cementing gels such as MSH, MAH, CSH, and CAH improved the UCS values of the stabilized samples, as these gels occupied the voids and improved the bonding between the soil particles. As a result, the treated soils' UCS values were improved.

Supplementary Materials: The following are available online at https://www.mdpi.com/article/10.3390/app11146618/s1, Figure S1. Particle size distribution (PSD) of RHA and GGBS used in this study.

Author Contributions: K.I.A.E. and A.M.A.E. designed and performed the experimental program; K.I.A.E. and M.A.M.I. collected and analyzed the experimental data; A.M.A.E. wrote, reviewed, and edited the manuscript; and M.A.M.I. provided supervision of the manuscript writing. All authors have read and agreed to the published version of the manuscript.

Funding: This work was funded by Universiti Sains Malaysia, under grant No. 1001.PAWAM.8014090.

Institutional Review Board Statement: Not applicable.

Informed Consent Statement: Not applicable.

Data Availability Statement: Not applicable.

Acknowledgments: The authors would like to acknowledge the Universiti Sains Malaysia for its financial support.

Conflicts of Interest: No conflict of interest to declare.

References

1. Aldaood, A.; Bouasker, M. Mechanical Behavior of Gypseous Soil Treated with Lime. *Geotech. Geol. Eng.* **2020**, *39*, 719–733. [CrossRef]
2. Li, W.; Yi, Y.; Puppala, A.J. Suppressing Ettringite-Induced Swelling of Gypseous Soil by Using Magnesia-Activated Ground Granulated Blast-Furnace Slag. *J. Geotech. Geoenviron. Eng.* **2020**, *146*, 06020008. [CrossRef]
3. Kalipcilar, I.; Mardani-Aghabaglou, A.; Sezer, A.; Altun, S.; Inan Sezer, G. Sustainability of cement-stabilised clay: Sulfate resistance. *Proc. Inst. Civ. Eng. Eng. Sustain.* **2016**, *171*, 254–274. [CrossRef]
4. Adeleke, B.; Kinuthia, J.; Oti, J. Strength and Swell Performance of High-Sulphate Kaolinite Clay Soil. *Sustainability* **2020**, *12*, 10164. [CrossRef]
5. Yao, K.; Wang, W.; Li, N.; Zhang, C.; Wang, L. Investigation on strength and microstructure characteristics of nano-MgO admixed with cemented soft soil. *Constr. Build. Mater.* **2019**, *206*, 160–168. [CrossRef]

6. Seco, A.; del Castillo, J.M.; Espuelas, S.; Marcelino, S.; García, B. Sulphate soil stabilisation with magnesium binders for road subgrade construction. *Int. J. Pavement Eng.* **2020**. [CrossRef]
7. Raja, P.S.K.; Thyagaraj, T. Effect of compaction time delay on compaction and strength behavior of lime-treated expansive soil contacted with sulfate. *Innov. Infrastruct. Solut.* **2020**, *5*, 14. [CrossRef]
8. Chegenizadeh, A.; Keramatikerman, M.; Miceli, S.; Nikraz, H. Investigation on Recycled Sawdust in Controlling Sulphate Attack in Cemented Clay. *Appl. Sci.* **2020**, *10*, 1441. [CrossRef]
9. Yi, Y.; Liska, M.; Jin, F.; Al-tabbaa, A. Mechanism of reactive magnesia—ground granulated blastfurnace slag (GGBS) soil stabilization. *Can. Geotech. J.* **2016**, *55*, 773–782. [CrossRef]
10. Bazyar, M.H.; Ebrahimi, M.; Lenjani, M.Z.; Makarchian, M. The Effect of Rice Husk Ash on Mechanical Properties of Clayey Soils Stabilized with Lime in the Presence of Sulphate. *J. Eng. Geol.* **2017**, *11*, 23–52.
11. Abdi, M.R.; Askarian, A.; Safdari Seh Gonbad, M. Effects of sodium and calcium sulphates on volume stability and strength of lime-stabilized kaolinite. *Bull. Eng. Geol. Environ.* **2019**, *79*, 941–957. [CrossRef]
12. Eyo, E.U.; Abbey, S.J.; Ngambi, S.; Ganjian, E.; Coakley, E. Incorporation of a nanotechnology-based product in cementitious binders for sustainable mitigation of sulphate-induced heaving of stabilised soils. *Eng. Sci. Technol. Int. J.* **2020**, *24*, 436–448. [CrossRef]
13. Kampala, A.; Jitsangiam, P.; Pimraksa, K.; Chindaprasirt, P. An investigation of sulfate effects on compaction characteristics and strength development of cement-treated sulfate bearing clay subgrade. *Road Mater. Pavement Des.* **2020**. [CrossRef]
14. Khadka, S.D.; Jayawickrama, P.W.; Senadheera, S.; Segvic, B. Stabilization of highly expansive soils containing sulfate using metakaolin and fly ash based geopolymer modified with lime and gypsum. *Transp. Geotech.* **2020**, *23*, 100327. [CrossRef]
15. Seco, A.; Miqueleiz, L.; Prieto, E.; Marcelino, S.; García, B.; Urmeneta, P. Sulfate soils stabilization with magnesium-based binders. *Appl. Clay Sci.* **2017**, *135*, 457–464. [CrossRef]
16. Celik, E.; Nalbantoglu, Z. Effects of ground granulated blastfurnace slag (GGBS) on the swelling properties of lime-stabilized sulfate-bearing soils. *Eng. Geol.* **2013**, *163*, 20–25. [CrossRef]
17. Pal, S.C.; Mukherjee, A.; Pathak, S.R. Investigation of hydraulic activity of ground granulated blast furnace slag in concrete. *Cem. Concr. Res.* **2003**, *33*, 1481–1486. [CrossRef]
18. McCarthy, M.J.; Csetenyi, L.J.; Sachdeva, A.; Dhir, R.K. Identifying the role of fly ash properties for minimizing sulfate-heave in lime-stabilized soils. *Fuel* **2012**, *92*, 27–36. [CrossRef]
19. Indraratna, B.; Qi, Y.; Tawk, M.; Heitor, A.; Rujikiatkamjorn, C.; Navaratnarajah, S.K. Advances in Ground Improvement Using Waste Materials for Transportation Infrastructure. *Proc. Inst. Civ. Eng. Ground Improv.* **2020**, 1–20. [CrossRef]
20. Yu, B.W.; Du, Y.J.; Jin, F.; Liu, C.Y. Multiscale study of sodium sulfate soaking durability of low plastic clay stabilized by reactive magnesia-activated ground granulated blast-furnace slag. *J. Mater. Civ. Eng.* **2016**, *28*, 04016016. [CrossRef]
21. Václavík, V.; Dirner, V.; Dvorský, T.; Daxner, J. The use of blast furnace slag. *Metalurgija* **2012**, *51*, 461–464.
22. Jin, F.; Al-Tabbaa, A. Evaluation of novel reactive MgO activated slag binder for the immobilisation of lead and zinc. *Chemosphere* **2014**, *117*, 285–294. [CrossRef]
23. Cai, G.H.; Liu, S.Y.; Zheng, X. Influence of drying-wetting cycles on engineering properties of carbonated silt admixed with reactive MgO. *Constr. Build. Mater.* **2019**, *204*, 84–93. [CrossRef]
24. Yi, Y.; Liska, M.; Al-Tabbaa, A. Initial investigation into the use of GGBS-MgO in soil stabilisation. *Grouting Deep Mix.* **2012**, *2012*, 444–453.
25. YI, Y.; LI, C.; LIU, S.; AL-TABBAA, A. Resistance of MgO–GGBS and CS–GGBS stabilised marine soft clays to sodium sulfate attack. *Géotechnique* **2014**, *64*, 673–679. [CrossRef]
26. Yi, Y.; Liska, M.; Al-Tabbaa, A. Properties of Two Model Soils Stabilized with Different Blends and Contents of GGBS, MgO, Lime, and PC. *J. Mater. Civ. Eng.* **2014**, *26*, 267–274. [CrossRef]
27. Yi, Y.; Gu, L.; Liu, S.; Jin, F. Magnesia reactivity on activating efficacy for ground granulated blastfurnace slag for soft clay stabilisation. *Appl. Clay Sci.* **2016**, *126*, 57–62. [CrossRef]
28. Sarkkinen, M.; Kujala, K.; Gehör, S. Efficiency of MgO activated GGBFS and OPC in the stabilization of highly sulfidic mine tailings. *J. Sustain. Min.* **2019**, *18*, 115–126. [CrossRef]
29. Tawk, M.; Qi, Y.; Indraratna, B.; Rujikiatkamjorn, C.; Heitor, A. Behaviour of a Mixture of Coal Wash and Rubber Crumbs under Cyclic Loading. *J. Mater. Civ. Eng.* **2021**, *33*, 04021054. [CrossRef]
30. Nidzam, R.M.; Kinuthia, J.M. Sustainable soil stabilisation with blastfurnace slag—A review. *Proc. Inst. Civ. Eng. Constr. Mater.* **2010**, *163*, 157–165. [CrossRef]
31. Pai, R.R.; Patel, S.; Bakare, M.D. Applicability of Utilizing Stabilized Native Soil as a Subbase Course in Flexible Pavement. *Indian Geotech. J.* **2020**, *50*, 289–299. [CrossRef]
32. Truong, S.B.; Thi, N.N.; Thanh, D.N. An Experimental Study on Unconfined Compressive Strength of Soft Soil-Cement Mixtures with or without GGBFS in the Coastal Area of Vietnam. *Adv. Civ. Eng.* **2020**, *2020*, 20–25.
33. Ghorbani, A.; Salimzadehshooiili, M. Dynamic Characterization of Sand Stabilized with Cement and RHA and Reinforced with Polypropylene Fiber. *J. Mater. Civ. Eng.* **2019**, *31*, 1–24. [CrossRef]
34. British Standards Institution. *BS EN ISO 17892-4:2016: GGeotechnical Investigation and Testing. Laboratory Testing of Soil Part. 4: Determination of Particle Size Distribution*; BSI Standards Ltd.: London, UK, 2014.

35. Phanikumar, B.R.; Nagaraju, T.V. Effect of Fly Ash and Rice Husk Ash on Index and Engineering Properties of Expansive Clays. *Geotech. Geol. Eng.* **2018**, *36*, 3425–3436. [CrossRef]
36. Yatawara, M.; Athukorala, S. Potential of replacing clay soil by rice husk ash (RHA) in enhancing the properties of compressed earth blocks (CEBs). *Environ. Dev. Sustain.* **2020**, *23*, 3474–3486. [CrossRef]
37. Nayel, I.H.; Burhan, S.K.; Nasr, M.S. Characterisation of prepared rice husk ash and its effects on strength development in recycled aggregate concrete. In Proceedings of the 2nd International Conference on Engineering Sciences, Kerbala, Iraq, 26–28 March 2018.
38. He, J.; Li, Z.X.; Wang, X.Q.; Shi, X.K. Durability of Soft Soil Treated with Soda Residue and Ground Granulated Blast Furnace Slag in a Soaking Environment. *J. Mater. Civ. Eng.* **2020**, *32*, 06019018. [CrossRef]
39. Gu, K.; Jin, F.; Al-Tabbaa, A.; Shi, B.; Liu, C.; Gao, L. Incorporation of reactive magnesia and quicklime in sustainable binders for soil stabilisation. *Eng. Geol.* **2015**, *195*, 53–62. [CrossRef]
40. National Academies of Sciences, E.A.M. *Recommended Practice for Stabilization of Sulfate-Rich Subgrade Soils*; The National Academies Press: Washington, DC, USA, 2009; ISBN 9780309429825.
41. Puppala, A.J.; Talluri, N.; Chittoori, B.C.S. Calcium-based stabiliser treatment of sulfate-bearing soils. *Proc. Inst. Civ. Eng. Ground Improv.* **2014**, *167*, 162–172. [CrossRef]
42. Talluri, N.; Puppala, A.J.; Congress, S.S.C.; Banerjee, A. Experimental Studies and Modeling of High Sulfate Soil Stabilization. *J. Geotech. Geoenviron. Eng.* **2020**, *146*, 1–16, in press. [CrossRef]
43. British Standards Institution. *BS EN 13286-2:2010: Unbound and Hydraulically Bound. Mixtures Part. 2: Test. Methods for Laboratory Reference Density and Water Content—Proctor Compaction*; BSI Standards Ltd.: London, UK, 2012.
44. British Standards Institution. *BS EN 13286-49:2004: Unbound and Hydraulically Bound. Mixtures Accelerated Swelling Test. for Soil Treated by Lime and/or Hydraulic Binder*; BSI Standards Ltd.: London, UK, 2004.
45. British Standards Institution. *BS EN ISO 17892-7:2018: Geotechnical Investigation and Testing. Laboratory Testing of Soil Part. 7: Unconfined Compression Test. Strength*; BSI Standards Ltd.: London, UK, 2018.
46. Yong, L.L.; Namal Jayasanka Perera, S.; Syamsir, A.; Emmanuel, E.; Paul, S.C.; Anggraini, V. Stabilization of a residual soil using calcium and magnesium hydroxide nanoparticles: A quick precipitation method. *Appl. Sci.* **2019**, *9*, 4325. [CrossRef]
47. Estabragh, A.R.; Jahani, A.; Javadi, A.A.; Babalar, M. Assessment of different agents for stabilisation of a clay soil. *Int. J. Pavement Eng.* **2020**. [CrossRef]

applied sciences

MDPI

Article

Reducing Soil Permeability Using Bacteria-Produced Biopolymer

Amanda Mendonça [1], Paula V. Morais [2,*], Ana Cecília Pires [2], Ana Paula Chung [2] and Paulo J. Venda Oliveira [3,*]

1 Department of Mechanical Engineering, University of Coimbra, 3030-788 Coimbra, Portugal; amaandamendonca@hotmail.com
2 Centre for Mechanical Engineering, Materials and Processes, Department of Life Sciences, Calçada Martim de Freitas, University of Coimbra, 3000-456 Coimbra, Portugal; accp@ua.pt (A.C.P.); apchung@gmail.com (A.P.C.)
3 ISISE, Department of Civil Engineering, University of Coimbra, 3030-788 Coimbra, Portugal
* Correspondence: pvmorais@ci.uc.pt (P.V.M.); pjvo@dec.uc.pt (P.J.V.O.)

Abstract: The building of civil engineering structures on some soils requires their stabilisation. Although Portland cement is the most used substance to stabilise soils, it is associated with a lot of environmental concerns. Therefore, it is very pertinent to study more sustainable alternative methodologies to replace the use of cement. Thus, this work analyses the ability of the more sustainable xanthan-like biopolymer, produced by *Stenotrophomonas maltophilia* Faro439 strain (LabXLG), to reduce the permeability of a sandy soil. Additionally, the effectiveness of this LabXLG is compared with the use of a commercial xanthan gum (XG) and cement for various hydraulic gradients and curing times. The results show that a treatment with either type of XG can be used to replace the cement over the short term (curing time less than 14 days), although a greater level of effectiveness is obtained with the use of the commercial XG, due to its higher level of purity. The soil treatment with LabXLG creates a network of fibres that link the soil particles, while the commercial XG fills the voids with a homogeneous paste.

Keywords: biopolymer; soil stabilisation xanthan gum; sandy soil

Citation: Mendonça, A.; Morais, P.V.; Pires, A.C.; Chung, A.P.; Oliveira, P.J.V. Reducing Soil Permeability Using Bacteria-Produced Biopolymer. *Appl. Sci.* **2021**, *11*, 7278. https://doi.org/10.3390/app11167278

Academic Editors: Daniel Dias and José A. González-Pérez

Received: 17 May 2021
Accepted: 4 August 2021
Published: 7 August 2021

Publisher's Note: MDPI stays neutral with regard to jurisdictional claims in published maps and institutional affiliations.

Copyright: © 2021 by the authors. Licensee MDPI, Basel, Switzerland. This article is an open access article distributed under the terms and conditions of the Creative Commons Attribution (CC BY) license (https://creativecommons.org/licenses/by/4.0/).

1. Introduction

The growing development and consequent increase in construction in the main urban centres around the world results in a need for infrastructures over and/or inside geotechnical formations lacking the most favourable characteristics. In the past, these formations were frequently considered inappropriate for civil engineering. The main limitations of these types of soils are usually related to insufficient mechanical properties and permeability characteristics that are inadequate for most building practices. Some ground improvement techniques are frequently used to enhance the properties of these soils to overcome these limitations and to allow the safe construction of infrastructures. Currently, chemical stabilisation using cement-based binders is one of the techniques most used [1]. However, the application of cement-based materials raises some concerns from an environmental point of view, in particular: (i) the activity associated with the production of cement creates significant CO_2 emissions; in fact, in 2014, this activity contributed to about 8–10% of the global CO_2 emissions [2–5]; (ii) the use of cement-based materials for soil stabilisation creates an irreversible composite material [4]; (iii) the use of cement-based binders for soil stabilisation induces an increase in the pH of the soil, with detrimental effects on the vegetation and microbial communities [4].

In order to decrease the ecological footprint caused by the cement-based materials, some alternative bio-based materials that promote the enhancement of the soil's properties have been investigated during the last few years. These biomaterials can be obtained from the activity of microorganisms and/or the use of substances derived from them. Bacteria-producing materials [6–8] and enzymes [8–13] have been used to catalyse the hydrolysis of the urea in porous media to promote biocementation and biopolymers have been used to modify the properties of soils [5,14]. The effect of the biocementation process on such properties is usually due to the association of two factors, the creation of some bonds in the solid skeleton and the filling of the empty spaces between the soil particles with calcium carbonate crystals. Results published with this methodology using sandy soils have showed that the biocementation process promotes an improvement in the mechanical behaviour [7,9–11,15–17] and a decrease in the coefficient of permeability of the biocemented soil [16,18]. During the last few years, several studies have investigated different biopolymers for soil improvement, specifically in terms of the: (i) improvement of the mechanical behaviour with the use of Guar gum [19], β-glucan [20,21], and xanthan gum [22–26]; (ii) change in the compressibility characteristics with the use of Guar gum [27], cationic e-polylysine [23], xanthan gum [19,23,26], gellan, and agar gums [4,21,28]; (iii) change in the plasticity properties with the use of Guar gum [29], cationic e-polylysine [23], and xanthan gum [22,23]; (iv) reduced permeability coefficient with the use of Guar gum [30], chitosan gum [31], and xanthan gum [20,25]; (v) use of Chitosan gum to remediate contaminated soils [4,32]; (vi) reduce the erosion of the soil with the use of xanthan gum [21] and casein and sodium caseinate salt [31]. Recent research concerning the effect of xanthan gum (XG) on the hydraulic conductivity of XG-soil mixtures can be summarised as follows:

(i) The biostabilisation of sandy and silty soils with XG induces the filling of the pore spaces of the soil with hydrogels, which obstruct the water flow (i.e., pore-clogging) and consequently reduce the permeability coefficient of the XG-soil mixture. This effect tends to be more significant with an increase in the XG content of the mixture [19,20,25,30,33,34]. In fact, the experimental results of Ayeldeen et al. [34] show that treatment of sand and silty soil with a content of 2% of XG leads to a reduction in the coefficient permeability to 10% of the value shown for the untreated soils.

(ii) The results of the mixture of XG (0.5%–3%) with sand-bentonite are not in line with the behaviour described previously for biostabilised silty and sandy soils since a slight increase in the permeability coefficient is observed with the increase in the XG content [35]. These results highlight the effect of the finer particles (silt or clay) in the soil type in terms of the effectiveness of the treatment with biopolymers. In fact, the increase in the XG content strengthens and makes more numerous the bonds between the clay particles (ionic/electrostatic, hydrogen, and Van der Waals). This promotes the creation of more and bigger-sized aggregates of clay particles. This leads to the widening of the water flow channels and a consequent slight increase in the coefficient of permeability, within one order of magnitude [35].

(iii) The experimental findings concerning the effect of the curing time on the permeability coefficient of treated soils show some contradictory results. Thus, while Khachatoorian et al. [33] show a decrease in the permeability coefficient of treated Ottawa sand (0.43–0.6 mm size sand) over 11 days of curing, the results obtained by Cabalar et al. [25] with treated Narli sand (0.075–1.0 mm size sand) do not show a clear tendency over the same range of curing time. On the other hand, for a longer curing time, most of the experimental results show an increase in the permeability coefficient with the increase in the curing time for mixtures of XG with sand and silt, which is justified by the dehydration of the XG expected over time [25,34]. Indeed, SEM images confirm the occurrence of shrinkage over time, which increases the size of the pore spaces, and consequently an increase in the soil's permeability [34].

(iv) The increase in the permeability coefficient observed over the curing time (in the long term) is more significant in treated sands than in treated silts since the finer particles of silt delay the evaporation of the water in the voids and the dehydration of the hydrogels deposited in the pore spaces [34].

(v) The gas permeability also decreases after the biostabilisation with XG; indeed, a mixture of 3% xanthan gum with kaolin clay decreases the gas permeability of the mixture in relation to the untreated clay by up to two orders of magnitude [36]. This effect is more significant for a higher water content since there is more water available for the hydration of the biopolymers, which promotes a more effective pore-clogging [36].

As the XG is a polysaccharide, there is an important concern, related to its long-term durability when it is used in geotechnical engineering since this biopolymer (like the generality of biopolymers) is biodegradable over time [5]. In spite of this issue, some works have revealed a satisfactory durability, for example: (i) a mixture of sand, gravel, and kaolin treated with XG shows a reasonable durability against water-induced deterioration [37]; (ii) a sand biostabilised with XG submitted to slake durability tests showed a better resistance to disintegration than with the use of Portland cement [38]; (iii) the long-term strength (until 730 days) of a sand treated with XG does not reveal a decrease in the strength, that is, traces of the biodegradability [39]. It should be mentioned that sometimes, the biodegradability of the XG can become an advantage, for instance, in temporary works performed in sensitive areas, from an environmental point of view and where, after the work, it is necessary to remove all traces of the intervention performed.

Although the use of XG in soil stabilization is more sustainable than the use of Portland cement, the industrial production of commercial xanthan gum is not entirely environmentally friendly, since it includes: processes of fermentation, filtration (or centrifugation), precipitation using non-solvents (such as isopropanol, ethanol, acetone), the use of salts, and pH adjustments [40]. In order to mitigate some of the environmental concerns related to the use of commercial XG, this work studies the ability of a xanthan-like gum obtained directly from the strain of *Stenotrophomonas maltophilia* Faro439 (LabXLG) to reduce the permeability of the soil treated. The aim of the present work is to compare the effectiveness of LabXLG with commercial XG and the traditional stabilisation with cement concerning the permeability coefficient of sandy soil. Additionally, the effect of the curing time (3, 7, 14, and 28 days), the hydraulic gradient (6.9, 8.2, 9.4, and 10.6), the XG content (0.16 and 0.33%), and cement (0.33 and 1.0%) are analysed. Scanning electron microscopy (SEM) with energy dispersive X-ray (EDX) was also used to study the microstructure and the chemical composition of the soil treated with LabXLG and commercial XG.

2. Materials and Methods

2.1. Soil Characterisation

The grain size distribution and the main characteristics of the soil used in the experiments are shown in Figure 1 and Table 1, respectively. This is a non organic and non-plastic silty sand, composed of 72.7% sand and 27.3% silt, classified as SM (ASTM D2487, 2017) [41]. The soil compaction characteristics reached by using the standard Proctor test (ASTM D698, 2012) Ref. [42] show an optimum water content (w_{opt}) of 14.3% and a maximum dry unit weight (γ_{dmax}) of 16.2 kN/m^3. The pH value of the non-treated soil is approximately 7.1 [43].

2.2. Production of the Xanthan Gum Biopolymer Compound

The production of the biopolymer enriched compound, illustrated in Figure 2, consisted of the following steps [43]: (i) *Stenotrophomonas maltophilia* strain Faro439 (UCCCB 127), obtained from the University of Coimbra Bacterial Culture Collection (https://ucccb.uc.pt/, accessed on 15 January 2020), was grown on R2A agar (Reasoner's 2A agar medium) plates for 3 days, at a temperature of 25 °C; (ii) the cells grown were used to inoculate the xanthan medium [44]. The cells were grown in a batch system incubated for 24 h in a shaker at 150 rpm. These cells were used as inoculum in a new batch system with xanthan medium incubated for 3 days in a shaker at 150 rpm before undergoing the centrifugation process to recover the biomass containing the biopolymer compound; (iii) the biomass was drained for 24 h to remove the excess of water and lyophilized for 24 h to obtain the powder form and to facilitate the mixing with the soil.

The use of this xanthan-like, enriched compound is justified by its simple production process. Indeed, the compound is obtained without the use of purification steps, solvents, or procedures, or any equipment that leads to consequences from the ecological point of view. Therefore, this xanthan gum-like compound production can be considered more environmentally friendly than the commercial XG. Its availability at reasonable prices in relation to other biopolymers [21] and the ability of this biopolymer to decrease the soil's permeability are other positive aspects.

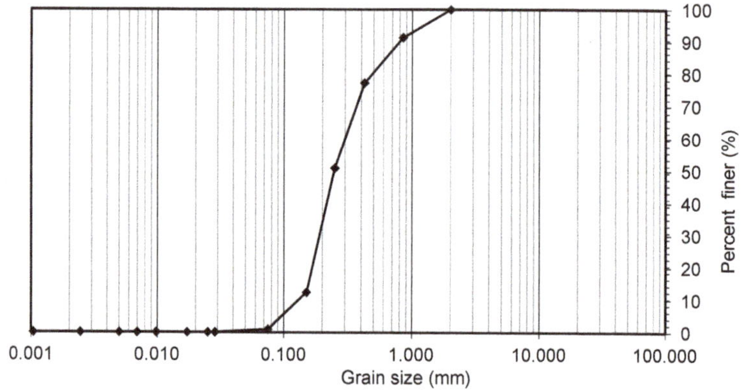

Figure 1. Grain size distribution of the soil used.

Table 1. Main characteristics of the soil used in the experiments.

Property	Value
Grain size distribution:	
Sand (%)	72.7
Silt (%)	27.3
Clay (%)	0
Specific gravity, G	2.66
Consistency limits	NP (**)
Standard Proctor test [ASTM D698]:	
Maximum dry unit weight, γ_{dmax}	16.2
Optimum water content, w_{opt}	14.3
Organic matter content, OM (%)	0.19
pH	7.1
Soil classification, USCS (*) [ASTM D2487]	SM

(*) Unified Soil Classification System; (**) Non plastic.

2.3. Commercial Xanthan Gum

The commercial XG used in this work is produced by Sigma-Aldrich(St. Louis, Missouri, EUA) Ref.: G1253), available in powder and is obtained from *Xanthomonas campestris* bacterium via the fermentation of glucose/sucrose. It is a heteropolysaccharide composed of units of glucose, mannose, and glucuronic acid [40]. This is a non-toxic hydrophilic colloid with pseudoplastic rheological properties [40] and is stable within a wide temperature range (10°–80°) and pH (1–13) values. Nowadays, XG is fundamentally used as a thickener and a stabilising agent for many applications, such as agriculture, food, the pharmaceutical industry, cosmetics, industrial lubrication of equipment, textile printing pastes, and explosives, among others [30,40].

2.4. Portland Cement

Some mixtures in this research (Table 2) were chemically stabilised with Portland cement Type I 42.5 R, which is composed of 45% of cement particles smaller than 45 µm. The cement is mainly composed of calcium oxide (CaO = 63.0%), which induces a spontaneous reaction with water, i.e., hydraulic properties.

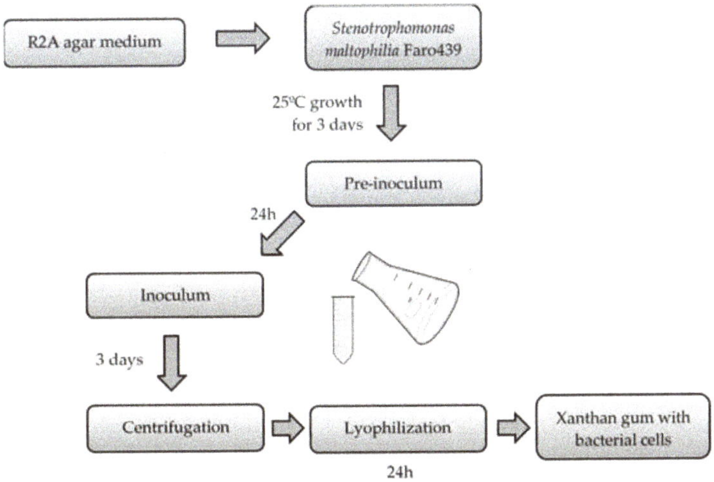

Figure 2. Diagram of the production of the *Stenotrophomonas maltophilia* Faro439 strain biopolymer.

2.5. Testing Procedure

The samples used in the permeability tests were produced in six different types: natural soil, a mixture of soil with the xanthan-like biopolymer produced by bacteria (Lab XG) (0.33%), a mixture of soil with the commercial biopolymer xanthan gum for the contents of 0.33% and 0.16%, and also a mixture of soil with cement for the contents of 0.33% and 1.0%. Each type of mixture was tested for different hydraulic gradients (6.9, 8.2, 9.4, and 10.6) and after four different curing times (3, 7, 14, and 28 days). The details of the testing programme are shown in Table 2. Although all the specimens were prepared equally, the permeability tests were repeated three times for each condition tested, to assure the quality and the reproducibility of the results obtained.

The stabilised soil specimens used in the permeability tests were prepared according to the following steps: (i) larger particles of the soil were removed with a sieve with a 2.0-mm mesh; (ii) the soil was dried in an oven at 105 °C for 24 h; (iii) the dry soil and powdered grout agent content (presented in Table 2) were mixed; (iv) the soil, the grout agent (LAB XG, commercial XG, and cement), and the water content of 14.3% (optimum water content for untreated soil using the standard Proctor test) were mixed until a homogeneous paste was obtained; (v) the paste was put inside PVC moulds (38 mm in diameter, 76 mm height) in 3 layers; (vi) each layer was compacted in order to obtain the dry unit weight of 16.2 kN/m^3 (optimum conditions for untreated soil using the standard Proctor test); (vii) the contact surfaces between two successive layers was scarified before the introduction of a new layer; (viii) the specimens were stored during the curing time (Table 2) in a humid chamber with a temperature of (20 ± 2 °C) and humidity (95 ± 5%) control; (viii) after the curing time, the specimens were assembled in the equipment used to carry out the permeability tests (Figure 3); (ix) a container (to measure the volume of the water that flowed through the specimen) was placed under each one; (x) the water level was adjusted in order to assure the hydraulic gradients stated in Table 2; (xi) the permeability tests were performed.

Although the optimum compaction conditions obtained from the standard Proctor test changed when Portland cement and the xanthan gum were added to the soil [26,45], all the specimens used in the work were prepared for the optimum water content for untreated soil, making it possible to compare all the results without including additional effects.

Table 2. Testing programme.

Grout Agents			Hydraulic Gradient (m/m)	Curing Time (Days)
Lab XLG (%) (*)	Commercial XG (%)	Cement (%)		
0 (*)	0 (*)	0 (**)	6.9 8.2 9.4 10.6	– – – –
–	0.16	–	6.9 8.2 9.4 10.6	3/7/14/28 3/7/14/28 3/7/14/28 3/7/14/28
–	0.33	–	6.9 8.2 9.4 10.6	3/7/14/28 3/7/14/28 3/7/14/28 3/7/14/28
0.33	–	–	6.9 8.2 9.4 10.6	3/7/14/28 3/7/14/28 3/7/14/28 3/7/14/28
–	–	0.33	6.9 8.2 9.4 10.6	3/7/14/28 3/7/14/28 3/7/14/28 3/7/14/28
–	–	1.0	6.9 8.2 9.4 10.6	3/7/14/28 3/7/14/28 3/7/14/28 3/7/14/28

(*) Percentage based on dry weight. (**) Reference tests (non-stabilised soil).

The permeability tests were performed according to the scheme illustrated in Figure 3, which consists of a constant head permeability test. This scheme allows us to test eight specimens simultaneously. An overflow drain in the upper container assures a constant level of water during the tests. As is usual in this type of test, the total volume of water that flows through the specimen is measured (the accumulated water inside the containers placed under each specimen was weighed on a precision balance). The system has a set of taps that make it possible to control the water flow for each specimen. The permeability coefficient (k) was calculated according to Darcy's law:

$$k = \frac{Q}{A \times i} = \frac{(\Delta V/\Delta t)}{A \times (\Delta h/L)} \quad (1)$$

where Q is the water flow rate, ΔV is the average of the accumulated volume of water that passed through each sample during three measurements, Δt is the period of time, A is the cross-sectional area of the specimen, i is the hydraulic gradient, L is the height of the specimen, and Δh is the sum of the L/2 and the water column height.

SEM (scanning electron microscope) with EDX (energy dispersive X-ray) were also performed, with the objective of analysing the microscopic structure of the soil-biopolymer mixtures and the chemical elements present in each mixture. The samples were prepared by fixing the soil samples with 2.5% of glutaraldehyde, following by dehydration by a grade of ethanol incubations (70%, 80%, 90%, 95%, and 100%). Previous to observation, a thin layer of gold coating was sputtered on the cracked surface of a thin section of the specimen tested and the SEM/EDS tests were performed in accordance with Pansu and Gautheyrou [46].

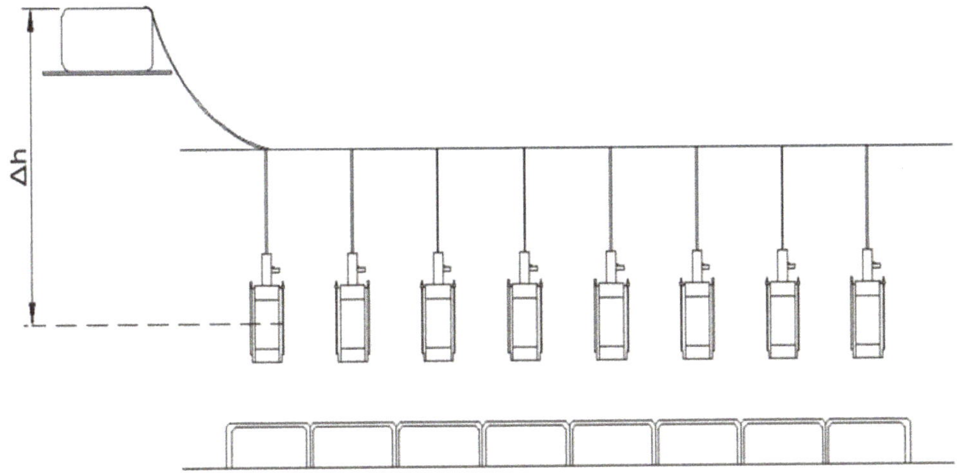

Figure 3. Set-up used for the permeability tests.

3. Results and Discussion

The permeability coefficients obtained experimentally with the different mixtures and hydraulic gradients are presented in Figure 4a–d for a curing time of 3, 7, 14, and 28 days, respectively. The results from the natural soil are used as reference values to evaluate the impact of each grouting agent on the permeability coefficient of the soil treated. Additionally, the inclusion of the samples chemically stabilised with cement in this work is intended to compare the effectiveness of the two XGs (commercial and the one produced from bacteria) to reduce the soil's permeability, with the most usual grouting agent (cement) in geotechnical engineering.

For a better analysis of the results, the average value, the maximum, and minimum errors for each condition tested are presented. As expected, and as is usual in this type of mixture [7,19], there is a high scattering of the results, which is more significant for the natural soil and the soil treated with Lab XG. In fact, these results reflect some heterogeneity of the natural soil, the lower level of purity of the Lab XG, and the difficulty in producing homogeneous specimens when a low content of XG is used. Indeed, an increase in the content of the commercial XG from 0.16% to 0.33% significantly decreases the scattering of the results.

3.1. *Effect of Type and Content of the Stabilising Agent*

Independently of the curing time and the hydraulic gradient, the results depicted in Figure 4 clearly show that the stabilisation of the soil with both types of XG induces a significant decrease in the coefficient of permeability. This behaviour is due to the development of some links and viscous hydrogels due to the contact of the XG with the water (i.e., hydration of the hydrogels) that fill part of the soil's voids [23,30], inducing a decrease in the soil's permeability. The results also reveal that the commercial XG is more effective than the Lab XG to reduce the soil's permeability; in fact, the use of 0.33% of the LabXLG in the mixtures has much less of an effect than the use of 0.33% of the commercial XG and even with the use of 0.16% of the commercial XG, mainly for higher curing times. This greater effectiveness of the commercial XG is consistent with its greater purity when compared with the LabXLG. Indeed, it is important to state that the LabXLG used in the tests is composed of the lyophilized biomass formed by *S. maltophilia* Faro439, containing the biopolymer and the lyophilized cells that were responsible for producing the biopolymer, consequently a low level of purity is obtained.

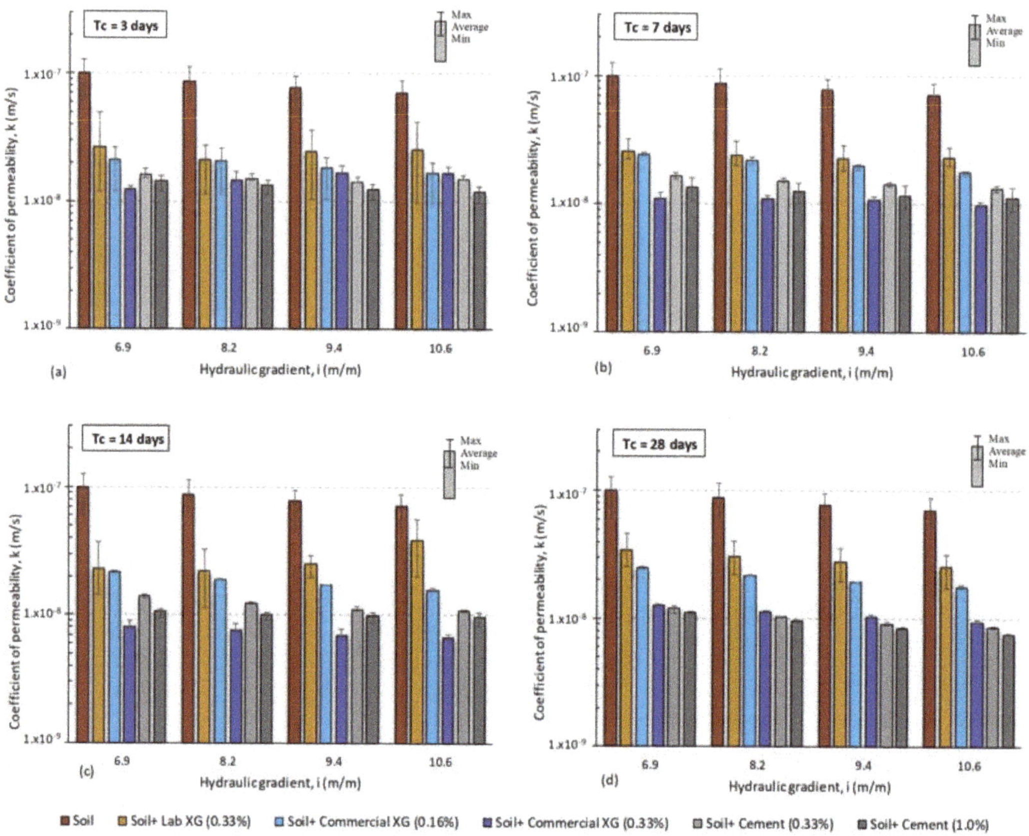

Figure 4. Permeability coefficient of the natural soil and stabilised soil for different hydraulic gradients; (**a**) curing time of 3 days; (**b**) curing time of 7 days; (**c**) curing time of 14 days; (**d**) curing time of 28 days.

The comparison of the use of XG with the use of Portland cement (the most usual binder for soil stabilisation) shows that the use of 0.33% commercial XG is more effective than the use of 1% cement, namely for a curing time of 7 and 14 days. Even for a curing time of 28 days, the effectiveness of 0.33% of commercial XG is similar to the use of 0.33% of cement. Thus, these results indicate that XG can be used to replace the use of cement, mainly in the short term (for instance in temporary works).

Considering the efficiency of both biopolymers and that the production process of the commercial XG is associated with a complex industrial process with more environmental concerns (as discussed in Section 1) than the production of the Lab XG, the use of the Lab XG emerges as a viable alternative to reduce a soil's permeability.

3.2. Effect of the Hydraulic Gradient

With some exceptions (for instance, the mixture of the soil with the LabXLG for a curing time of 14 days), most of the results show a slight decrease in the permeability coefficient with the increase in the hydraulic gradient, mainly for the natural soil and the treated soil with 0.16% of the commercial XG. Although, this tendency is not in line with most of the results obtained by Cabalar et al. [25] for a range of hydraulic gradient from 5 to 100; in some cases, the increase in the hydraulic gradient from 5 and 10 (similar to the range considered in this work) also shows a decrease in the soil's permeability [25].

The differences previously observed may be due to the methodology used in the tests to evaluate the permeability coefficient. As shown in Figure 3 in the present work, the permeability coefficient was measured with a descendent water flow. Thus, the increase in the hydraulic gradient tends to increase the vertical effective stress with the consequent decreases in the void ratio, which tends to decrease the soil's permeability, as observed in most of the results in this work. In the case of the specimens being treated with the Lab and the commercial XG, the results also seem to indicate that there may be a movement of some hydrated hydrogels towards the voids in the soil, which tends to restrict the water flow, thereby inducing the decreases in the permeability coefficient.

The results also show that the specimens with a high level of stabilisation (i.e., with a low permeability coefficient) show a lower impact on the variation of the hydraulic gradient. Indeed, the stronger solid skeleton promoted by the stabilisation with a high content of the commercial XG (0.33%) and cement (0.33% and 1%), restricts the eventual movement of the soil particles and hydrogels, which leads to a lower influence of the hydraulic gradient.

3.3. Effect of Curing Time

Figure 5 highlights the effect of the curing time on the evolution of the average permeability coefficient over the curing time, for the mixtures with a grout content of 0.33% and subjected to a hydraulic gradient of 6.9. Independently of the process of production of the XG, a sharp decrease in the soil's permeability during the first 3 days of curing is observed, followed by a slight decrease over time to a minimum permeability coefficient for a curing time of 14 days. Although these results indicate that the hydration of the hydrogels tends to occur fundamentally over the short term (<3 days), they also suggest that there is still some hydration of the hydrogels during the first 14 days of curing time. In fact, as the hydration starts from the outside of the hydrogels, this seems to hinder the water penetrating to the inner part of the hydrogels, which leads to a slow and gradual hydration of this inner zone. After 14 days of curing, a slight increase in the soil's permeability is observed with the use of both types of XG, which is probably due to some biodegradation of the XG [14] and/or the dehydration of the hydrogels with the consequent shrinkage of the XG. These two factors increase the pore spaces in the soil over time, which favours the flow of water: consequently, an increase in the soil's permeability is observed. The decrease in the soil's permeability over the short term (fewer than 14 days) is in line with the results of Khachatoorian et al. [33] for a similar range of curing times, while the increase in the permeability coefficient over the long term (more than 14 days) matches the findings of Ayelden et al. [34] and Cabalar et al. [25].

As expected, the specimens chemically stabilised with cement show a reduction in the permeability coefficient with curing time, which is consistent with the development of the pozzolanic reactions that occur in the long term. Thus, for a curing time of 28 days, the coefficient of permeability of the soil treated with cement and the commercial XG with the same content (0.33%) show similar effectiveness, while in the short term, the commercial XG induces a greater reduction in the soil's permeability. Figure 5 also emphasizes the better effectiveness of the commercial XG in relation to the LabXLG, which can be fundamentally attributed to its higher level of purity.

3.4. Analysis of the Structures Formed in the Biopolymer-Sol Mixture and Chemical Analysis

The SEM images of the natural soil (Figure 6) show an "open" structure, with a great void ratio that is associated with its high permeability coefficient. Figures 7 and 8 show the SEM images of the structure of the solid skeleton of the soil treated with both types of XG. The use of the LabXLG (Figure 7) seems to induce the creation of structures similar to a network of fibres that link the soil particles [5], which decreases the voids in the soil, thus reducing its permeability. Chang et al. [14] also observed that XG creates bridges between soil particles. However, the use of the commercial XG (Figure 8) seems to produce a homogeneous paste, probably due to the hydration of the hydrogels, which is responsible for the reduction in the permeability coefficient. The differences in the structures of the solid skeleton induced by both types of XG seem to be related to the differences in the effectiveness of the two XGs to decrease the soil's permeability. In fact, the more open structure promoted by the LabXLG is associated with a higher soil permeability, while the homogeneous paste induced by the use of the commercial XG restricts the water flow, decreasing the soil's permeability in relation to the use of the LabXLG.

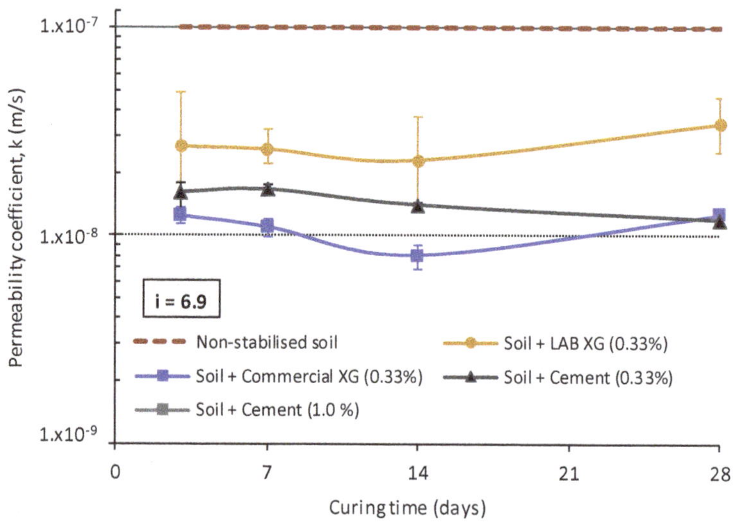

Figure 5. Effect of curing time on the permeability coefficient for a hydraulic gradient of 6.9.

Figure 9 shows the chemical analysis carried out on a fibre of the LabXLG, while Figure 9b shows all the chemical elements present in the sample. It is worth mentioning that the gold (Au) must be disregarded because it was used to prepare the specimens for the SEM test. The results obtained show a significant content of carbon (C, 30.8%) and oxygen (O, 22.7%), which clearly indicate the presence of organic fibres [40]. This finding supports the conclusion that biological fibres can be related with the soil permeability reduction.

Figure 6. SEM image of the natural soil at 150×.

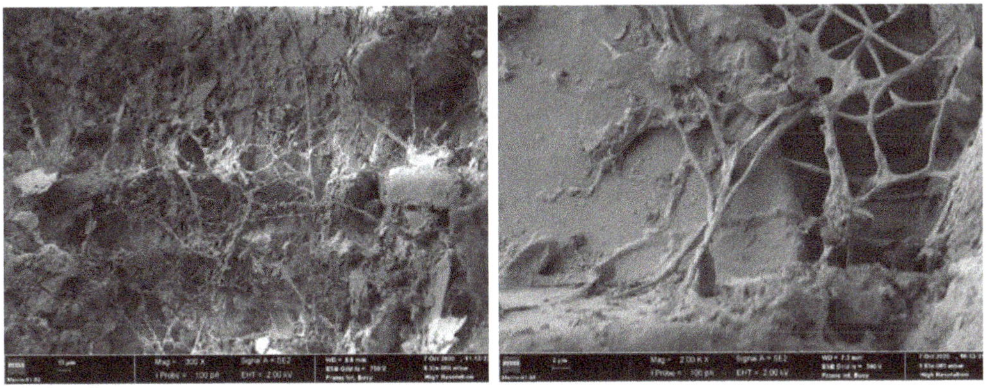

(**a**) (**b**)

Figure 7. (**a**) SEM image of the mixture Lab XG-soil at 300×. (**b**) Scale-up of Figure 7a to 2000×.

Figure 8. SEM image of the mixture produced with the commercial XG at 300×.

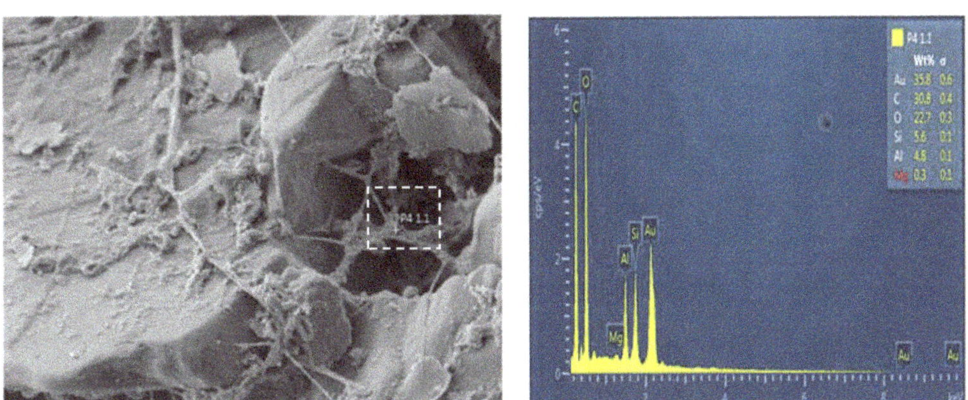

(**a**) (**b**)

Figure 9. (**a**) Area targeted (P4 1.1) for the chemical analysis. (**b**) Chemical analysis of the LabXLG-soil specimen.

4. Conclusions

This work compares the ability of the xanthan-like gum obtained directly from a strain of *S. maltophilia* Faro439 species (LabXLG), with commercial xanthan gum (XG) and Portland cement to decrease the permeability coefficient of a sandy soil. In parallel, the effect of the curing time (3, 7, 14, and 28 days), the hydraulic gradient (6.9, 8.2, 9.4, and 10.6), and the XG content (0.16 and 0.33%) and cement (0.33 and 1.0%) are also analysed. From the experimental testing programme, the following conclusions may be drawn:

(i) The treatment of the soil with both types of XG induces a significant decrease in the coefficient of permeability compared to the untreated soil.
(ii) The comparison of the two types of XG shows that the LabXLG is less effective than the commercial XG, which is related to its lower level of purity. The lyophilized biomass formed by *S. maltophilia* Faro439 also contains lyophilized cells as well as the biopolymer.
(iii) The use of a 0.33% content of commercial XG is more effective to reduce the coefficient of permeability of the treated soil than the use of 1% cement, namely for a curing time of 7 and 14 days. These results indicate that XG can replace the use of cement in the short term, for instance in temporary works.
(iv) The increase in the hydraulic gradient induces a slight decrease in the permeability coefficient. This is consistent with the descendent water flow used in the permeability tests since the increase in the hydraulic gradient tends to increase the vertical effective stress with the consequent decreases in the void ratio and the soil's permeability.
(v) The treatment with both types of XG demonstrates a slight decrease in the permeability coefficient during the first 14 days of curing, followed by an increase for longer times. The initial decrease in the soil's permeability is associated with the hydration of the hydrogels, while the increase in the soil's permeability over time probably reflects the existence of some biodegradation of the XG and/or the eventual dehydration of the hydrogels associated with the consequent shrinkage.
(vi) The microstructure of the treated soil depends on the type of XG, which is linked with each one's effectiveness to reduce the soil's permeability. Thus, the LabXLG creates a network of fibres that links the soil particles and decreases the voids in the soil, while the commercial XG induces the partial filling of the pore spaces with a homogeneous paste, probably due to the hydration of the hydrogels.

Finally, it should be mentioned that taking the production process and its effectiveness in reducing the permeability coefficient into consideration, the experimental results of this work show that the xanthan gum produced using the *S. maltophilia* Faro439 strain emerges as an interesting alternative for use in the stabilisation of sandy soils.

Although *Stenotrophomonas maltophilia* is included in the BSL-2 facultative pathogen group, there are no regulations available for the release of this type of microorganism into the environment, and the number of cells detected in the assays with soil is very low. Therefore, future studies are still needed to ensure that this material is fully safe.

Author Contributions: Conceptualization, A.C.P., P.V.M. and A.P.C.; writing—original draft preparation, A.M., A.C.P. and P.V.M.; writing—review and editing, A.M., P.V.M., P.J.V.O. and A.P.C.; funding acquisition, P.V.M. and P.J.V.O. All authors have read and agreed to the published version of the manuscript.

Funding: This work was supported by the projects ERAMIN2 REVIVING funded by Fundação para a Ciência e Tecnologia (FCT), project PTDC/CTA-AMB/31820/2017 and FCT UIDB/00285/2020, BIORECOVER H2020 grant agreement 821096, project POCI-01-0145-FEDER-028382 and the R&D Unit Institute for Sustainability and Innovation in Structural Engineering (ISISE), under reference UIDB/04029/2020.

Institutional Review Board Statement: Not applicable.

Informed Consent Statement: Not applicable.

Data Availability Statement: No new data were created or analysed in this study. Data sharing is not applicable to this article.

Acknowledgments: This work had the support of the lab technician J.A. Lopes, Department of Civil Engineering, University of Coimbra.

Conflicts of Interest: The authors declare no conflict of interest.

References

1. Correia, A.A.S.; Casaleiro, P.D.F.; Rasteiro, M.G. Applying multiwall carbon nanotubes for soil stabilization. *Procedia Eng.* **2015**, *102*, 1766–1775. [CrossRef]
2. Rapier, R. Global Carbon Dioxide Emissions-Facts and Figures. *Consum. Energy Repor.* **2012**, *102*, 1766–1775.
3. Oss, H.G.V. *Cement Statistics and Information*; US Geological Survey: Reston, VA, USA, 2014.
4. Chang, I.; Im, J.; Cho, G.C. Introduction of microbial biopolymers in soil treatment for future environmentally friendly and sustainable geotechnical engineering. *Sustainability* **2016**, *8*, 251. [CrossRef]
5. Mendonça, A.; Morais, P.V.; Pires, A.C.; Chung, A.P.; Oliveira, P.V. A Review on the Importance of Microbial Biopolymers Such as Xanthan Gum to Improve Soil Properties. *Appl. Sci.* **2021**, *11*, 170. [CrossRef]
6. Whiffin, V.S.; van Paassen, L.A.; Harkes, M.P. Microbial carbonate precipitation as a soil improvement technique. *Geomicrobiol. J.* **2007**, *24*, 417–423. [CrossRef]
7. Venda Oliveira, P.J.; Costa, M.S.; Costa, J.N.P.; Nobre, M.F. Comparison of the ability of two bacteria to improve the behaviour of a sandy soil. *J. Mater. Civ. Eng.* **2015**, *27*, 06014025. [CrossRef]
8. Jiang, N.J.; Soga, K. The applicability of microbially induced calcite precipitation (MICP) for internal erosion control in gravel–sand mixtures. *Géotechnique* **2017**, *67*, 42–55. [CrossRef]
9. Venda Oliveira, P.J.; Freitas, L.D.; Carmona, J.P.S.F. Effect of Soil Type on the Enzymatic Calcium Carbonate Precipitation Process Used for Soil Improvement. *J. Mater. Civ. Eng.* **2016**, *29*, 04016263. [CrossRef]
10. Venda Oliveira, P.J.; Neves, J.P.G. Effect of Organic Matter Content on Enzymatic Biocementation Process Applied to Coarse-Grained Soils. *J. Mater. Civ. Eng.* **2019**, *31*, 04019121. [CrossRef]
11. Venda Oliveira, P.J.; Rosa, J.A.O. Confined and unconfined behavior of a silty sand improved by the enzymatic biocementation method. *Transp. Geotech.* **2020**, *24*, 100400. [CrossRef]
12. Carmona, J.P.S.F.; Venda Oliveira, P.J.; Lemos, L.J.L.; Pedro, A.M.G. Improvement of a sandy soil by enzymatic $CaCO_3$ precipitation. *Proc. Inst. Civ. Eng.-Geotech. Eng.* **2018**, *171*, 3–15. [CrossRef]
13. Yasuhara, H.; Neupane, D.; Hayashi, K.; Okamura, M. Experiments and predictions of physical properties of sand cemented by enzymatically-induced carbonate precipitation. *Soils Found.* **2012**, *52*, 539–549. [CrossRef]
14. Jang, J. A Review of the Application of Biopolymers on Geotechnical Engineering and the Strengthening Mechanisms between Typical Biopolymers and Soils. *Adv. Mater. Sci. Eng.* **2020**, *2020*, 1465709. [CrossRef]
15. van Paassen, L.A.; Ghose, R.; van der Linden, T.J.M.; van der Star, W.R.L.; van Loosdrecht, M.C.M. Quantifying biomediated ground improvement by ureolysis: Large scale biogrout experiment. *J. Geotech. Geoenviron. Eng.* **2010**, *136*, 1721–1728. [CrossRef]
16. Al Qabany, A.; Soga, K. Effect of chemical treatment used in MICP on engineering properties of cemented soils. *Géotechnique* **2013**, *63*, 331–339. [CrossRef]
17. Lin, H.; Suleiman, M.T.; Brown, D.G.; Kavazanjian, E., Jr. Mechanical behavior of sands treated by microbially induced carbonate precipitation. *J. Geotech. Geoenviron. Eng.* **2016**, *142*, 04015066. [CrossRef]
18. Chou, C.W.; Seagren, E.A.; Aydilek, A.H.; Lai, M. Biocalcification of sand through ureolysis. *J. Geotech. Geoenviron. Eng.* **2011**, *137*, 1179–1189. [CrossRef]
19. Venda Oliveira, P.J.; Cabral, D.J.R. Behaviour of a silty sand stabilized with xanthan gum under unconfined and confined conditions. *Proc. Inst. Civ. Eng.-Ground Improv.* **2021**. [CrossRef]
20. Dehghan, H.; Tabarsa, A.; Latifi, N.; Bagheri, Y. Use of Xanthan and guar gums in soil strengthening. *Clean Technol. Environ. Policy* **2019**, *21*, 155–165. [CrossRef]
21. Chang, I.; Prasidhi, A.K.; Im, J.; Shin, H.D.; Cho, G.C. Soil treatment using microbial biopolymers for anti-desertification purposes. *Geoderma* **2015**, *253-254*, 39–47. [CrossRef]
22. Chang, I.; Kwon, Y.-M.; Im, J.; Cho, G.-C. Soil consistency and interparticle characteristics of Xanthan gum biopolymer–containing soils with pore-fluid variation. *Can. Geotech. J.* **2019**, *56*, 1206–1213. [CrossRef]
23. Kwon, Y.-M.; Chang, I.; Lee, M.; Cho, G.-C. Geotechnical engineering behavior of biopolymer-treated soft marine soil. *Geomech. Eng.* **2019**, *17*, 453–464.
24. Lee, S.; Chung, M.; Park, H.M.; Song, K.-I.; Chang, I. Xanthan gum Biopolymer as Soil-Stabilization Binder for Road Construction Using Local Soil in Sri Lanka. *J. Mater. Civ. Eng.* **2019**, *31*, 06019012. [CrossRef]
25. Cabalar, A.F.; Wiszniewski, M.; Skutnik, Z. Effects of Xanthan Gum Biopolymer on the Permeability, Odometer, Unconfined Compressive and Triaxial Shear Behavior of a Sand. *Soil Mech. Found. Eng.* **2017**, *54*, 356–361. [CrossRef]
26. Antonette, J.; Donza, J.; Jaradat, K.A.; Darbari, Z.; Abdelaziz, S.L. Filler-Stabilized Xanthan Gum for Soil Improvement. In Proceedings of the Eighth International Conference on Case Histories in Geotechnical Engineering, Philadelphia, PA, USA, 24–27 March 2019. [CrossRef]

27. Ayeldeen, M.; Negm, A.; El-Sawwaf, M.; Kitazume, M. Enhancing mechanical behaviors of collapsible soil using two biopolymers. *J. Rock Mech. Geotech. Eng.* **2017**, *9*, 329–339. [CrossRef]
28. Chang, I.; Jeon, M.; Cho, G.C. Application of microbial biopolymers as an alternative construction binder for earth buildings in underdeveloped countries. *Int. J. Polym. Sci.* **2015**, *2015*, 326745. [CrossRef]
29. Nugent, R.; Zhang, G.; Gambrell, R. Effect of exopolymers on the liquid limit of clays and its engineering implications. *Transp. Res. Rec.* **2010**, *2101*, 34–43. [CrossRef]
30. Bouazza, A.; Gates, W.P.; Ranjith, P.G. Hydraulic conductivity of biopolymer-treated silty sand. *Géotechnique* **2009**, *59*, 71–72. [CrossRef]
31. Fatehi, H.; Abtahi, S.M.; Hashemolhosseini, H.; Hejazi, S.M. A novel study on using protein based biopolymers in soil strengthening. *Constr. Build. Mater.* **2018**, *167*, 813–821. [CrossRef]
32. Yong, S.K.; Shrivastava, M.; Srivastava, P.; Kunhikrishnan, A.; Bolan, N. Environmental applications of chitosan and its derivatives. *Rev. Environ. Contam. Toxicol. Vol.* **2015**, *233*, 1–43.
33. Khachatoorian, R.; Petrisor, I.G.; Kwan, C.-C.; Yen, T.F. Biopolymer plugging effect: Laboratory-pressurized pumping flow studies. *J. Pet. Sci. Eng.* **2003**, *38*, 13–21. [CrossRef]
34. Ayeldeen, K.M.; Abdelazim, M.N.; Mostafa, A.E.S. Evaluating the physical characteristics of biopolymer/soil mixtures. *Arab. J. Geosci.* **2016**, *9*, 371. [CrossRef]
35. Biju, M.S.; Arnepalli, D.N. Effect of biopolymers on permeability of sand-bentonite mixtures. *J. Rock Mech. Geotech. Eng.* **2020**, *12*, 1093–1102. [CrossRef]
36. Ng, C.W.W.; So, P.S.; Lau, S.Y.; Zhou, C.; Coo, J.L.; Ni, J.J. Influence of biopolymer on gas permeability in compacted clay at different densities and water contents. *Eng. Geol.* **2020**, *272*, 105631. [CrossRef]
37. Muguda, S.; Booth, S.J.; Hughes, P.N.; Augarde, C.E.; Perlot, C.; Bruno, A.W.; Gallipoli, D. Mechanical properties of biopolymer-stabilised soil-based construction materials. *Géotech. Lett.* **2017**, *7*, 309–314. [CrossRef]
38. Qureshi, M.U.; Chang, I.; Al-Sadarani, K. Strength and durability characteristics of biopolymer-treated desert sand. *Geomech. Eng.* **2017**, *12*, 785–801. [CrossRef]
39. Chang, I.; Im, J.; Prasidhi, A.K.; Cho, G.C. Effects of Xanthan gum biopolymer on soil strengthening. *Constr. Build. Mater.* **2015**, *74*, 65–72. [CrossRef]
40. García-Ochoa, F.; Santos, V.E.; Casas, J.A.; Gómez, E. Xanthan gum: Production, recovery, and properties. *Biotechnol. Adv.* **2020**, *18*, 549–579. [CrossRef]
41. ASTM D2487-17e1. *Standard Practice for Classification of Soils for Engineering Purposes (Unified Soil Classification System)*; American Society for Testing and Materials: West Conshohocken, PA, USA, 2017.
42. ASTM D698-12e2. *Standard Test Methods for Laboratory Compaction Characteristics of Soil Using Standard Effort (12.400 ft-lbf/ft^3 (600 kN-m/m^3)*; American Society for Testing and Materials: West Conshohocken, PA, USA, 2012.
43. Mendonça, A.C.S. Use of Microbial Biopolymer to Decrease Soil Permeability by Bioclogging. Master's Thesis, University of Coimbra, Coimbra, Portugal, 2020.
44. Carignatto, C.R.; Oliveira, K.S.; Lima, V.M.; Neto, P.O. New Culture Medium to Xanthan Production by Xanthomonas campestris pv. campestris. *Indian J Microbiol.* **2011**, *51*, 283–288. [CrossRef]
45. Joga, J.R.; Varaprasad, B.J.S. Sustainable Improvement of Expansive Clays Using Xanthan Gum as a Biopolymer. *Civ. Eng. J.* **2019**, *5*, 1893–1903. [CrossRef]
46. Pansu, M.; Gautheyrou, J. *Handbook of Soil Analysis—Mineralogical, Organic and Inorganic Methods, English Version*; Springer: Berlin/Heidelberg, Germany, 2006.

applied sciences

MDPI

Article

Solidification/Stabilization of Contaminated Soil in a South Station of the Khurmala Oil Field in Kurdistan Region, Iraq

Sazan Nariman Abdulhamid *, Ahmed Mohammed Hasan and Shuokr Qarani Aziz

Civil Engineering Department, College of Engineering, Salahaddin University-Erbil, Erbil 44001, Kurdistan Region, Iraq; ahmed.hasan@su.edu.krd (A.M.H.); shuokr.aziz@su.edu.krd (S.Q.A.)
* Correspondence: sazan.abdulhamid@su.edu.krd; Tel.: +964-750-463-0741

Abstract: Currently, the primary source of pollution is crude oil production. Crude oil production has dramatic consequences for farmlands, communities, and in terms of the construction materials required for earthworks. The main aims of the present study were to reduce the level of pollution caused by oil production in the Khurmala soil and then reuse it as a construction material. Soil remediation using the solidification/stabilization method was applied in the field using Portland limestone cement (CEM II). The performance of using CEM II in the remediation process was then investigated in the laboratory by taking the natural, contaminated, and treated soils from the Khurmala site. Furthermore, the results of the soils were compared with their corresponding soil samples using ordinary Portland cement (OPC). The comparison was performed by investigating the physical, chemical, and mechanical properties of the soils. The discussion was supported using the scanning electron microscopy (SEM) results. Chemical and SEM results revealed that there were fourfold and tenfold decreases in the percentage of oil and grease using OPC and CEM II, respectively, confirming the higher performance of using CEM II over OPC. The values of the coefficient of permeability, shear strength parameters, and California bearing ratio of the treated soils were significantly improved, compared to those of the contaminated soils.

Keywords: cement; soil contamination; oil field; SEM; treatment

Citation: Abdulhamid, S.N.; Hasan, A.M.; Aziz, S.Q. Solidification/Stabilization of Contaminated Soil in a South Station of the Khurmala Oil Field in Kurdistan Region, Iraq. *Appl. Sci.* **2021**, *11*, 7474. https://doi.org/10.3390/app11167474

Academic Editor: Paulo José da Venda Oliveira

Received: 24 July 2021
Accepted: 12 August 2021
Published: 14 August 2021

Publisher's Note: MDPI stays neutral with regard to jurisdictional claims in published maps and institutional affiliations.

Copyright: © 2021 by the authors. Licensee MDPI, Basel, Switzerland. This article is an open access article distributed under the terms and conditions of the Creative Commons Attribution (CC BY) license (https://creativecommons.org/licenses/by/4.0/).

1. Introduction

Kurdistan is an autonomous region rich in natural resources in the northern part of Iraq, including crude oil. In the Republic of Iraq, there are several oil fields, some of which are located in the Kurdistan Region. The area has 13 petroleum fields, one of which is the Khurmala oil field. It is located in the southwest of Erbil City, where 64 crude oil wells are operated. The oil sector is a crucial contributor to the Iraqi Kurdish economy.

The activities and stages of crude oil production in Khurmala, including discovery, loading/unloading stations, and storage facilities, have an adverse effect on all modes of life and ecosystems [1,2]. The environmental impact of these processes cannot be overlooked or disregarded. Among the impacts, the soil pollution which can alter soil engineering properties is considered to be the most worrying, due to its negative impact on civil engineering infrastructure protection [3–5]. Crude oil contaminated soil is possibly the result of oil being released from gas, liquid, or solid components; compounds; or mixtures, leading to changes in the soil's physical or chemical composition [2]. Crude petroleum is regarded as the most dangerous source of soil pollution. If soil has been contaminated with crude oil, it becomes inappropriate for engineering purposes due to the effect of crude oil on shear strength parameters, resulting in a lack of bearings and immoderate settlement and resulting in the extreme cracking of existing foundations and structures [6]. Nevertheless, it should be noted that the majority of soil pollution occurred in the past, although it continues today through regular industrial and agricultural activity [7,8]. Moreover, soil pollutant outcrops can result from agricultural activity, leaking from aboveground or underground storage tanks and accidental discharges [9,10].

Several researchers, including Akinwumi et al. [11], Wang et al. [12], Oluremi and Osuolale [13], Khamehchiyan et al. [9], and Kermani and Ebadi [14], investigated the geotechnical properties of oil-contaminated soils, indicating decreased soil strength and increased plasticity due to oil contamination. The permeability of soils also decreased significantly. Furthermore, quartz sand completely saturated with engine oil can lead to a substantial reduction in soil friction angles and a dramatic increase in soil volumetric strain [15]. In addition, Alfach and Wilkinson [16] reported that the contamination of soil by crude oil had an adverse effect on the base of the pile regarding geotechnical behavior degradation. Moreover, Nasehi et al. [17] and Khosravi et al. [18] investigated the impact of the contamination of gas oil on fine and coarse-grained soil's geotechnical properties; a decrease in the MDD and the optimum humidity levels was also observed with the increase in Atterberg's clay and silt limits.

Various methods have been used in recent years to remediate crude-oil-polluted soil. The solvent/surfactant soil-washing technique shows that petroleum pollution soils can cause solubility and extract crude oil soil components [1]. Although biosurfactant solutions have a considerable capacity to extract crude oil from polluted soil by washing conditions, the results showed that the washing-temperature efficiency of crude oil removal from contaminated soil was the most significant factor, compared with the least influential factor which was washing time [2]. In turn, the bioremediation of crude oil polluted soil was achieved by isolating strains of the most efficient biodegradable material in the laboratory; this study demonstrates that many aromatic and saturation hydrocarbons with a chemical composition that is similar to that of crude oil were extracted successfully by the strain [19]. With different remediation approaches, the active degradation of crude oil contaminated saline soil can be achieved by using nitrogen additions, the inoculation of arbuscular mycorrhizas, and the cultivation of Suaeda salsa [20]. In other studies, soil was remediated through pyrolytic treatment. Compared to the reaction time, the pyrolytic efficacy was more affected by the working temperature [21]. Almost all studies in the literature, as mentioned earlier, were focused on agriculture, soil science, and the climate. Hence, it is crucial to analyze these research results for the aim of engineering applications in order to promote practical soil remediation. Thus, oil-contaminated soils must be cured efficiently with methods to enhance the mechanical and geotechnical properties of the soil [22,23]. Furthermore, the solidification/stabilization method, which is accomplished by incorporating cement [24,25], lime [26], fly ash [27], as well as other bonding products into a mixture which is used to impale the contaminants in the polluted medium and ensure long-term safety, is the most effective technique. Solidification describes a process that converts contaminated media into a homogenous solid material with strong structural integrity through its encapsulation in order to change its physical properties. Stabilization describes a process that minimizes contaminated soil's hazardous potential by limiting its solubility, mobility, or toxicity. Therefore, satisfactory results can be achieved using this technology. For example, Akinwumi et al. [28] and Yu et al. [29] stated that an improvement in crude oil soil achieved with a different proportion of Portland cement increased its strength and reduced its permeability and plasticity, making the soil more suitable after the cement treatment. Similarly, Shah et al. [3] reported better results of soil geotechnical properties with the utilization of various additives, such as cement, lime, and fly ash to stabilize contaminated soils with crude oil.

In further experimental work, Wang et al. [30] indicated that the results of the geotechnical properties presented a notable increase in undrained shear strength, solid content (water content), and Atterberg limit values of the soil, achieved through using different curing times and various doses of cement after stabilizing the mature fine tailings. Additional research carried out by Nasr [31] examined the sand's strength behavior when contaminated with oil by utilizing the cement kiln dust (CKD) to determine the stabilized soil's engineering properties for use in rural road construction. Results showed that with the addition of CKD, the unconfined compressive strength and California bearing ratio (CBR) values of the oil-contaminated sand were increased. The stability of polluted sand decreases with the increase in oil percentage. Consequently, Al-Rawas et al. [32] concluded that oil-contaminated land could be reused when stabilized with cement or cement by-pass dust, due to the enhanced geotechnical properties of construction and engineering applications, offering practical, safe, and cost-effective solutions.

This study investigated the effectiveness of using two different types of cements to stabilize oil-contaminated soil. Additionally, the physical, mechanical, and chemical behavior of both polluted crude-oil soil and stabilized soil has been studied in order to enable their reuse as earth construction materials. To date, this kind of research has not been carried out in the oil fields in Kurdistan Region, Iraq.

2. Scope

The Khurmala oil and gas field is spread from 30 km southwest of Erbil City, and is 22 km long and 3 km wide. The crude oil produced in the field wells that are spread across the Khurmala dome was collected at stations (1—North: collection from a network of 29 oil wells; 2—South: collection from a network of 26 oil wells; and 3—Middle: collection from a network of 20 oil wells) through various 150 mm flow lines and then sent to the Central Process Station (CPS-1 and CPS-2) through 500 mm trunk pipelines. After processing, the crude oil was pumped for export.

The historical activities of the Khurmala dome began in 1935 with the drilling of the first well. In 1935–1977, 12 wells were drilled. Development then began in 1988 and the first oil production occurred in 2009.

Crude oil is among the leading causes of terrestrial pollution due to its superior ability to spread, interact, and penetrate the soil in many forms and various means through its dependence on biological, physical, and air variables. There are several different sources of hazards in crude oil contaminated soil in the Khurmala oil field, including exploration and application processes. Therefore, a goal was set to minimize the amount of pollution in the field and remediate the contaminated soil by constructing specific concrete containers for collecting waste crude oil and mixing the contaminated soil with an appropriate cement type, respectively.

The above activities create significant soil pollution due to inappropriate disposal, oil spills, tank leakage, and pipeline breakage [29]. In the Khurmala oil field, the main sources of soil pollution are as follows:

1. Burning pit: This is a pit that is prepared to collect the crude oil that is tested and drained during oil well testing through a special pipe called the burning pipe. A check is necessary to determine the quantity and quality of crude oil, utilizing a test point and a flow meter attached to the burning pipe, as shown in Figure 1a.
2. Random pit: If the pipeline is not accessible for a particular location, the alternative is unloading. The oil in the tankers must be tested. The tested oil then has to be randomly handled. The tested oil is dealt with by disposal in a designated pit called the random pit, as shown in Figure 1b, from which oil can leak into the soil from the older pits. Therefore, these old pits must be remedied. Fortunately, in the Khurmala oil field, a particular separator system is currently used. An oil–water separator system is designed to isolate total quantities of oil and suspended solids from the oil refinery wastewater effluents. This system is based on preventing any leakage into the surrounding and underground soil under Health Safety and Environmental regulations (HSE), as shown in Figure 1c.
3. The absence of a closed drain system in the facilities (including pumps, equipment, pipes, and valves) frequently causes various oil leakage accidents, which can cause severe pollution to the surrounding soil, as shown in Figure 1d. These problems can be controlled through the use of close drain systems linked to a piping system connected to a particular basin for this leakage. Unfortunately, this system is not currently in use at the Khurmala oil station.

4. Flow lines under or above the ground that transport crude oil from the well to stations are subjected to corrosion due to H_2S if a corrosion inhibiter is not used, leading to holes in these pipes, causing oil leakage and then soil contamination.

Figure 1. Most sources of polluted soils at Khurmala oil field: (**a**) burning pits, (**b**) random pits, (**c**) separator system, and (**d**) oil leakage from the facilities.

3. Materials and Methods

3.1. Materials

Soil, Portland limestone cement (CEM II), ordinary Portland cement (OPC), and crude-oil were the primary materials used in this work. In this section, the physical, mechanical, and chemical properties of these materials are described as follows:

3.1.1. Soil

This research was carried out on natural, contaminated, and treated soils. All soils were obtained from an oil pit at the south station's Khurmala oil field treatment area (latitude: 39.0424; longitude: 39.76083).

Figure 2 shows the grain size distribution curve for the natural soil. The soil is classified under the Unified Soil Classification System (USCS) as silty sandy soil (SM). These characteristics designated according to the American Standard of Testing Materials (ASTM). Table 1 shows the geotechnical properties of the soil which were obtained by performing the tests in the Geotechnical Laboratory, Civil Engineering Department, College of Engineering, Salahaddin University-Erbil, Erbil, Kurdistan Region, Iraq, while Table 2 shows the chemical characteristics of the natural soil that were obtained by performing the tests in the Kurd Central Research Facilities (KCRF) laboratory in the Soran District, Erbil City.

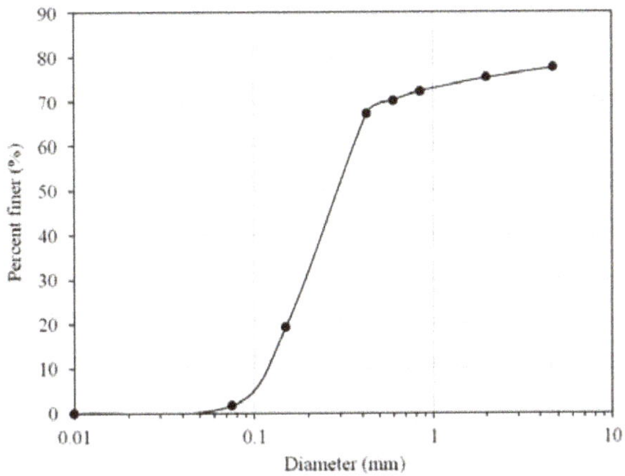

Figure 2. The grain size distribution curve of the natural soil.

Table 1. The geotechnical properties of the natural soil.

Soil Properties		Natural Soil	Standard
Natural moisture content	w (%)	2.4	ASTM D2216 [33]
Specific Gravity	Gs	2.67	ASTM D854 [33]
Grain size	Gravel (%)	19.51	ASTM D421-85(2007) [33] ASTM D2217-85 R98 [33]
	Sand (%)	66.05	
	Fines (%)	14.45	
	C_u	3.47	
	C_c	1.06	
Soil classification	USCS	SM	ASTM D2487 [33]
	AASHTO	A-2-4	
Maximum dry unit weight	$\gamma_{d\ max}$ (kN/m^3)	17.6	ASTM D698 [33]
Optimum moisture content	(%)	12.6	
Angle of internal friction	Φ	28.56°	ASTM D3080 [33]
Cohesion	C (kPa)	34.5	
Coefficient of permeability	k (cm/s)	3.47×10^{-5}	ASTM D2434 [33] ASTM D5084 [33]
CBR	Unsoaked CBR %	41.883	ASTM D1883 [33]
	Soaked CBR %	25.257	

Table 2. The chemical characteristics of the natural soil.

Parameter	Unit	Value
pH		7.7
Electrical conductivity	µmho/cm	703
Alkalinity	mg/L	39
Carbonate	mg/L	0
Bicarbonate	mg/L	39
Sulfate	mg/L	104
Chloride	mg/L	132
Total organic carbon	%	0.43
Oil and Grease	%	0.52

3.1.2. Crude Oil

The petroleum specimen was obtained from the Khurmala Station in Iraqi Kurdistan, run by the Kar Group Petroleum Production Company. A description of the fundamental crude oil properties is shown in Table 3. The Khurmala Block refinery authorities provided the values of the crude oil properties, which have a value of American Petroleum Institute (API) gravity equal to 32.29 at 15.6 °C, and a value of specific gravity equal to 0.8639 at 15.6 °C.

Table 3. The physical properties of the Khurmala crude oil.

Test	H_2S (ppm)	BS&W * (%)	Total Sulphur (%)	Salt (ptb) **	Density (kg/m³)	API Gravity (at 15.6 °C)	Viscosity (mm²/s)
Standard	UOP 163	ASTM D4007 [34]	ASTM D4294 [35]	ASTM D3230 [36]	ASTM D1298 [37]	ASTM D1298 [37]	ASTM D7042 [38]
Results	41.3	0.6	2.22	229	863.1	32.29	12.8

* Basic sediment and water content of crude oils. ** Ptb = pounds of salt per thousand barrels of crude oil.

3.1.3. Cement

In this study, CEM II is available in the local market and used in the field (according to BS EN 196—Methods of testing cement). Simultaneously, a locally produced OPC is available in the Erbil market and used in the laboratory study. The cements' chemical composition is presented in Table 4.

Table 4. The physical and chemical properties of the CEM II and OPC.

Chemical Analysis	Results (%)	
	CEM II	OPC
SiO_2	20.04	20.17
CaO	61.84	63.11
Al_2O_3	4.37	4.22
Fe_2O_3	3.71	3.78
MgO	3.48	3.82
SO_3	2.67	2.08
Insoluble Material	0.32	0.59
Loss of Ignition	3.05	1.55
Lime Saturation Factor	0.87	0.96
C_3A	5.3	4.79
C_3S	42.09	63.94
C_2S	25.72	9.79
C_4AF	11.28	11.5

3.2. The Solidification/Stabilization Process of Pollutant Soil in the Field

Solidification/stabilization requires the immobilization of the polluted soil constituents through a process of chemical modification into insoluble substances or by encapsulating the solid. Mixing the polluted soil with cement results in this process. The treated soil in the site underwent a solidification/stabilization process at the Khurmala oil field treatment area. The soil was mixed with CEM II (1 ton cement/7 m³ soil) at approximately 8–9% by weight of the soil with a water–cement ratio of 40%, and the treated soil was then left as a construction earth material for two months in order to gain an equilibrium between cement and soil, before being reused in the area. The main goals in this process were the following:

- Improve soil handling and physical characteristics;
- Minimize available surface area for the movement or loss of pollutants and limit fluid movement by the total hard matrix volume;
- Minimize the solubility of the contaminant into the amount of contaminated soil.

The project was initiated on 16 January 2019 and lasted until 19 November 2019. The method of treatment included the following:

1. International drilling fluids and engineering services, a Qmax solutions company, waste management division, provided services of remediating and encapsulating all the pit's wastes, such as oily sludge and contaminated soil waste.
2. The remediation involves remediating the contaminated oil pit, reserve pit (Figure 3a), and overflow pit at the south production station, located at Khurmala Site, in the Erbil City.
3. The equipment and machinery were mobilized to the southern production plant site on 15 January 2019.
4. The contaminated soil in the oil pit was first treated by digging and treating all the soil contaminated with the existing crude oil (Figure 3b). The total volume of treated soil reached 2980 m^3.
5. The treated soil was stored beside the oil pit (Figure 3c) to be backfilled, after reconstructing and lining the pit, and prepared as a construction earth material.
6. When the test results of the oil pit's bottom and sides showed that it was cleaned of contaminant, the pit was reconstructed and lined up with a geosynthetic clay layer with high-density polyethylene liner to be backfilled with remediated soil (Figure 3d).
7. The oil pit was backfilled with treated soil, covered with a GCL liner on the top (Figure 3e), then backfilled to 3.5 m of fresh soil from the area around the pit, leveled, and compacted to the natural ground level.
8. The site underwent general clean-up and restoration. The procedures of the treatment are illustrated in Figure 3.

3.3. The Solidification/Stabilization Process of Pollutant Soil in the Laboratory

The main objectives of the laboratory tests were as follows: (1) to check whether the process of the solidification/stabilization of pollutant soil in the field was performed effectively in the field and (2) to emphasize that the CEM II is a suitable type of cement used in the process. To achieve this, natural, contaminated, and treated soil samples were collected from the Khurmala site. Then, all samples were transported in closed, labeled plastic bags to the Geotechnical Laboratory, Civil Engineering Department, College of Engineering, Salahaddin University-Erbil, Erbil, Kurdistan Region, Iraq, in order to study their physical, mechanical, and chemical properties. In addition, the impact of stabilizing crude oil polluted soil treated by OPC was studied and compared to the treated sample with CEM II from the field.

The treated (CEM II and OCP) soil specimens in the laboratory were prepared by mixing the contaminated specimens (at oil content 14%). The specimens were mixed with 8.7% of ordinary Portland cement by weight with a water–cement ratio of 0.4 in order to match the field conditions. The mixed samples were placed into closed containers for two weeks, allowing possible reactions between the soil and cement.

The clean soils were taken as reference samples. These were obtained from a location that was ensured, through the detection of vision and color, to be uncontaminated. The site had similar geological conditions to the contaminated site. The clean samples were taken 10 cm from the earth's surface.

3.4. Laboratory Test Program

Laboratory work was designed to obtain parameters, including the specific gravity, compaction, coefficient of permeability, un-soaked CBR and soaked CBR, direct shear, and scanning electron microscopy (SEM) tests for the natural soil, soils polluted with crude oil, and contaminated soil samples stabilized with CEM II and OPC. The laboratory investigation was performed to explore the impact of different types of cements on the geotechnical properties of oil-contaminated soils. On average, three specimens were used to avoid any uncertainty and scattering in data.

The soils' compaction characteristics were studied by conducting a standard compaction test following ASTM [34]. The MDD and the OMC were obtained from the compaction curve for all the soils.

Figure 3. The treatment process of pollutant soil: (**a**) the oil pit before treatment; (**b**,**c**) the treatment of contaminated soil in the oil pit; (**d**) reconstruction and lining of the oil pit; and (**e**) backfilling of the pit with treated soil and leveling.

The shear strength parameters of the soils are essential to consider, as they influence the design of many geotechnical engineering projects, such as embankments, soil slopes stability, and foundations. Direct shear tests were performed according to the method recommended by ASTM [33]. The samples were tested at their MDD and OMC. Soaked and un-soaked CBR tests were performed on oil-contaminated soil samples with and without cement and clean soil samples as described in ASTM [33]. The falling head equipment was used to determine the permeability coefficient. The test was performed on all soils. The technique, used to assess permeability via the falling head method, was compatible with the work of Head [39].

4. Results and Discussion

The findings show that the values of specific gravity of both the natural and treated CEM II and OPC were 2.67, 2.68, and 2.38, respectively. The untreated soil had the lowest value of specific gravity of 2.35. This could be attributed to the high oil content (which was up to 14%).

4.1. Compaction Test Results

The compaction test results are shown in Figure 4 in the form of dry density versus water content. Generally, the compaction curves of the contaminated soil and both treated soils moved below the natural soil curve. The MDD of the contaminated soil substantially decreased with a 14% oil content to a low value of MDD (1.625 g/cm^3) due to the oil content in contaminated soils. This reduction is attributed to the effect of the specific gravity value of crude oil on the soil. Moreover, with silty sandy soils polluted with crude oil, the particles separated as the voids filled with the oil and coated the granules. Therefore, a decline in dry density was observed as the soil transited into a loose material state. Similar results are reported by Al-sanad et al. [40], Meegoda et al. [7], and Nasr [31]. Nevertheless, these findings differ from other studies by Khamechiyan et al. [9], Al-Rawas et al. [32], and Nasehi et al. [17]. At the same time, no discernible change in the OMC was noticed between the natural soils and polluted soil by crude oil.

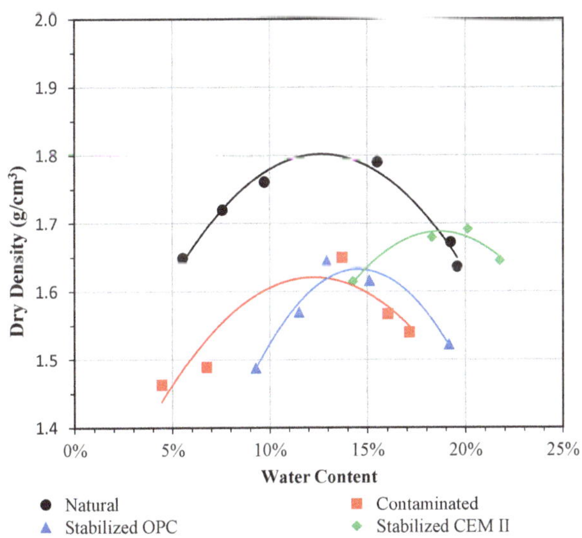

Figure 4. Compaction curves of natural, contaminated, and treated (CEM II and OPC) soils.

The MDD of the treatment soil slightly increased when the soil was solidified with CEM II and OPC, reaching a peak at 1.69 g/cm^3 and 1.635 g/cm^3. A high increase in OMC could be observed compared to the OMC of natural soil, particularly in relation to the soil treated with CEM II. In comparison, the value of MDD of the natural soil was much higher than that for treated soils. By adding the cement to the polluted soil, the MDD of the stabilized contaminated soil increased due to the specific gravity of cement (commonly 3.15) compared with the contaminated soil (2.35). Meanwhile, the OMC increased since cement has a better absorption potential for water.

4.2. Direct Shear Test Results

The results from the direct shear tests are presented in Figure 5, which are illustrated in the form of shear stress against normal stress. It can be seen that all soils had almost the same trend: a noticeable increase in the treated soil was observed compared with other soils.

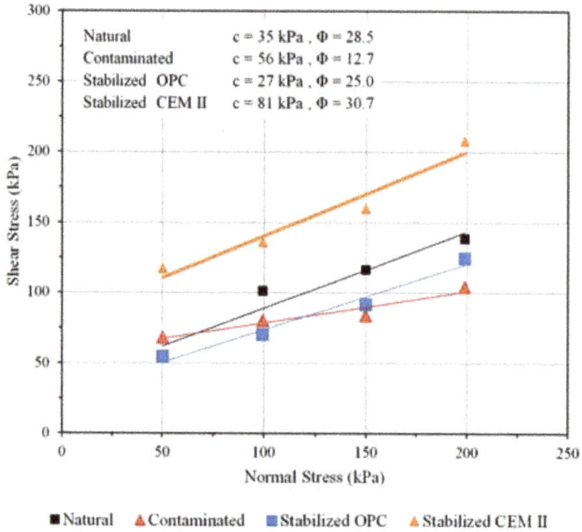

Figure 5. Direct shear test results of natural, contaminated, and treated (CEM II and OPC) soils.

The direct shear test results showed that the internal friction angle (Φ) decreased drastically from 28.56° to 12.7° for natural, treated (CEM II and OPC), and contaminated soils, respectively. Shin et al. [41] stated that oil contamination causes a noticeable decrease in the Φ value. Ghaly [42], Khamehchiyan et al. [10], and Nasehi et al. [17] reported that in the presence of a high crude oil content, the friction angle decreases. This inverse correlation might be explained by the coating of soil particles with crude oil, which acts as a lubricant that decreases the inter-granular contact force between the sand particles. However, Abousnina et al. [43] reported that, for samples containing 2% to 20% oil, no significant difference in the frictional angle of the sand was detected, which indicates that the sand particles were totally covered with crude oil at a level of more than 2%, and their frictional angle remained unchanged. For the stabilized soils, the treated soils indicated an increment in the angle of internal friction by stabilizing with CEM II and OPC to 30.7° and 25.0°, respectively, as shown in Figure 5. This increment could be related to the cement action that increases the agglomeration between grains and minimizes lubrication, increasing the contact force between particles. However, these results are different from the findings of Al-Rawas et al. [32].

The cohesion of the natural soil was 35 kPa. Crude oil contamination led to an increase in the cohesion value of this soil to 56 kPa. These findings match the results of Nasehi et al. [17], but are incompatible with the findings of Khamehchiyan et al. [9]. It is clear that crude oil's ability to resist shear force is greater than water, since its viscosity is more. Therefore, during the application of shear force to the contaminated specimen, crude oil resists a portion of that shear force besides the soil particles and, in turn, increases the soil's apparent cohesion.

In soil stabilized with CEM II, the rise in cohesion was dramatic, changing to 81 kPa, while the cohesion in soil treatment with OPC reached a low value of 27 kPa. This is due to the increase in the material's cohesiveness as a result of the cementing action caused by the hydration process. This is in line with the finding that CEM II hydrated more quickly and provided a higher strength than OPC. Since the C_2S is responsible for the subsequent rise after the first week in the strength of the cement's hydraulic components, its value is 25.72% in CEM II, compared to 9.79% in OPC.

4.3. Permeability Tests

Table 5 shows the permeability test results for the natural, contaminated and treated soils. As expected, the permeability has a reverse correlation with oil content. The coefficient of permeability (k) for contaminated soil (7.34 $\times$ 10^{-6} cm/s) was lower than for natural soil (3.47 $\times$ 10^{-5} cm/s). However, even at 14% crude oil, the decrease in the value of k is not as high as expected. It is clear that oil-contaminated soil decreases the k due to the occupation of the crude oil for the pore spaces, which causes a reduction in the flow rate through the soil by minimizing the volume of the pores responsible for facilitating the movement of fluids within the soil. Similar results are presented by Khamehchiyan et al. [9] and Abousnina et al. [43].

Table 5. Permeability test results.

Soil Properties		Natural Soil	Contaminated Soil	Treatment Soil with CEM II	Treatment Soil with OPC
Coefficient of permeability	k (cm/s)	3.47 $\times$ 10^{-5}	7.34 $\times$ 10^{-6}	4.55 $\times$ 10^{-8}	4.87 $\times$ 10^{-6}

The results for the treated soils indicated a decrease in the value of k, compared to the natural and contaminated ones. By adding 8.7% cement, the permeability of CEM II and OPC decreased to 4.55 $\times$ 10^{-8} cm/s and 4.87 $\times$ 10^{-6} cm/s, respectively. With the addition of cement to the content, a cement product, such as a bonding gel, was produced, which reduced the porosity that binds the soil particles together and hindered the passage of water into the soil. Consequently, the permeability coefficient was reduced. Similar results were found by Al-Rawas et al. [32].

4.4. CBR Tests

The CBR is a test usually performed to assess the strength of subgrade soils and base course materials in pavement work. As summarized in Table 6 and Figure 6, the CBR values of the crude oil contaminated soils under un-soaked conditions significantly decreased compared with the natural ones. This reduction is probably due to the combination of excessive oil presence and the low maximum density of the contaminated soil. These results are consistent with those of Al-Sanad et al. [40] and Nasr [31]. In contrast, the values of CBR for natural and contaminated soils under soaked conditions were similar.

Table 6. CBR test results for natural, contaminated, and treated (CEM II and OPC) soil samples.

Soil Identification	Maximum Dry Unit Weight (kN/m^3)	Optimum Moisture Content %	CBR %	
			Unsoaked	Soaked
Natural Soil	17.60	12.6	41.88	25.25
Contaminated Soil	15.89	12.2	26.24	23.72
Treatment Soil with CEM II	16.55	18.7	75.16	38.26
Treatment Soil with OPC	16.01	14.5	60.35	34.68

Figure 6 also shows that, for both soaked and un-soaked treated (CEM II and OPC) soils, the values of CBR were significantly improved in comparison to those for the natural soil. Based on the review of Wang [44], cement contains hydration products that increase therapy strength and performance. The enhancement in un-soaked and soaked CBR values is due to the production of cementitious components, such as calcium silicate hydrates and calcium aluminate hydrates in the contaminated soil stabilized/solidified by cement [45].

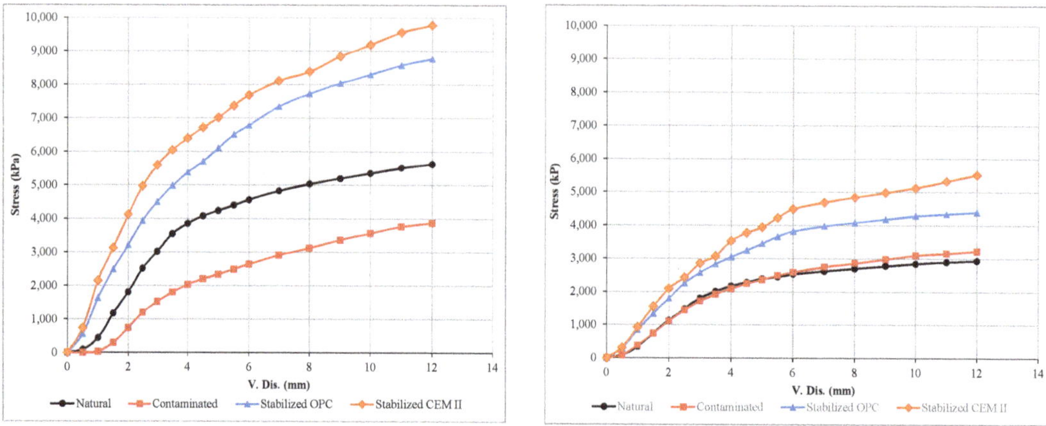

(a) (b)

Figure 6. Stress versus vertical displacement for (**a**) un-soaked and (**b**) soaked samples.

4.5. SEM Analysis

SEM is a technique that provides many magnified images and explains differences that soil enhancement produces in physical, chemical, and mechanical behavior, including shape, size, composition, and crystallography properties [46]. SEM was used in this study to investigate the microstructure particles for the natural, contaminated, and treated soils, in order to detect the structure of the bonding between sand particles in the previous cases.

Figures 7a–d and 8a–d illustrate the geometric arrangement for the natural, contaminated, and treated soils, respectively. In natural soil fabric, the diameter in the singular grains can be observed. However, it is not possible to distinguish individual floccules in these micrographs, as indicated in Figure 7a. Subsequently, the morphological shape of the natural soil, as shown in Figure 8a, indicated the appearance of burrs in the soil grains, confirming its non-coated properties.

Crude oil firmly coated the singular soil particles via hydrogen bonding and van der Waals forces. As a result, it was shown in the form of a dense-packed structure with almost no visible voids, as shown in Figure 7b, since the lining oils created a water-resistant layer that blocked the voids, causing a reduction in permeability. Moreover, in the photomicrograph of Figure 8b, the surface of the contaminated soil appeared as one flock with no distinct pore spaces, indicating that it was filled with oil.

Figure 7c illustrates a significant improvement in the soil treatment with CEM II. The similarity of the microscopic surface of the treated soil to the natural soil was obvious, in addition to agglomerated morphology of the soil sample. The change in color from dark to light in the samples signifies that the crude oil was removed from the soil in a satisfactory proportion. These results are in agreement with the chemical results shown in Table 7, which indicate a decline in total organic carbon from 11.7% to 0.8% and in oil and grease from 14% to 0.96%, simultaneously, which is a significant performance. However, the structural features for the soil sample in Figure 8c showed a small proportion of oil covering some of the grains with the presence of apparent voids in the surface element.

In comparison, no considerable improvement was noticed in the soil that was stabilized with OPC, as shown in Figure 7d, compared to CEM II. The microscopic surface of the treated soil was more similar to the contaminated soil than it was to the natural soil. The crude oil still coats the soil particles. If we combine the impact of the oil and cement, the influence of oil is still dominant. This result is consistent with the chemical results; soil treated with OPC had 8.2% total organic carbon and 9.8% oil and grease. Although the soil grains became aggregated, they were formed in the shape of flocks coated with oil, as shown in the microscopic image in Figure 8c.

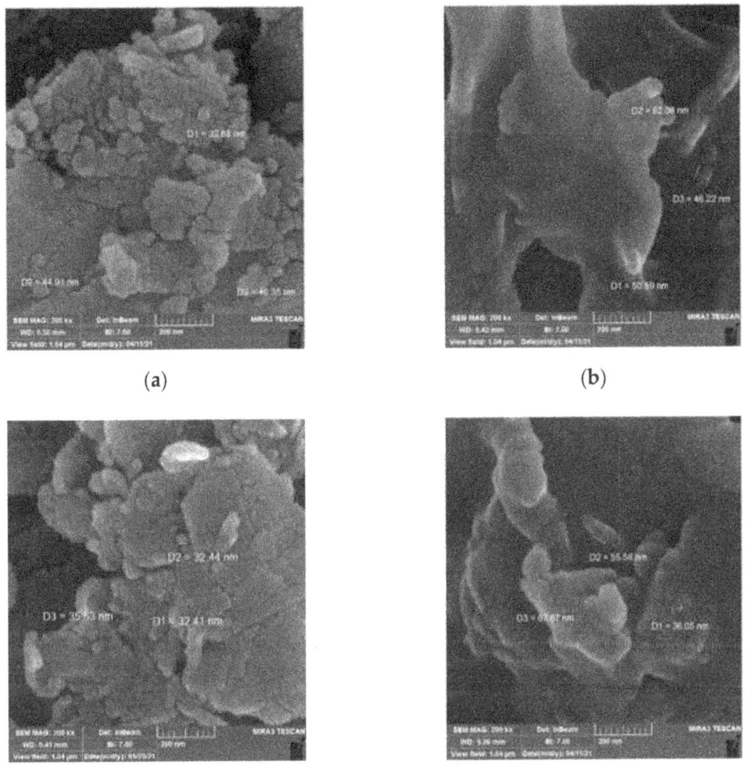

(**a**) (**b**) (**c**) (**d**)

Figure 7. SEM photograph of 200 nm: (**a**) natural soil; (**b**) contaminated soil; (**c**) soil stabilized with CEM II; and (**d**) soil stabilized with OPC.

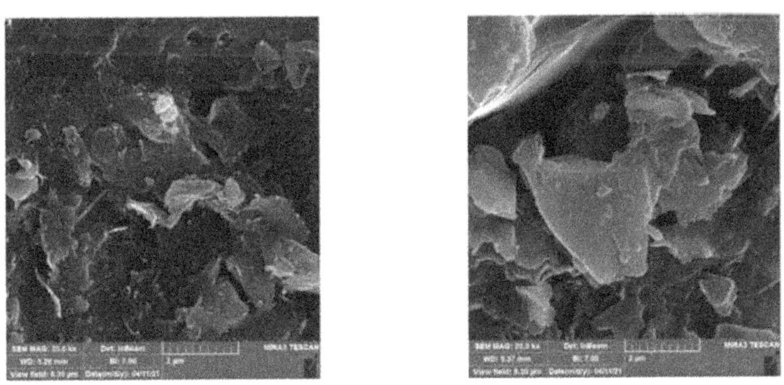

(**a**) (**b**)

Figure 8. *Cont.*

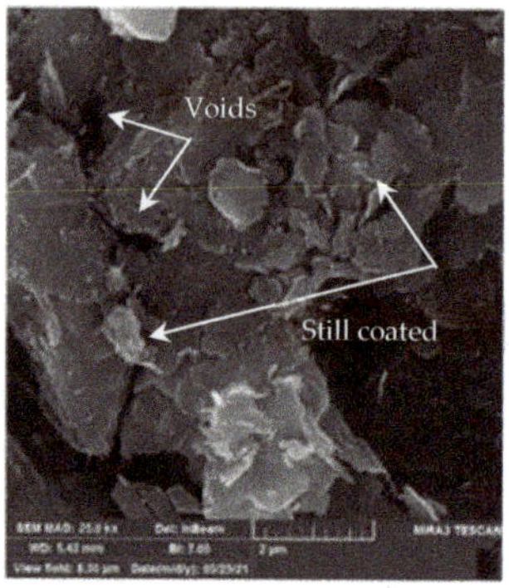

(**c**)

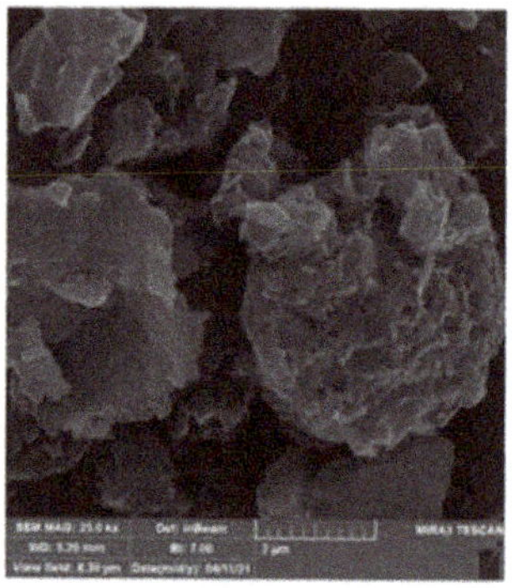

(**d**)

Figure 8. SEM photograph of 2 µm: (**a**) natural soil; (**b**) contaminated soil; (**c**) soil stabilized with CEM II; and (**d**) soil stabilized with OPC.

Table 7. Chemical test results of the clean, contaminated, and tread soil samples.

Parameter	Unit	Results			
		Natural	Contaminated	Stabilized with CEM II	Stabilized with OPC
pH		7.7	7.9	7.1	12.74
Electrical conductivity	µmho/cm	703	599	3160	4710
Alkalinity	mg/L	39	87	106	87
Carbonate	mg/L	0	0	0	87
Bicarbonate	mg/L	39	87	106	0
Sulfate	mg/L	104	II52	1320	20
Chloride	mg/L	132	104	65	52
Total organic carbon	%	0.43	11.7	0.8	8.2
Oil and Grease	%	0.52	14	0.96	9.8

4.6. Chemical Tests

Chemical tests for the clean, contaminated, and tread soil samples are shown in Table 7. It is clear that the values of pH, alkalinity, bicarbonate, sulfate, total organic matter, and oil and grease in the polluted samples were higher than for the clean sample. It is clear that oil disposal caused the contamination of the soil at the Khurmala oil field. The research conducted by Ergozhin et al. [47], Kuany et al. [48], Wang et al. [49], Trejos-Delgado et al. [50], and Jabbarov et al. [51] confirms the obtained results. All pH values were in the normal range, except the pH value for the soil sample treated with OPC, which is classified in the high alkaline range (pH = 12.74). Limited variation in the alkalinity, carbonate, and bicarbonate values for normal, polluted, and treated soil samples was reported.

Treatment using CEM II led to a decrease in the pH, chloride, total organic matter, oil and grease in the treated soil sample, while treatment via OPC resulted in a decrease in pollutants, such as alkalinity, bicarbonate, sulfate, chloride, total organic matter, and oil and grease in the processed soil sample. Generally, treatment with CEM II and OPC caused a decrease in contaminates, especially chloride, total organic matter, and oil and grease. A fluctuation in pH, electrical conductivity, alkalinity, and sulfate values was observed; this may be due to the chemical reactions between pollutants and the components of the treatment materials. Sulfate and chloride figures after treatment became lower than those of the normal soil sample. Results revealed that the application of CEM II for the treatment of the polluted soil samples was often superior to that of the OPC.

4.7. XRD Tests

XRD is a powerful nondestructive method for symbolizing crystalline materials. It offers information on the structures, stages, preferred crystal locations (texture), and other structural factors. XRD peaks are formed by the productive interference of a monochromatic beam of X-rays distributed at definite angles from each set of lattice planes in an illustration. The highest strengths are found using the atomic positions within the lattice planes. Accordingly, the XRD pattern is the print of periodic atomic arrangements in a specified material [52].

XRD test results for the soil samples are illustrated in Table 8 and Figure 9. The pollutants changed the shape of Figure 9b, when compared with Figure 9a; silicon oxide and calcium carbonate values were increased in the polluted soil sample, when compared with the normal soil sample (Table 8). Furthermore, calcium aluminum silicate also increased, while silicon oxide and calcium carbonate decreased after both treatment methods. Additionally, treatment using CEM II is shown in Figure 9c. Using OPC for the treatment of the polluted soil sample affected the soil components, as shown in Figure 9d. Values of aluminum calcium silicon, magnesium aluminum silicate, iron silicate hydroxide, magnesium dialuminium disilicide-U1, and sodium aluminum silicate hydrate increased in the treated samples (Table 8). The obtained results shown in Table 7 are in coincidence with the illustration of XRD results. The variation of values in Figure 9a–d agrees with the obtained results in Table 8. The present results agree with the published work of Aziz [53].

Table 8. XRD test results of the natural, contaminated, and tread soil samples.

Soil Type	Compound Name and Chemical Formula								
	Silicon Oxide	Calcium Carbonate	Albite Low	Calcium Aluminum Silicate	Aluminum Calcium Silicon	Magnesium Aluminum Silicate	Iron Silicate Hydroxide	Magnesium Dialuminium Disilicide–U1	Sodium Aluminum Silicate Hydrate
	SiO_2	$CaCO_3$	Al1 NaO_8 Si_3	Ca Al_2 Si_2 O_8	Al_2 Ca_3 Si_2	Mg_2 Al_4 Si_5 O_{18}	Fe_3 Si_2 O_5 $(OH)_4$	Al_2 Mg Si_2	Na_3 Al_3 Si_3 O_{12} $(H_2O)_{1.8}$
Natural	76	73	66						
Contaminated	82	74		61					
Stabilized OPC	66	-			53	41			
Stabilized CEM II	71	69					54	69	66

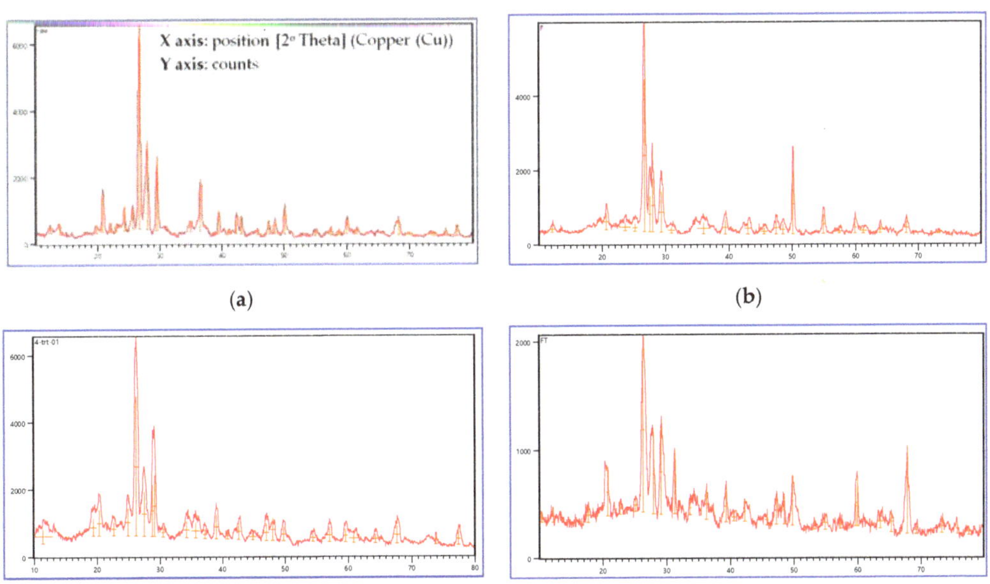

(a) (b)

(c) (d)

Figure 9. XRD test for the soil sample: (**a**) natural soil; (**b**) contaminated soil; (**c**) soil stabilized with CEM II; and (**d**) soil stabilized with OPC.

5. Conclusions

According to the study's results presented above, the following conclusions can be drawn:

- The disposal of crude oil resulted in soil contamination at the Khurmala oil field.
- Compaction characteristics and CBR values deteriorated with the presence of crude oil content. At the same time, when the contaminated soil was treated with a stabilization agent containing both types of cement (i.e., CEM II and OPC), an increase in the MDD and OMC and CBR values was observed, but the best result was achieved with CEM II.
- The greatest improvement in the shear strength parameters (c' and Φ') was achieved when the contaminated soil was treated using CEM II.
- Generally, the contamination of sandy soil with crude oil induced a permeability reduction, and a further decrease in permeability was detected as the soil solidified with cement.
- A substantial reduction in the oil and grease of the treated soil was achieved using CEM II, compared to soils treated with OPC. The SEM results confirm this.
- The solidification/stabilization (S/S) method provides an effective remediation method for processing waste to produce a safe, dry material acceptable for onsite burial. The application of the S/S process via utilizing cement has an influential role in strengthening the geotechnical characteristics for the contamination of soils with crude oil.
- The remediation of contaminated soil with crude oil utilizing CEM II resulted in a larger improvement compared to when using OPC.

Author Contributions: Conceptualization, S.N.A. and A.M.H.; methodology, A.M.H. and S.N.A.; formal analysis, S.N.A., A.M.H. and S.Q.A.; site visiting and soil sampling, A.M.H.; writing—original draft preparation, S.N.A.; writing—review and editing, A.M.H. and S.Q.A.; supervision, A.M.H. and S.N.A. All authors have read and agreed to the published version of the manuscript.

Funding: This research received no external funding.

Institutional Review Board Statement: Not applicable.

Informed Consent Statement: Not applicable.

Data Availability Statement: MDPI Research Data Policies.

Conflicts of Interest: The authors declare no conflict of interest.

References

1. Wang, M.; Zhang, B.; Li, G.; Wu, T.; Sun, D. Efficient Remediation of Crude Oil-Contaminated Soil Using a Solvent/Surfactant System. *RSC Adv.* **2019**, *9*, 2402–2411. [CrossRef]
2. Urum, K.; Pekdemir, T.; Çopur, M. Screening of Biosurfactants for Crude Oil Contaminated Soil Washing. *J. Environ. Eng. Sci.* **2005**, *4*, 487–496. [CrossRef]
3. Shah, S.J.; Shroff, A.V.; Patel, J.V.; Tiwari, K.C.; Ramakrishnan, D. Stabilization of Fuel Oil Contaminated Soil—A Case Study. *Geotech. Geol. Eng.* **2003**, *21*, 415–427. [CrossRef]
4. Cheng, L.; Shahin, M.A. Stabilisation of Oil-Contaminated Soils Using Microbially Induced Calcite Crystals by Bacterial Flocs. *Geotech. Lett.* **2017**, *7*, 146–151. [CrossRef]
5. Rao, S.M.; Rao, S.S. Ground Heave from Caustic Soda Solution Spillage—A Case Studt. *Soils Found.* **1994**, *34*, 13–18. [CrossRef]
6. Rahman, Z.A.; Hamzah, U.; Taha, R.; Ithnain, N.S.; Ahmad, N. Influence of Oil Contamination on Geotechnical Properties of Basaltic Residual Soil. *Am. J. Appl. Sci.* **2010**, *7*, 954–961. [CrossRef]
7. Meegoda, J.N.; Chen, B.; Gunasekera, S.D.; Pederson, P. Compaction Characteristics of Contaminated Soils-Reuse as a Road Base Material. *Geotech. Spec. Publ.* **1998**, *79*, 195–209.
8. Ratnaweera, P.; Meegoda, J.N. Shear Strength and Stress-Strain Behavior of Contaminated Soils. *Geotech. Test. J.* **2006**, *29*, 133–140. [CrossRef]
9. Khamehchiyan, M.; Hossein Charkhabi, A.; Tajik, M. Effects of Crude Oil Contamination on Geotechnical Properties of Clayey and Sandy Soils. *Eng. Geol.* **2007**, *89*, 220–229. [CrossRef]
10. Estabragh, A.R.; Beytolahpour, I.; Moradi, M.; Javadi, A.A. Consolidation Behavior of Two Fine-Grained Soils Contaminated by Glycerol and Ethanol. *Eng. Geol.* **2014**, *178*, 102–108. [CrossRef]

11. Akinwumi, I.I.; Diwa, D.; Obianigwe, N. Effects of Crude Oil Contamination on the Index Properties, Strength and Permeability of Lateritic Clay. *Int. J. Appl. Sci. Eng. Res.* **2014**, *3*, 816–824. [CrossRef]
12. Wang, Y.; Feng, J.; Lin, Q.; Lyu, X.; Wang, X.; Wang, G. Effects of Crude Oil Contamination on Soil Physical and Chemical Properties in Momoge Wetland of China. *Chin. Geogr. Sci.* **2013**, *23*, 708–715. [CrossRef]
13. Oluremi, J.; Osuolale, O. Oil Contaminated Soil as Potential Applicable Material in Civil Engineering Construction. *J. Environ. Earth Sci. Int. J.* **2014**, *4*, 87–100.
14. Kermani, M.; Ebadi, T. The Effect of Oil Contamination on the Geotechnical Properties of Fine-Grained Soils. *Soil Sediment Contam.* **2012**, *21*, 655–671. [CrossRef]
15. Evgin, E.; Das, B.M. Mechanical Behavior of an Oil Contaminated Sand. In Proceedings of the Mediterranean Conference on Environmental Geotechnology, Izmir, Turkey, 25–27 May 1992; pp. 101–108.
16. Alfach, M.T.; Wilkinson, S. Effect of Crude-Oil-Contaminated Soil on the Geotechnical Behaviour of Piles Foundation. *Geotech. Res.* **2020**, *7*, 76–89. [CrossRef]
17. Nasehi, S.A.; Uromeihy, A.; Nikudel, M.R.; Morsali, A. Influence of Gas Oil Contamination on Geotechnical Properties of Fine and Coarse-Grained Soils. *Geotech. Geol. Eng.* **2016**, *34*, 333–345. [CrossRef]
18. Khosravi, E.; Ghasemzadeh, H.; Sabour, M.R.; Yazdani, H. Geotechnical Properties of Gas Oil-Contaminated Kaolinite. *Eng. Geol.* **2013**, *166*, 11–16. [CrossRef]
19. Yan, G.; Cai, B.; Chen, C.; Yue, Y.; Wang, Q.; Deng, H.; Liu, S.; Guo, S. Bioremediation of Crude Oil Contaminated Soil. *Pet. Sci. Technol.* **2015**, *33*, 717–723. [CrossRef]
20. Gao, Y.C.; Guo, S.H.; Wang, J.N.; Li, D.; Wang, H.; Zeng, D.H. Effects of Different Remediation Treatments on Crude Oil Contaminated Saline Soil. *Chemosphere* **2014**, *117*, 486–493. [CrossRef]
21. Kang, C.U.; Kim, D.H.; Khan, M.A.; Kumar, R.; Ji, S.E.; Choi, K.W.; Paeng, K.J.; Park, S.; Jeon, B.H. Pyrolytic Remediation of Crude Oil-Contaminated Soil. *Sci. Total Environ.* **2020**, *713*, 136498. [CrossRef]
22. Russell, D.L. *Remediation Manual for Petroleum Contaminated Sites*; CRC Press, Taylor & Francis Group: Lancaster, UK, 1992; ISBN 9780877628767.
23. Balba, M.T.; Al-Awadhi, N.; Al-Daher, R. Bioremediation of Oil-Contaminated Soil: Microbiological Methods for Feasibility Assessment and Field Evaluation. *J. Microbiol. Methods* **1998**, *32*, 155–164. [CrossRef]
24. Moon, D.H.; Grubb, D.G.; Reilly, T.L. Stabilization/Solidification of Selenium-Impacted Soils Using Portland Cement and Cement Kiln Dust. *J. Hazard. Mater.* **2009**, *168*, 944–951. [CrossRef] [PubMed]
25. Estabragh, A.R.; Khatibi, M.; Javadi, A.A. Effect of Cement on Treatment of a Clay Soil Contaminated with Glycerol. *J. Mater. Civ. Eng.* **2016**, *28*, 04015157. [CrossRef]
26. Alpaslan, B.; Yukselen, M.A. Remediation of Lead Contaminated Soils by Stabilization/Solidification. *Water Air Soil Pollut.* **2002**, *133*, 253–263. [CrossRef]
27. Dermatas, D.; Meng, X. Utilization of Fly Ash for Stabilization/Solidification of Heavy Metal Contaminated Soils. *Eng. Geol.* **2003**, *70*, 377–394. [CrossRef]
28. Akinwumi, I.I.; Booth, C.A.; Diwa, D.; Mills, P. Cement Stabilisation of Crude-Oil-Contaminated Soil. *Proc. Inst. Civ. Eng. Geotech. Eng.* **2016**, *169*, 336–345. [CrossRef]
29. Yu, C.; Liao, R.; Zhu, C.; Cai, X.; Ma, J. Test on the Stabilization of Oil-Contaminated Wenzhou Clay by Cement. *Adv. Civ. Eng.* **2018**, *2018*, 9675479. [CrossRef]
30. Wang, J.; Guo, Y.; Shang, J.Q.A. Portland Cement Stabilization of Canadian Mature Fine Oil Sands Tailings. *Environ. Geotech.* **2021**, *8*, 1–43. [CrossRef]
31. Taylor, P.; Nasr, A.M.A. Utilisation of Oil-Contaminated Sand Stabilised with Cement Kiln Dust in the Construction of Rural Roads. *Int. J. Pavement Eng.* **2014**, *15*, 889–905. [CrossRef]
32. Graettinger, A.J.; Johnson, P.W.; Sunkari, P.; Duke, M.C.; Al-rawas, A.; Hassan, H.F.; Taha, R. Stabilization of Oil-Contaminated Soils Using Cement and Cement by-Pass Dust. *Manag. Environ. Qual. Int. J.* **2005**, *16*, 670–680. [CrossRef]
33. ASTM. *Annual Book of ASTM Soil and Rock*; ASTM: West Consh, PA, USA, 2017.
34. ASTM D4007-11e1. *Standard Test Method for Water and Sediment in Crude Oil by the Centrifuge Method (Laboratory Procedure)*; ASTM Int.: West Consh, PA, USA, 2016.
35. ASTM D4294-16e1. *Standard Test Method for Sulfur in Petroleum and Petroleum Products by Energy Dispersive X-Ray Fluorescence Spectrometry*; ASTM Int.: West Consh, PA, USA, 2016.
36. ASTM D3230-19. *Standard Test Method for Salts in Crude Oil (Electrometric Method)*; ASTM Int.: West Consh, PA, USA, 2019.
37. ASTM D1298-12b. *Standard Test Method for Density, Relative Density, or API Gravity of Crude Petroleum and Liquid Petroleum Products by Hydrometer Method*; ASTM Int.: West Consh, PA, USA, 2017.
38. ASTM D7042-21. *Standard Test Method for Dynamic Viscosity and Density of Liquids by Stabinger Viscometer (and the Calculation of Kinematic Viscosity)*; ASTM Int.: West Consh, PA, USA, 2021.
39. Head, K.H.; Epps, R.J. *Manual of Soil Laboratory Testing*, 3rd ed.; Whittles Publishing, CRC Press, Taylor & Francis Group: Boca Raton, FL, USA, 2011.
40. Al-Sanad, H.A.; Eid, W.K.; Ismael, N.F. Closure to "Geotechnical Properties of Oil-Contaminated Kuwaiti Sand" by Hasan A. Al-Sanad, Walid K. Eid, and Nabil F. Ismael. *J. Geotech. Eng.* **1996**, *122*, 787–788. [CrossRef]

41. Shin, E.C.; Omar, M.T.; Tahmaz, A.A.; Das, B.M. Shear Strength and Hydraulic Conductivity of Oil-Contaminated Sand. In *the Fourth International Congress on Environmental Geotechnics*; de Mello, L.G., Almeida, M., Eds.; Balkema Publishers: Rio de Janeiro, Brazil, 2002; pp. 9–13.
42. Ghaly, A.M. Strength Remediation of Oil Contaminated Sands. In *The 17th International Conference on Solid Waste Technology and Management*; Zandi, I., Mersky, R.L., Shieh, W.K., Eds.; Widener University: Philadelphia, PA, USA, 2001; pp. 289–298.
43. Abousnina, R.M.; Manalo, A.; Shiau, J.; Lokuge, W. Effects of Light Crude Oil Contamination on the Physical and Mechanical Properties of Fine Sand. *Soil Sediment Contam.* **2015**, *24*, 833–845. [CrossRef]
44. Wang, L. Cementitious Stabilization of Soils in the Presence of Sulfate. Ph.D. Thesis, Louisiana State University, Baton Rouge, LA, USA, 2002.
45. Prusinski, J.R.; Bhattacharja, S. Effectiveness of Portland Cement and Lime in Stabilizing Clay Soils. *Transp. Res. Rec.* **1998**, *1652*, 215–227. [CrossRef]
46. Ural, N. The Significance of Scanning Electron Microscopy (SEM) Analysis on the Microstructure of Improved Clay: An Overview. *Open Geosci.* **2021**, *13*, 197–218. [CrossRef]
47. Ergozhin, Y.; Dzhusipbekov, U.; Teltayev, B.; Nurgalieva, G.; Shakirova, A.; Khudaibergenova, K.; Izmailova, G.; Yelshibayev, N. Crude Oil Contaminated Soil: Its Neutralization and Use. *Sustainability* **2020**, *12*, 3087. [CrossRef]
48. Bol, P.; Kuany, G.; Zhou, X.; Abdelhafeez, I.A.; Abdelhafez, A.A. Oil Contaminated Soil, Global Environmental Impact (Overview). *Int. J. Curr. Sci. Eng.* **2019**, *1*, 124–129.
49. Wang, S.; Xu, Y.; Lin, Z.; Zhang, J.; Norbu, N.; Liu, W. The Harm of Petroleum-Polluted Soil and Its Remediation Research. *AIP Conf. Proc.* **2017**, *1864*, 020222. [CrossRef]
50. Trejos-Delgado, C.; Cadavid-Restrepo, G.E.; Hormaza-Anaguano, A.; Agudelo, E.A.; Barrios-Ziolo, L.; Loaiza-Usuga, J.C.; Cardona-Gallo, S.A. Oil Bioremediation in a Tropical Contaminated Soil Using a Reactor. *An. Acad. Bras. Cienc.* **2020**, *92*, 1–18. [CrossRef]
51. Jabbarov, Z.; Abdrakhmanov, T.; Pulatov, A.; Kováčik, P.; Pirmatov, K. Change in the Parameters of Soils Contaminated by Oil and Oil Products. *Agriculture* **2019**, *65*, 88–98. [CrossRef]
52. Kohli, R.; Mittal, K.L. *Developments in Surface Contamination and Cleaning*; Elsevier: Amsterdam, The Netherlands, 2019; Chapter three; pp. 23–105, ISBN 9780128160817.
53. Aziz, S.Q. Landfill Leachate Treatment Using Powdered Activated Carbon Augmented Sequencing Batch Reactor (SBR) Process. Ph.D. Thesis, Universiti Sains Malaysia, Gelugor, Malaysia, 2011.

applied sciences

MDPI

Article

Soil-Cement Mixtures Reinforced with Fibers: A Data-Driven Approach for Mechanical Properties Prediction

Joaquim Tinoco [1,*], António Alberto S. Correia [2] and Paulo J. Venda Oliveira [3]

1 ISISE, Department of Civil Engineering, University of Minho, 4710-057 Braga, Portugal
2 CIEPQPF-Chemical Process Engineering and Forest Products Research Centre, Department of Civil Engineering, University of Coimbra, 3004-531 Coimbra, Portugal; aalberto@dec.uc.pt
3 ISISE, Department of Civil Engineering, University of Coimbra, 3004-531 Coimbra, Portugal; pjvo@dec.uc.pt
* Correspondence: jtinoco@civil.uminh.pt; Tel.: +351-253-510-200

Abstract: The reinforcement of stabilized soils with fibers arises as an interesting technique to overcome the two main limitations of the stabilized soils: the weak tensile/flexural strength and the higher brittleness of the behavior. These types of mixtures require extensive laboratory characterization since they entail the study of a great number of parameters, which consumes time and resources. Thus, this work presents an alternative approach to predict the unconfined compressive strength (UCS) and the tensile strength of soil-binder-water mixtures reinforced with short fibers, following a Machine Learning (ML) approach. Four ML algorithms (Artificial Neural Networks, Support Vector Machines, Random Forest and Multiple Regression) are explored for mechanical prediction of reinforced soil-binder-water mixtures with fibers. The proposed models are supported on representative databases with approximately 100 records for each type of test (UCS and splitting tensile strength tests) and on the consideration of sixteen properties of the composite material (soil, fibers and binder). The predictive models provide an accurate estimation (R^2 higher than 0.95 for Artificial Neuronal Networks algorithm) of the compressive and the tensile strength of the soil-water-binder-fiber mixtures. Additionally, the results of the proposed models are in line with the main experimental findings, i.e., the great effect of the binder content in compressive and tensile strength, and the significant effect of the type and the fiber properties in the assessment of the tensile strength.

Keywords: soil-cement mixtures; fibers; mechanical properties; machine learning; artificial neural networks

Citation: Tinoco, J.; Correia, A.A.S.; Venda Oliveira, P.J. Soil-Cement Mixtures Reinforced with Fibers: A Data-Driven Approach for Mechanical Properties Prediction . *Appl. Sci.* **2021**, *11*, 8099. https://doi.org/10.3390/app11178099

Academic Editor: Dario De Domenico

Received: 2 August 2021
Accepted: 28 August 2021
Published: 31 August 2021

Publisher's Note: MDPI stays neutral with regard to jurisdictional claims in published maps and institutional affiliations.

Copyright: © 2021 by the authors. Licensee MDPI, Basel, Switzerland. This article is an open access article distributed under the terms and conditions of the Creative Commons Attribution (CC BY) license (https://creativecommons.org/licenses/by/4.0/).

1. Introduction

In the last two decades, soil stabilization using chemical binders has been spreading rapidly around the world. This technique is used to improve the properties of problematic soils, mainly when the soils show a low shear strength and high compressibility to support in safe conditions the loads applied by several works, such as: foundations of buildings and/or embankments, slope reinforcement, deep retaining walls [1,2], stabilization of contaminated soils [3], among others. The main constraint of this methodology is related to the weak tensile strength of the stabilized soil, which restrains its use in works where a non-negligible tensile strength is required, namely in the case of structures subject to horizontal vibrations (e.g., induced by heavy machinery, traffic, wind, sea waves, explosives and earthquakes) or horizontal loading/displacement (e.g., deep mixing columns used in slope stabilization or installed in the lateral of embankments, retaining walls [4]). The tensile/flexural strength of the soil-binder-water mixtures can be increased through the inclusion of short fibers [5,6] or by the installation of steel H-beams inside deep mixing columns. In fact, this approach of including fibers to improve the mechanical behavior of the mixtures has been adopted in other similar industries [7–9].

The reinforcement of soil-binder-water mixtures with short fibers, addressed in several works, induces an increase in the ductility, post-peak strength and tensile/flexural strength [6,10–18]. However, the experimental results also show that the impact of the reinforcement changes with the type of soil, type and content of fiber, the amount of binder and the mechanism induced by the test used to characterize the tensile strength [6,10,11]. In fact, the reinforcement with synthetic fibers in soil-binder-water mixtures for a binder content lower than 10% induces an increase in the compressive strength [16–18], while a higher amount of binder originates an opposite tendency [6,10,19]. Moreover, the effect of the reinforcement with fibers on the tensile strength depends on the strain level imposed at failure by each type of test [10]. Thus, when the tests originate a reduced strain at failure (as the direct tensile strength tests), which is insufficient to mobilize the tensile strength of the fibers, the effect of the reinforcement is less expressive or even detrimental. On the other hand, when the failure is associated with a deformation high enough to mobilize the tensile strength of the fibers (as in the case of the flexural strength and the split tensile strength tests), an increase in the tensile strength is observed with the reinforcement with fibers.

As previously described, the evaluation of the mechanical characteristics of soil-fiber-binder-water mixtures depends on a great number of factors, requiring the execution of specific tests for each of the desired properties. Additionally, the specimens should be prepared in conditions to replicate as possible the field conditions, mainly the soil and water content, which increases the costs, especially when dealing with natural materials rich in heterogeneities as soils are. Thus, the use of tools to predict the mechanical characteristics of soil-fiber-binder-water mixtures can be very useful, particularly in the pre-design stage of a work allowing to minimize the associated costs. Keeping this in mind, this work followed a data-driven approach by exploring the capabilities of four Machine Learning (ML) algorithms. In particular, Artificial Neural Networks (ANNs) [20], Support Vector Machines (SVMs) [21] and Random Forest (RF) [22] have been explored for mechanical prediction of reinforced soil-binder-water mixtures with fibers. As a baseline comparison, a Multiple Regression (MR) was also implemented. These advanced algorithms have been widely applied in different knowledge domains [23,24] with very promising results and taking advantage of a consolidated experience. In the field of Civil Engineering, several successful applications of these tools can be found [25–27], including solving complex geotechnical problems related to slopes stability assessment [28,29]. These algorithms have also been applied in the study of mechanical properties of soil-binder-water mixtures as reported on Tinoco et al. [30], which underline the non-linear learning capabilities of these algorithms. Thus, considering its past application on unconfined compressive strength [30,31] estimation of non-reinforced soil-water-cement mixtures, the focus and main novelty of this work is the prediction of the unconfined compressive strength and, mainly, the tensile strength of stabilized soils reinforced with some types of short fibers.

2. Methodology

2.1. Modeling

For both mechanical property's prediction of reinforced soil-binder-water mixtures with fibers, a data-driven approach was adopted. Thus, four different ML algorithms were fitted to each one of the databases previously compiled and prepared that contained unconfined compression strength tests results and indirect tensile strength tests results related to laboratory mixtures, as well as a set of input variables related to the soil, binder and fibers characteristics used to prepare the mixtures. Particularly, Artificial Neural Networks (ANNs) [20,32,33], Support Vector Machines (SVMs) [21,34–37] and Random Forests (RF) [22,27,38–40] were trained for Unconfined Compressive Strength (UCS) and Indirect Tensile Strength (ITS) estimation of reinforced soil-binder-water mixtures with fibers. In addition, as a baseline comparison, also a Multiple Regression (MR) [41] algorithm was implemented.

For a detailed overview of each one of the adopted ML algorithms, the readers are advised to check the literature, namely the above-indicated references. Concerning the definitions and hyperparameters of each algorithm, Figure 1 summarizes the adopted parameters.

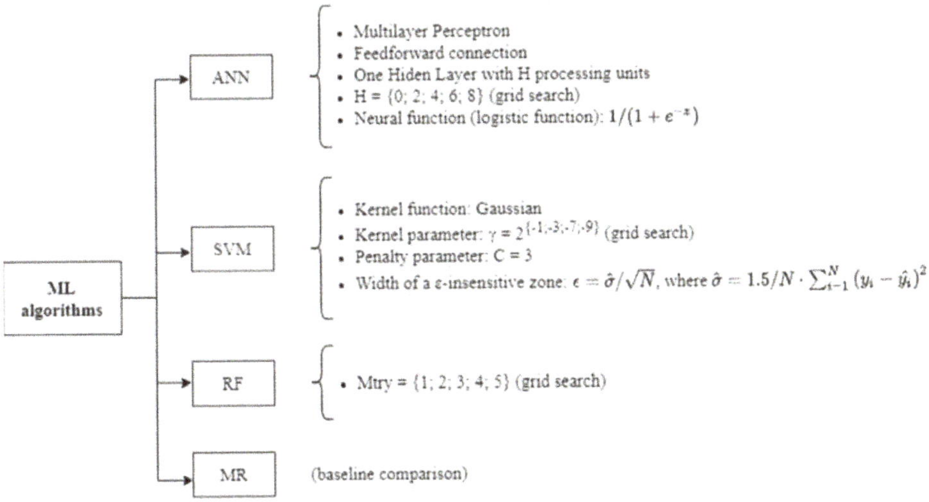

Figure 1. Definitions and hyperparameter adopted for each ML algorithm.

All experiments were conducted using the R statistical environment [42] and supported through the rminer package [43], which facilitates the implementation of several DM algorithms, including ANNs, SVMs and RF algorithms, as well as different validation schemas such as the cross-validation adopted in this work.

2.2. Models Evaluation

Models' accuracy and interpretability are two important steps for a deeper understanding and assessment of the proposed models.

Concerning the models' comparison and accuracy measurement, three distinct metrics were calculated [44]: Mean Absolute Error (MAE), Root Mean Square Error (RMSE) and Coefficient of correlation (R^2). On a perfect model, the first two metrics (MAE and RMSE) should present a value close to zero and R^2 equal to one. Although similar, MAE and RMSE allow a model's assessment under distinct and complementary perspectives. When compared with MAE, RMSE penalizes more heavily a model that in a few cases produces high errors since it uses the square of the distance between the real and predicted values [26,31]. In addition, the Regression Error Characteristic (REC) curve proposed by Bi and Bennett [45] was also adopted. An REC curve plots the error tolerance on the x-axis versus the percentage of points predicted within the tolerance on the y-axis, allowing quick and easy comparison of different regression models.

Generalization capacity is also a key point for the model's assessment. For this purpose, in this work, a 5-run under cross-validation (k-fold = 10) approach [44] was implemented. A k-fold validation evaluates the data across the entire training set, but it does so by dividing the training set into k folds (or subsections, where k is a positive integer) and then training the model k times, each time leaving a different fold out of the training data and using it instead as a validation set. In the end, the performance metric is averaged across all k tests. Lastly, as before, once the best parameter combination has been found, the model is retrained on the full data.

From an engineering point of view, the model's interpretability is a key aspect to take into account. Due to the high complexity of most ML algorithms, namely SVMs or ANNs that rely on complex statistical analysis and are frequently referred to as "black boxes", it is fundamental to find a way to "open" such models in order to understand what was learnt by them. With this purpose, Cortez and Embrechts [46] proposed a novel visualization approach based on sensitivity analysis (SA), which is used in this work. SA is a simple method that is applied after the training phase and measures the model responses when a given input is changed, allowing the quantification of the relative importance of each attribute as well as its average effect on the target variable. In particular, it was applied the Global Sensitivity Analysis (GSA) method [46], which is able to detect interactions among input variables. This is achieved by performing a simultaneous variation of F inputs. Each input is varied through its range with L levels and the remaining inputs fixed to a given baseline value. In this work, the average input variable value as a baseline was adopted and set to L = 12, which allows an interesting detail level under a reasonable amount of computational effort.

With the sensitivity response of the GSA, different visualization techniques can be computed. In this work, it is calculated the input importance barplot, which shows the relative influence (R_a) of each input variable in the model (from 0 to 100%). The rationale of GSA is that the higher the changes produced in the output, the more important is the input. To measure this effect, first, the gradient metric (g_a) for all inputs was calculated. After that, the relative influence was computed according to the following equation:

$$R_a = g_a/\sum_{i=1}^{I} g_i \cdot 100(\%) \quad where, \quad g_a = \sum_{j=2}^{L} \left|\hat{y}_{a,j} - \hat{y}_{a,j-1}\right| / (L-1), \tag{1}$$

where a denotes the input variable under analysis, and $\hat{y}_{a,j}$ is the sensitivity response for $x_{a,j}$.

2.3. Database

For models training and testing, two independent databases were compiled, respectively, for UCS and ITS studies, containing 121 records in the first case and 94 in the second. All samples were prepared under a controlled environment in the framework of a laboratory testing program developed at the University of Coimbra. This program aimed to characterize the compression and tensile behavior of soil-binder-water mixtures reinforced with fibers through unconfined compressive strength tests and indirect tensile strength tests (the later ones also called split tensile strength tests). Soils characteristics (grain size composition, organic matter content, water content, Atterberg limits), binder content, curing time and fibers characteristics (changing origin, length, fiber content and mechanical properties) were parameters considered in the study [4,6,10–12,47,48].

The soils used in the preparation of the laboratory samples comprise natural soils (collected in the Mondego river lower valley area and in a gravel-silty pit) and laboratory-made soils (starting from natural soils a specific property was varied, e.g., organic matter content and sand content), ranging from cohesive to cohesionless soils, organic to nonorganic soils, presenting different geotechnical properties. In all cases, soils were chemically stabilized with Portland cement, the most widely used binder in soils stabilization [49], applied in different amounts ranging from 75 to 500 kg/m^3. Concerning the fibers, four distinct types have been used trying to encompass all the types of fibers usually applied in soils stabilization. Thus, it was selected a natural fiber (Sisal) and three artificial fibers, a synthetic one (polypropylene), and two metallic fibers (Dramix and Wiremix, varying the fibers anchorage conditions), characterized by different mechanical properties, namely stiffness and tensile strength. The fibers length changed from 12 to 30 mm, and they were applied in different amounts ranging from 2 to 150 kg/m^3. A detailed description of all materials may be found in [4,6,10–12,47,48].

As models input, a set of 16 variables were selected. Among all variables available in the framework of the study, these 16 features are identified in the literature as influents on mechanical properties behavior [30,50–53]. Moreover, from a statistical point of view, they were also identified as relevant, as shown in the correlation matrix depicted in Figure 2, which relates to the UCS study. Considering that the formulations prepared for both studies (UCS and ITS) are similar, the equivalent representation for ITS is also similar. For that reason, it was not included in the paper. In addition, the selection of the variables was also supported on a try and error procedure using the evaluation metrics described above. Below, all 16 input variables considered in this study are listed on both mechanical properties' prediction of reinforced soil-binder-water mixtures with fibers:

- Soil sand content (%)—$\%Sand$
- Soil silt content (%)—$\%Silt$
- Soil clay content (%)—$\%Clay$
- Soil organic matter content (%)—$\%OM$
- Liquid limit—W_{LL}
- Plastic limit—W_{PL}
- Water content (%)—ω_0
- Cement content (%)—a_w
- Cement dosage (kg/m^3)—D_{Kg/m^3}
- Ratio between water and cement contents—ω_0/a_w
- Age of the mixture (days)—t
- Length of the fiber (mm) —L_{fiber}
- Fiber content (%)—T_{fiber}
- Fiber dosage (kg/m^3)—F_{Kg/m^3}
- Tensile strength of the fiber (MPa)—f_{ct_fiber}
- Deformability modulus of the fiber (GPa)—E_{fiber}

Table 1 summarizes the main statistics of all 16 inputs variables, as well as of the output variables (UCS and ITS), showing the wide range of binder and fiber contents.

Table 1. A summary of the main statistics of the input and output variables used in mechanical properties prediction of reinforced soil-binder-water mixtures with fibers.

Variable	Minimum		Maximum		Mean		Std. Deviation	
	UCS	ITS	UCS	ITS	UCS	ITS	UCS	ITS
$\%Sand$	14.00	14.00	100.00	97.82	36.41	37.83	33.23	35.41
$\%Silt$	0.00	1.77	61.00	61.00	45.49	44.26	23.56	25.07
$\%Clay$	0.00	0.41	25.00	25.00	18.10	17.91	10.09	10.34
$\%OM$	0.00	0.24	13.05	13.05	8.01	7.79	5.12	4.93
W_{LL}	0.00	0.00	80.00	80.00	54.68	55.49	32.10	33.07
W_{LP}	0.00	0.00	48.80	42.90	32.97	31.61	19.13	18.61
ω_0	14.20	14.20	113.00	80.87	67.05	63.85	27.41	29.23
a_w	7.52	7.52	73.98	71.50	25.91	22.34	22.03	21.22
$D_{kg.m^3}$	75.00	75.00	500.00	500.00	236.78	221.81	116.86	113.19
ω_0/a_w	1.09	1.13	8.85	8.85	4.27	4.72	3.20	3.47
t	3.00	3.00	28.00	28.00	25.02	24.17	7.36	8.16
L_{fiber}	12.00	12.00	30.00	30.00	19.72	22.51	8.87	8.82
T_{fiber}	0.19	0.33	13.96	13.96	2.41	2.85	2.70	2.89
F_{kg/m^3}	2.00	10.00	150.00	150.00	29.62	35.43	27.45	28.17
f_{ct_fiber}	250.00	250.00	1345.00	1345.00	684.69	838.70	468.65	456.06
E_{fiber}	3.70	3.70	210.00	210.00	92.36	124.31	98.61	96.97
Output	6.00	1.40	5172.30	676.89	1451.15	251.90	1391.01	232.12

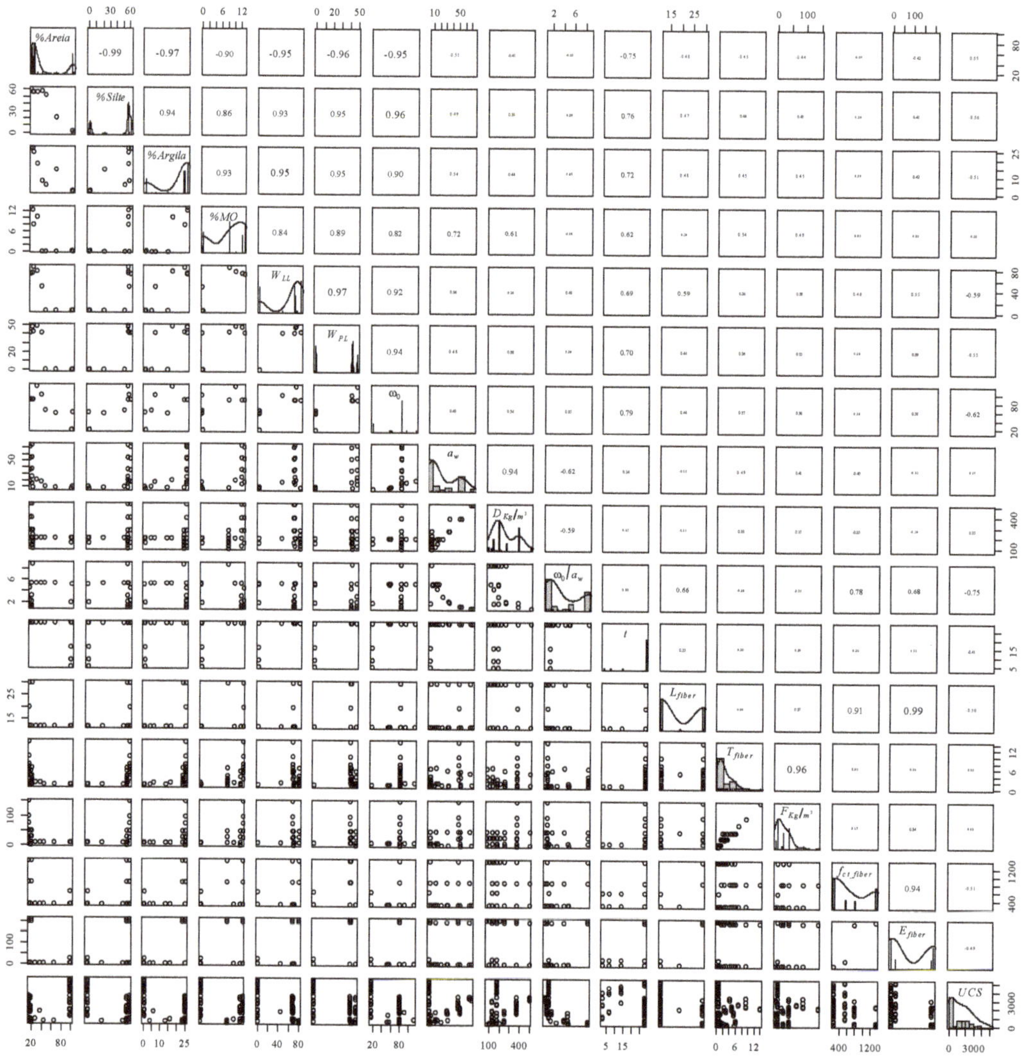

Figure 2. A correlation matrix of all variables considered in the study of UCS prediction of reinforced soil-binder-water mixtures with fibers (scatter plot of matrices (SPLOM), with bivariate scatter plots below the diagonal, histograms on the diagonal, and the Pearson correlation above the diagonal).

3. Results and Discussion

This section summarizes the main achievements of the study. Thus, the main achievements concerning the UCS prediction are presented and discussed in Section 3.1, followed by ITS results in Section 3.2. In both sections, after an overall comparison of all four ML algorithms trained, a more in-depth analysis is presented for ANN and RF algorithms, which achieved an overall superior performance. For simplification purposes, the following notation is adopted for the models' names: ML algorithm (ANN, SVM, RF or MR) dot followed by the prediction type (UCS or ITS). For example, ANN.UCS refers to the developed model for UCS prediction based on the ANN algorithm.

The average hyperparameters and fitting time values (and respective 95% level confidence intervals according to t-student distribution) of the four ML algorithms trained for both mechanical properties prediction of soil-binder-water mixtures reinforced with fibers (i.e., ANN, SVM, RF and MR) are summarized in Table 2. The slowest one is the RF on UCS modelling, which takes an average of 6 s over the five runs. If excluding MR, SVM was the fastest one taking on average around 2 s over the five runs, followed by ANN with more than 4.7 s. As expected, MR was very fast to model UCS and ITS, taking less than 0.50 s. It should be noted that these computational times are related to the time that each algorithm took to fit the training data. In the future, when the proposed models (namely the ANN and RF models) are applied to predict new cases, the time required is very close to zero (the computation is almost instantaneous). In terms of hyperparameter, and particularly for the ANN, the optimized number of neurons in the hidden layer was 6 and 5, respectively, for UCS and ITS prediction.

Table 2. Hyperparameters and computation time of each DM model.

Model	Hyperparameters		Time (s)	
	UCS	ITS	UCS	ITS
ANN	$H = 6 \pm 1$	$H = 5 \pm 1$	5.18 ± 0.18	4.79 ± 0.23
SVM	$\gamma = 0.23 \pm 0.05$ $\varepsilon = 0.03 \pm 0.01$	$\gamma = 0.17 \pm 0.08$ $\varepsilon = 0.03 \pm 0.01$	2.12 ± 0.07	2.32 ± 0.05
RF	$Mtry = 9 \pm 1$	$Mtry = 9 \pm 1$	6.21 ± 0.11	4.16 ± 0.11
MR	-	-	0.35 ± 0.04	0.38 ± 0.12

Table 3 compares the performance of the four ML algorithms in both UCS and ITS prediction of soil-binder-water mixtures reinforced with fibers based on MAE, RMSE and R^2 metrics (mean value and respective 95% level confidence intervals according to t-student distribution). Apart from MR, all other three algorithms present a particularly good and similar performance in both mechanical properties' prediction of soil-binder-water mixtures reinforced with fibers. Taken R^2 as a reference, all three algorithms (ANN, SVM and RF) achieved, on average, a value close to 0.95.

Table 3. A comparison of the models' performances based on the metrics MAE, RMSE and R^2 (best values in bold).

Model	MAE		RMSE		R^2	
	UCS	ITS	UCS	ITS	UCS	ITS
ANN	**158.19 ± 46.73**	**23.62 ± 4.32**	310.26 ± 159.03	**42.00 ± 11.23**	**0.95 ± 0.05**	**0.97 ± 0.02**
SVM	201.06 ± 37.68	33.17 ± 2.74	355.70 ± 85.68	54.58 ± 5.01	0.93 ± 0.03	0.94 ± 0.01
RF	197.06 ± 8.59	31.80 ± 2.74	**302.78 ± 12.56**	50.94 ± 7.61	**0.95 ± 0.00**	0.95 ± 0.02
MR	472.99 ± 52.27	66.03 ± 52.27	672.27 ± 187.19	88.26 ± 21.67	0.78 ± 0.10	0.86 ± 0.06

A detailed analysis shows that ANN achieved an overall superior performance on both mechanical properties' prediction (best values in bold in Table 3 as described in Section 2.2), followed by RF and SVM. As expected, the lower performance is observed for MR, which evidenced clear difficulties in modelling UCS and ITS efficiently, which can be explained by the characteristic non-linear behavior of soil-binder-water mixtures reinforced with fibers.

3.1. Uniaxial Compressive Strength

Concerning the UCS study, Figure 3 compares REC curves of all four ML algorithms, confirming the lower performances of MR and the superior response of ANN. In a REC representation, a high performance corresponds to an accuracy of one (y-axis) achieved for as low as possible absolute deviation (x-axis). Thus, taken ANN.UCS as a reference, one can observe that ANN.UCS achieved accuracy close to one for an absolute deviation of

750 kPa. On the opposite side, and for the same absolute deviation, the MR.UCS accuracy is around 25% lower. SVM.UCS and RF.UCS have similar performances, although the first one shows a better response for lower absolute deviations.

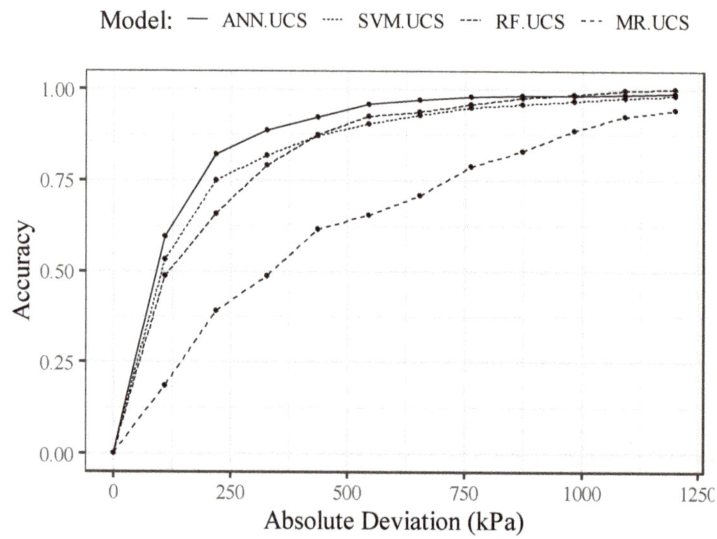

Figure 3. A comparison of ANN.UCS, SVM.UCS, RF.UCS and MR.UCS models performance in UCS prediction of soil-binder-water mixtures reinforced with fibers based on REC curves.

Figure 4 depicts the relation between observed and predicted UCS values (scatterplot) according to ANN.UCS (Figure 4a) and RF.UCS (Figure 4b) models. From its analysis, a very interesting fit can be observed (all points are very close to the diagonal line), which corroborates the metrics values above summarized in Table 3 and discussed.

As important as the model's accuracy is its interpretability, particularly from an engineering point of view. Accordingly, in this study, a detailed sensitivity analysis was applied, aiming to measure the relative importance of each model attribute and, this way, understand what has been learnt by the algorithms and compare it with the empirical knowledge. Figure 5 plots the relative importance of each one of the sixteen attributes considered in the UCS prediction of soil-binder-water mixtures reinforced with fibers, according to the four ML algorithms implemented in this study. Taken ANN.UCS model as reference, which achieved the best overall performance as above shown, in the ranking of the first four key variables, it may be found the influence of the binder dosage (D_{Kg/m^3}= *13.5%*), soil characteristics (ω_0= *12.8%*, *%Clay*= *8.5%*) and fiber type (T_{fiber}= *8.0%*). These variables are indeed some of the most important parameters controlling the behavior of soil-binder-water mixtures reinforced with fibers, as observed in some experimental studies [4–9,14,16,19,54–56]. Additionally, according to the SVM.UCS model, a similar distribution is observed. Concerning the RF.UCS model, although has achieved the second-best overall performance on UCS prediction of soil-binder-water mixtures reinforced with fibers, in terms of relative importance distribution, the influence of ω_0/a_w, seems too high (40%). However, it should be noted that based on previous studies [30] related to soil-cement mixtures, this ratio has been identified as one of the most influential variables on mechanical properties development.

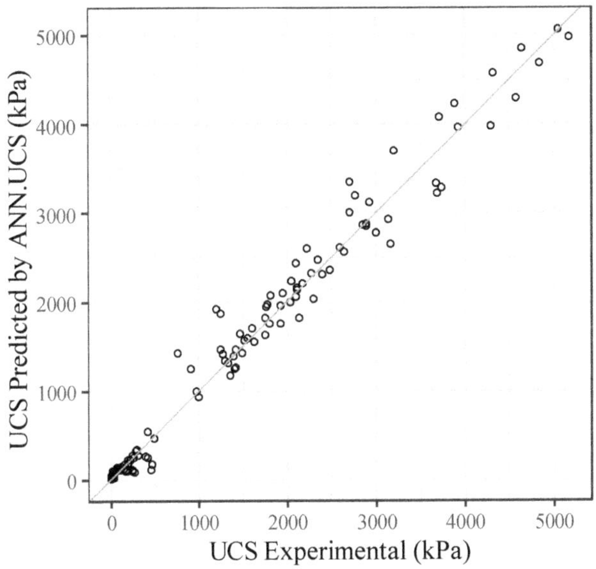

(**a**)

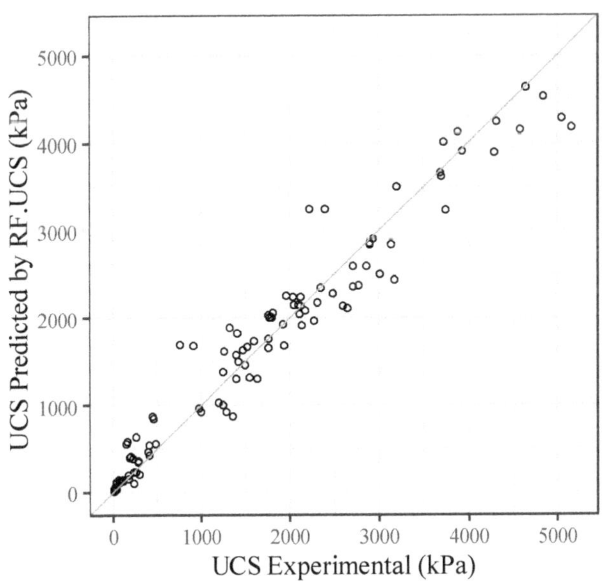

(**b**)

Figure 4. The relationship between UCS experimental versus predicted values of soil-binder-water mixtures reinforced with fibers according to: (**a**) the ANN.UCS model; (**b**) the RF.UCS model.

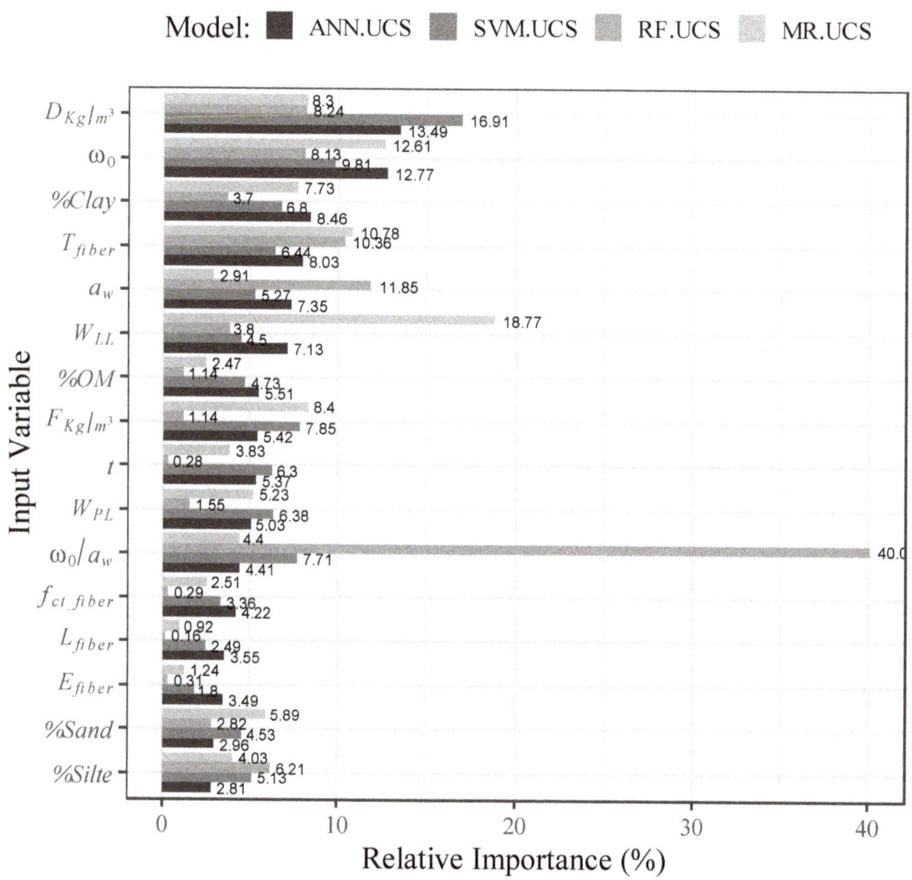

Figure 5. A comparison of the relative importance of each input variable based on a GSA in the UCS prediction of soil-binder-water mixtures reinforced with fibers.

3.2. Indirect Tensile Strength

Following the same procedure adopted in the UCS study, the performance of all four algorithms in ITS prediction was compared based on REC curves, as depicted in Figure 6. As previously discussed and shown in Table 3, it is also clear here that the superior performance of ANN algorithm on ITS prediction and the weak response of a linear approach (MR.ITS model). Concerning RF.ITS and SVM.ITS, both present a very similar response on ITS prediction.

Looking in detail to ANN.ITS model, it is observed that around 96% of all records can be predicted with an absolute deviation lower than 100 kPa. Moreover, even for a tighter tolerance, such as an absolute deviation around 50 kPa, ANN.ITS presents an accuracy higher than 85%, showing its good performance.

Figure 7 validates the high performance of both ANN.ITS (Figure 7a) and RF.ITS (Figure 7b) models on ITS prediction. As shown, particularly according to the ANN.ITS model, all predictions are close to the experimental values (diagonal line).

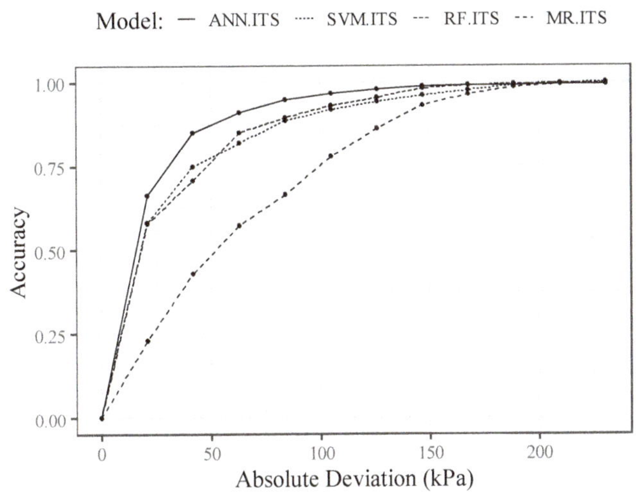

Figure 6. A comparison of ANN.ITS, SVM.ITS, RF.ITS and MR.ITS models performance in ITS prediction of soil-binder-water mixtures reinforced with fibers based on REC curves.

Concerning to model's interpretability, Figure 8 compares the relative importance of each model attribute. As for the UCS study, with the ANN.ITS model taken as a reference, the dosage of the binder ($D_{Kg/m3}$) was identified as the most relevant variable in ITS prediction with a relative influence close to 16%. A higher influence of the fibers is also observed, which was considered by the ANN.ITS model through E_{fiber} (7.2%), $F_{Kg/m3}$ (7.1%) and T_{fibre} (7.0%), which ranks in the five most relevant variables. This higher influence of the fibers on the ITS prediction, when compared to the UCS study, is in agreement with some empirical studies [10,14,16]. In fact, when the composite material is subject to indirect tensile through a splitting failure mechanism, there is an effective mobilization of the tensile strength of the fibers that cross the vertical failure plane imposed by the ITS test, and consequently, the tensile strength is directly related to the fibers' characteristics. According to RF.ITS, once again, an influence above 40% is observed for ω_U/a_w, which demonstrates the coherence of the algorithm.

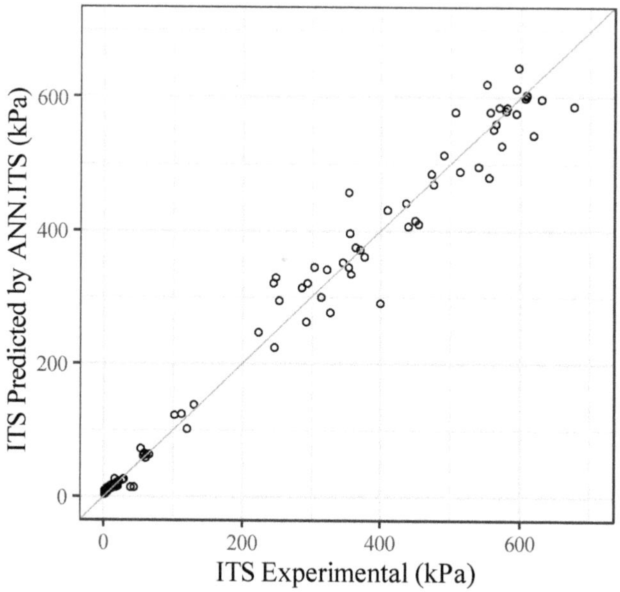

(a)

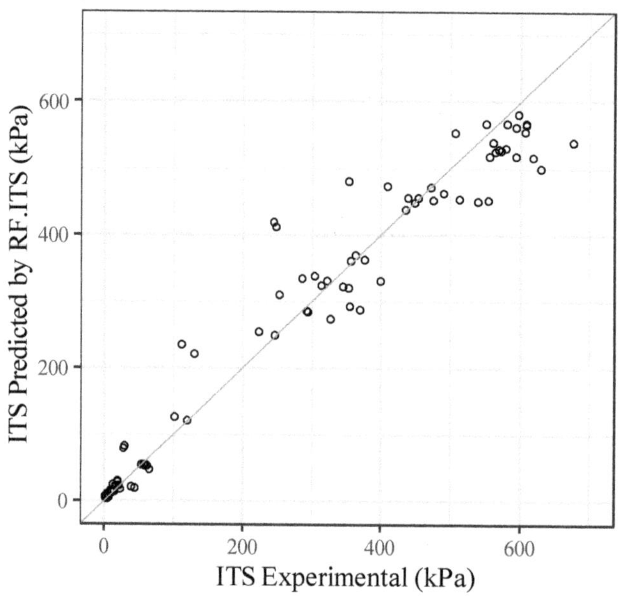

(b)

Figure 7. The relationship between ITS experimental versus predicted values of soil-binder-water mixtures reinforced with fibers according to: (**a**) the ANN.ITS model; (**b**) the RF.ITS model.

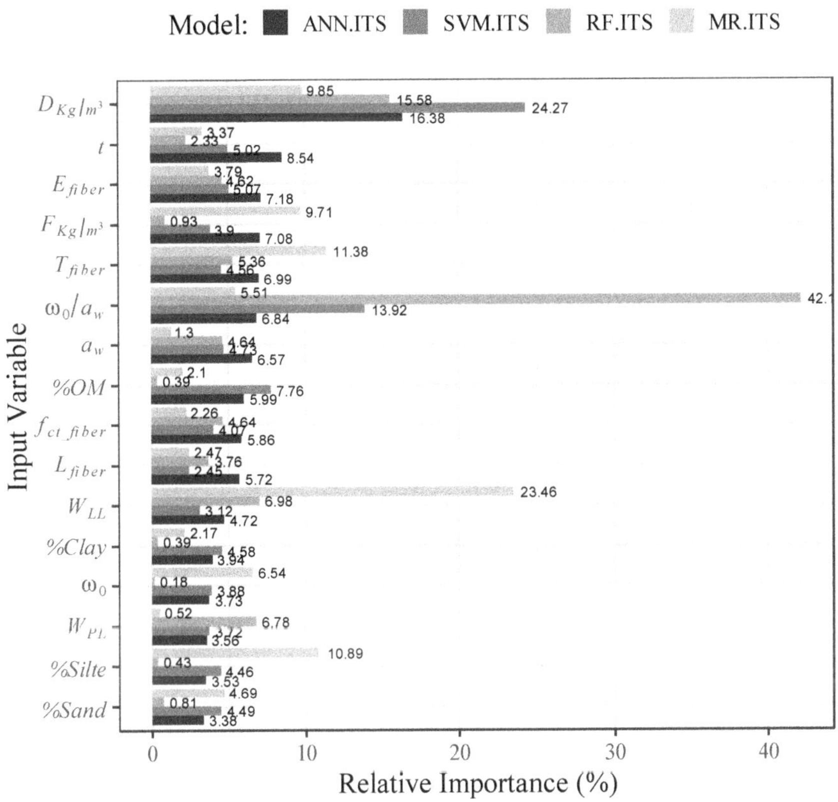

Figure 8. A comparison of the relative importance of each input variable based on a GSA in ITS prediction of soil-binder-water mixtures reinforced with fibers.

4. Conclusions

This work explored four Machine Learning (ML) algorithms to predict the mechanical properties of soil-binder-water mixtures reinforced with fibers. Thus, Artificial Neuronal Networks (ANNs), Support Vector Machines (SVMs), Random Forest (RF) and Multiple Regression (MR), which were used as a baseline comparison, were implemented to predict Unconfined Compressive Strength (UCS) and Indirect Tensile Strength (ITS) development. The proposed models, supported on representative databases comprising around 100 records each, were able to catch both mechanical properties behavior with a promising performance (R^2 higher than 0.95), particularly those based on ANNs. For that, sixteen variables covering information about the three main components involved in these types of mixtures (i.e., soil, fibers and binder) have been considered.

By addressing a global sensitivity analysis, a deeper understanding of the proposed models was extracted, showing that the binder content is one the most influential variable in both UCS and ITS prediction. Moreover, it was observed that in the ITS study, the type and characteristics of the fibers are more relevant than in the UCS study, which corroborates some experimental findings.

In conclusion, the proposed models can be used as an important tool for design purposes, allowing a very accurate estimation of the final properties of soil-binder-water mixtures reinforced with fibers by considering only information available without preparing/testing any sample. Moreover, it was shown once again the advantages of implementing a data-driven approach to explore complex geotechnical problems.

Author Contributions: Conceptualization, J.T., A.A.S.C. and P.J.V.O.; Software, J.T.; Investigation, J.T., A.A.S.C. and P.J.V.O.; Methodology, J.T., A.A.S.C. and P.J.V.O.; Validation, J.T., A.A.S.C. and P.J.V.O.; Writing—original draft, J.T.; Writing—review & editing, J.T., A.A.S.C. and P.J.V.O.; Project administration, P.J.V.O.; Funding acquisition, A.A.S.C. and P.J.V.O. All authors have contributed for all parts of the work. All authors have read and agreed to the published version of the manuscript.

Funding: This research was partly financed by FCT/MCTES through national funds (PIDDAC) under the R&D Unit Institute for Sustainability and Innovation in Engineering Structures (ISISE), under reference UIDB/04029/2020, the R&D Unit Chemical Process Engineering and Forest Products Research Centre (CIEPQPF) under reference UIDB/00102/2020, and under the project PTDC/ECI-CON/28382/2017.

Institutional Review Board Statement: Not applicable.

Informed Consent Statement: Not applicable.

Data Availability Statement: The data used in the present study are under privacy issues and cannot be shared.

Acknowledgments: This work was partly financed by FCT/MCTES through national funds (PIDDAC) under the R&D Unit Institute for Sustainability and Innovation in Engineering Structures (ISISE), under reference UIDB/04029/2020, and the R&D Unit Chemical Process Engineering and Forest Products Research Centre (CIEPQPF) under reference UIDB/00102/2020. A special thanks also to ACIV for the financial support.

Conflicts of Interest: The authors declare no conflict of interest.

References

1. Holm, G. Keynote Lecture: Towards a sustainable society—Recent advances in deep mixing. In Proceedings of the International Conference on Deep Mixing—Best Practice and Recent Advances, Stockholm, Sweden, 23–25 May 2005; pp. k13–k24.
2. Terashi, M.; Kitazume, M. Keynote Lecture: Current practice and future perspective of QA/QC for deep-mixed ground. In Proceedings of the International Symposium of Deep Mixing & Admixture Stabilization, Okinawa, Japan, 19–21 May 2009.
3. Al-Tabbaa, A.; Barker, P.; Evans, C.W. Keynote Lecture: Design of deep mixing in infrastructure applications. In Proceedings of the International Symposium of Deep Mixing & Admixture Stabilization, Okinawa, Japan, 19–21 May 2009.
4. Oliveira, P.J.V.; Correia, A.A.; Cajada, J.C. Effect of the type of soil on the cyclic behaviour of chemically stabilised soils unreinforced and reinforced with polypropylene fibres. *Soil Dyn. Earthq. Eng.* **2018**, *115*, 336–343. [CrossRef]
5. Sukontasukkul, P.; Jamsawang, P. Use of steel and polypropylene fibres to improve flexural performance of deep soil-cement column. *Constr. Build. Mater.* **2012**, *29*, 201–205. [CrossRef]
6. Oliveira, P.J.V.; Correia, A.A.S.; Teles, J.M.N.P.C.; Custódio, D.G. Effect of fibre type on the compressive and tensile strength of a soft soil chemically stabilised. *Geosynth. Int.* **2016**, *23*, 171–182. [CrossRef]
7. Koksal, F.; Yıldırım, M.S.; Benli, A.; Gencel, O. Hybrid effect of micro-steel and basalt fibers on physico-mechanical properties and durability of mortars with silica fume. *Case Stud. Constr. Mater.* **2021**, *15*, e00649. [CrossRef]
8. Horňáková, M.; Lehner, P. Relationship of Surface and Bulk Resistivity in the Case of Mechanically Damaged Fibre Reinforced Red Ceramic Waste Aggregate Concrete. *Materials* **2020**, *13*, 5501. [CrossRef] [PubMed]
9. Shelar, A.B.; Mahindrakar, A.B.; Neeraja, D. Sustainable alternatives in concrete along with the use of medicinal plant Sapindus Mukorossi as a green workability agent. *Innov. Infrastruct. Solut.* **2021**, *6*, 228. [CrossRef]
10. Correia, A.A.S.; Oliveira, P.J.V.; Custódio, D.G. Effect of polypropylene fibres on the compressive and tensile strength of a soft soil, artificially stabilised with binders. *Geotext. Geomembr.* **2015**, *43*, 97–106. [CrossRef]
11. Correia, A.A.S.; Oliveira, P.J.V.; Teles, J.M.N.P.C.; Pedro, A.M.G. Strength of a stabilised soil reinforced with steel fibres. *Proc. Inst. Civ. Eng.-Geotech. Eng.* **2017**, *170*, 312–321. [CrossRef]
12. Correia, A.A.S.; Venda Oliveira, P.J.; Lemos, L.J.L. Strength assessment in chemically stabilized soft soils. *ICE—Geotech. Eng.* **2019**, *172*, 218–227. [CrossRef]
13. Güllü, H.; Khudir, A. Effect of freeze-thaw cycles on unconfined compressive strength of fine-grained soil treated with jute fibre, steel fibre and lime. *Cold Reg. Sci. Technol.* **2014**, *106–107*, 55–65. [CrossRef]
14. Consoli, N.C.; Moraes, R.R.; Festugato, L. Parameters controlling tensile and compressive strength of fibre-reinforced cemented soil. *J. Mater. Civ. Eng.* **2013**, *25*, 1568–1573. [CrossRef]
15. Consoli, N.C.; De Moraes, R.R.; Festugato, L. Variables controlling strength of fibre-reinforced cemented soils. *Proc. Inst. Civ. Eng.-Ground Improv.* **2013**, *166*, 221–232. [CrossRef]
16. Olgun, M. Effects of polypropylene fibre inclusion on the strength and volume change characteristics of cement-fly ash stabilized clay soil. *Geosynth. Int.* **2013**, *20*, 263–275. [CrossRef]
17. Park, S. Unconfined compressive strength and ductility of fibre-reinforced cemented sand. *Constr. Build. Mater.* **2011**, *25*, 1134–1138. [CrossRef]

18. Consoli, N.C.; Moraes, R.R.; Festugato, L. Split tensile strength of monofilament polypropylene fibre-reinforced cemented sandy soils. *Geosynth. Int.* **2011**, *18*, 57–62. [CrossRef]
19. Consoli, N.C.; Vendruscolo, M.A.; Fonini, A.; Rosa, F.D. Fiber reinforcement effects on sand considering a wide cementation range. *Geotext. Geomembr.* **2009**, *27*, 196–203. [CrossRef]
20. Torregrosa, A.; García-Cuevas, L.; Quintero, P.; Cremades, A. On the application of artificial neural network for the development of a nonlinear aeroelastic model. *Aerosp. Sci. Technol.* **2021**, *115*, 106845. [CrossRef]
21. Deiss, L.; Margenot, A.J.; Culman, S.W.; Demyan, M.S. Tuning support vector machines regression models improves prediction accuracy of soil properties in MIR spectroscopy. *Geoderma* **2020**, *365*, 114227. [CrossRef]
22. Genuer, R.; Poggi, J.-M.; Tuleau-Malot, C.; Villa-Vialaneix, N. Random Forests for Big Data. *Big Data Res.* **2017**, *9*, 28–46. [CrossRef]
23. Bhattacharya, S.; Roy, S.; Chowdhury, S. A neural network-based intelligent cognitive state recognizer for confidence-based e-learning system. *Neural Comput. Appl.* **2016**, *29*, 205–219. [CrossRef]
24. Murthy, V.M.S.R.; Kumar, A.; Sinha, P.K. Prediction of throw in bench blasting using neural networks: An approach. *Neural Comput. Appl.* **2016**, *29*, 143–156. [CrossRef]
25. Kang, F.; Li, J.-S.; Wang, Y.; Li, J. Extreme learning machine based surrogate model for analyzing system reliability of soil slopes. *Eur. J. Environ. Civ. Eng.* **2017**, *21*, 1341–1362. [CrossRef]
26. Tinoco, J.; Parente, M.; Correia, A.G.; Cortez, P.; Toll, D. Predictive and prescriptive analytics in transportation geotechnics: Three case studies. *Transp. Eng.* **2021**, *5*, 100074. [CrossRef]
27. Tinoco, J.; de Granrut, M.; Dias, D.; Miranda, T.; Simon, A.-G. Piezometric level prediction based on data mining techniques. *Neural Comput. Appl.* **2020**, *32*, 4009–4024. [CrossRef]
28. Tinoco, J.; Correia, A.G.; Cortez, P.; Toll, D.G. Stability Condition Identification of Rock and Soil Cutting Slopes Based on Soft Computing. *J. Comput. Civ. Eng.* **2018**, *32*, 04017088. [CrossRef]
29. Tinoco, J.; Correia, A.G.; Cortez, P.; Toll, D.G. Data-Driven Model for Stability Condition Prediction of Soil Embankments Based on Visual Data Features. *J. Comput. Civ. Eng.* **2018**, *32*, 04018027. [CrossRef]
30. Tinoco, J.; Alberto, A.; da Venda, P.; Correia, A.G.; Lemos, L. A novel approach based on soft computing techniques for unconfined compression strength prediction of soil cement mixtures. *Neural Comput. Appl.* **2020**, *32*, 8985–8991. [CrossRef]
31. Tinoco, J.; Correia, A.G.; Cortez, P. Support vector machines applied to uniaxial compressive strength prediction of jet grouting columns. *Comput. Geotech.* **2014**, *55*, 132–140. [CrossRef]
32. McCulloch, W.S.; Pitts, W. A logical calculus of the ideas immanent in nervous activity. *Bull. Math. Biol.* **1943**, *5*, 115–133. [CrossRef]
33. Werbos, P. Beyond Regression: New Tools for Prediction and Analysis in the Behavioral Sciences. Ph.D. Dissertation, Harvard University, Cambridge, MA, USA, 1974.
34. Cortes, C.; Vapnik, V. Support vector networks. *Mach. Learn.* **1995**, *20*, 273–297. [CrossRef]
35. Akbarzadeh, S.; Paap, A.; Ahderom, S.; Apopei, B.; Alameh, K. Plant discrimination by Support Vector Machine classifier based on spectral reflectance. *Comput. Electron. Agric.* **2018**, *148*, 250–258. [CrossRef]
36. Jiang, P.; Huang, Y.; Liu, X. Intermittent demand forecasting for spare parts in the heavy-duty vehicle industry: A support vector machine model. *Int. J. Prod. Res.* **2020**. [CrossRef]
37. Cherkassky, V.; Ma, Y. Practical selection of SVM parameters and noise estimation for SVM regression. *Neural Netw.* **2004**, *17*, 113–126. [CrossRef]
38. Breiman, L. Random forests. *Mach. Learn.* **2001**, *45*, 5–32. [CrossRef]
39. Berry, M.; Linoff, G. *Mastering Data Mining: The Art and Science of Customer Relationships Management*; Wiley: New York, NY, USA, 2000.
40. Breiman, L.; Friedman, J.H.; Olshen, R.A.; Stone, C.J. *Classification and Regression Trees*; CRC Press: Wadsworth, OH, USA; Belmont, Australian, 1984.
41. Deisenroth, M.P.; Faisal, A.A.; Ong, C.S. *Mathematics for Machine Learning*; Cambridge University Press (CUP): Cambridge, UK, 2020; p. 391.
42. Venables, W.N.; Smith, D.M.; R Development Core Team. *An Introduction to R.: Notes on R, A Programming Environment for Data Analysis and Graphics*; Network Theory Ltd: Godalming, UK, 2021.
43. Cortez, P. Data mining with neural networks and support vector machines using the r-rminer tool. In *Advances in Data Mining: Applications and Theoretical Aspects, Proceedings of the 10th Industrial Conference on Data Mining, Berlin, Germany, 12–14 July 2010*; Perner, P., Ed.; Springer: Berlin/Heidelberg, Germany, 2010; pp. 572–583.
44. Hastie, T.; Tibshirani, R.; Friedman, J. *The Elements of Statistical Learning: Data Mining, Inference, and Prediction*; Springer Science & Business Media: New York, NY, USA, 2009.
45. Bi, J.; Bennett, K.P. Regression error characteristic curves. In Proceedings of the Twentieth International Conference on Machine Learning, Washington, DC, USA, 21–24 August 2003; pp. 43–50.
46. Cortez, P.; Embrechts, M.J. Using sensitivity analysis and visualization techniques to open black box data mining models. *Inf. Sci.* **2013**, *225*, 1–17. [CrossRef]
47. Venda Oliveira, P.J.; Anunciação, G.R.; Correia, A.A.S. Effect of cyclic loading frequency on the behavior of a stabilized sand reinforced with polypropylene and sisal fibers. *J. Mater. Civ. Eng.* **2021**.

48. Oliveira, P.V.; Correia, A.; Teles, J.M.; Pedro, A. Effect of cyclic loading on the behaviour of a chemically stabilised soft soil reinforced with steel fibres. *Soil Dyn. Earthq. Eng.* **2017**, *92*, 122–125. [CrossRef]
49. Kitazume, M.; Terashi, M. *The Deep Mixing Methodd Principle, Design and Construction*; CRC Press/Balkema: Leiden, The Netherlands, 2002.
50. Sariosseiri, F.; Muhunthan, B. Effect of cement treatment on geotechnical properties of some Washington State soils. *Eng. Geol.* **2009**, *104*, 119–125. [CrossRef]
51. Lee, F.H.; Lee, Y.; Chew, S.-H.; Yong, K.-Y. Strength and Modulus of Marine Clay-Cement Mixes. *J. Geotech. Geoenviron. Eng.* **2005**, *131*, 178–186. [CrossRef]
52. Lorenzo, G.A.; Bergado, D.T. Fundamental Parameters of Cement-Admixed Clay—New Approach. *J. Geotech. Geoenviron. Eng.* **2004**, *130*, 1042–1050. [CrossRef]
53. Chen, H.; Wang, Q. The behaviour of organic matter in the process of soft soil stabilization using cement. *Bull. Int. Assoc. Eng. Geol.* **2005**, *65*, 445–448. [CrossRef]
54. Kaniraj, S.R.; Havanaji, V.J. Behavior of cement-stabilized fiber-reinforced fly ash-soil mixtures. *J. Geotech. Geoenviron. Eng.* **2001**, *127*. [CrossRef]
55. Khattak, M.J.; Alrashidi, M. Durability and mechanistic characteristics of fiber reinforced soil-cement mixtures. *Int. J. Pavement Eng.* **2006**, *7*, 53–62. [CrossRef]
56. Kumar, A.; Walia, B.S.; Bajaj, A. Influence of Fly Ash, Lime, and Polyester Fibers on Compaction and Strength Properties of Expansive Soil. *J. Mater. Civ. Eng.* **2007**, *19*, 242–248. [CrossRef]

applied sciences

MDPI

Article

Key-Parameters in Chemical Stabilization of Soils with Multiwall Carbon Nanotubes

António Alberto S. Correia [1,*], Pedro D. F. Casaleiro [2], Diogo T. R. Figueiredo [3], Marta S. M. R. Moura [3] and Maria Graça Rasteiro [4]

1 CIEPQPF—Chemical Process Engineering and Forest Products Research Centre, Department of Civil Engineering, University of Coimbra, 3030-788 Coimbra, Portugal
2 Department of Civil Engineering, University of Coimbra, 3030-788 Coimbra, Portugal; pedrocasaleiro@gmail.com
3 Department of Chemical Engineering, University of Coimbra, 3030-790 Coimbra, Portugal; cmolico@hotmail.com (D.T.R.F.); moura_marta@hotmail.com (M.S.M.R.M.)
4 CIEPQPF, Department of Chemical Engineering, University of Coimbra, 3030-790 Coimbra, Portugal; mgr@eq.uc.pt
* Correspondence: aalberto@dec.uc.pt; Tel.: +351-239-797-277

Abstract: Chemical stabilization is one of the most successful techniques that has been applied to improve the geomechanical behavior of soil. Several additives have been studied to be a sustainable alternative to traditional additives (Portland cement and lime) normally associated with high cost and carbon footprint. Nanomaterials are one of the most recent additives proposed. This work is focused on one type of nanomaterial, multiwall carbon nanotubes (MWCNTs) with unique characteristics, applied to chemical stabilization of soils and aiming to identify the key parameters affecting the stabilization improvement. It was found that a surfactant should be added in order to oppose the natural tendency of MWCNTs to aggregate with the consequent loss of benefits. The surfactant choice is not so dependent on the charge of the surfactant but rather on the balance between the concentration and the hydrodynamic diameter/molecular weight due to their impact on the geomechanical compression behavior. As time evolves from 7 to 28 days, there is a decrease in the geomechanical benefits associated with the presence of MWCNTs explained by the development of the cementitious matrix. MWCNTs applied in a proper concentration and enriched with a specific surfactant type may be a short-time valid alternative to the partial replacement of traditional additives.

Keywords: soil improvement; multiwall carbon nanotubes (MWCNTs); unconfined compressive strength tests; surfactant

Citation: Correia, A.A.S.; Casaleiro, P.D.F.; Figueiredo, D.T.R.; Moura, M.S.M.R.; Rasteiro, M.G. Key-Parameters in Chemical Stabilization of Soils with Multiwall Carbon Nanotubes. *Appl. Sci.* **2021**, *11*, 8754. https://doi.org/10.3390/app11188754

Academic Editor: Claudio De Pasquale

Received: 23 July 2021
Accepted: 17 September 2021
Published: 20 September 2021

Publisher's Note: MDPI stays neutral with regard to jurisdictional claims in published maps and institutional affiliations.

Copyright: © 2021 by the authors. Licensee MDPI, Basel, Switzerland. This article is an open access article distributed under the terms and conditions of the Creative Commons Attribution (CC BY) license (https://creativecommons.org/licenses/by/4.0/).

1. Introduction

Soil is the loose particulate natural material that covers the Earth's surface. Soil is a multiphase material containing an aqueous, gaseous and solid phase, each composed of inorganic and organic components. The interactions and relative proportions between the different components of the soil, and the arrangements, size and shape of the solid particles determine the soil's physical and chemical properties [1,2], and ultimately its geomechanical behavior (soil's response in terms of strength and deformability to external actions). In many cases, the soil does not meet the safety and stability requirements for construction, and ground improvement techniques are required [3–6]. This is the case of soft soils, characterized by exhibiting low strength and high deformability.

One of the most successful ground improvement techniques applied to soft soils is chemical stabilization with additives [7–10]. This ground improvement technique consists of in situ mixing additives to the soil, aiming to increase the soil's strength and decrease the soil's deformability. (There may be other objectives such as reducing permeability or immobilization of pollutants in the soil). The traditional additives most used in chemical stabilization of soft soils are Portland cement and lime, applied alone or in combination in percentages ranging from 5% to 20% *w*/*w* (additive/soil) [11–13]. These additives have high costs and high environmental impacts associated to their production, which encourages the development of new additives. Industrial byproducts (e.g., slag, fly ash, rice ash), pozzolanic materials (e.g., fly ash, natural pozzolana, silica fume), biobased products (e.g., polymers, enzymes) and nanomaterials (e.g., carbon nanotubes, carbon nanofibers, nano-ashes, nanoclays) are some examples of promising additives that may be used as a total or partial replacement of the traditional additives [11,13–20].

Additives consisting of extremely fine particles (nanomaterials) are particularly attractive for use as replacement of part of cementitious additives, resulting in environmental, technical and economic advantages [21–23]. Due to extraordinary properties of carbon nanotubes (CNTs) (fine structure, ultrahigh specific surface, very high strength and moduli of elasticity, elastic and ductile behavior [24–28]), they have a great potential to be used as an additive in chemical stabilization of soils, replacing part of the cementitious main additive. However, due to CNT morphology and very high aspect ratio, CNTs have a natural tendency to aggregate, resulting in the loss of their beneficial properties [29,30]. Different strategies have been proposed to minimize this problem, including mechanical methods (e.g., ultrasonic energy applied to disperse CNTs in suspension) and chemical methods (e.g., functionalization of CNT surface by the addition of surfactants/polymers to the system), among others. The introduction of surfactants has a double advantage since it allows the dispersion of the CNTs and other additive particles, while at the same time minimizes ultrasound energy requirements.

The application of CNTs to geotechnics is still at the laboratory development and proof-of-concept stage, with few studies published thus far, being possible to conclude the following: (i) The introduction of CNTs in a content of 0.2% up to 1% of soil's dry weight is able to increase slightly the specific gravity, dry density and pH [31]; increase the plasticity index; increase the compression and swelling indices and reduce the hydraulic conductivity of the soil [32]. (ii) Mixing a clayey sand with CNTs applied in a content of 0.05–3% by weight of the soil promotes an increase of the compressive strength of the composite material up to 120% when compared with the original clayey soil, and increases the cohesion while decreasing the friction angle [33]. (iii) The combination of Portland cement with CNTs in a content of 0.001% to 0.01% Portland cement's dry weight has the potential to increase the unconfined compressive strength and the Young's modulus of the composite material up to 77% and 155%, respectively [8,34]. These studies show that it is possible to conclude that CNT presence in a soil matrix have an effect on the physical structure (reducing the interparticle spacing and nanoreinforcing the soil), and on the chemical reaction development when a cementitious material is added, allowing the construction of a stronger and stiffer soil skeleton matrix, therefore improving the geomechanical behavior of the composite material [34].

Despite the research to date, the impact of CNTs in chemically stabilized soil matrixes has not been properly studied. Furthermore, the fundamental parameters and their effect on the geomechanical behavior of the composite material have not been clearly identified and quantified. Thus, the present work aims to identify and evaluate the impact of some of the most important parameters on the geomechanical behavior of a chemically stabilized soil that contains carbon nanotubes. The surfactant type and concentration parameters, time and CNT concentration are studied in this work, aiming to quantify their importance on the geomechanical behavior of the composite material. These parameters and their effects are the focus and main novelty of this work, advancing the existing knowledge of the composite materials.

2. Methodology

2.1. Testing Plan

In order to determine the key-parameters on the geomechanical behavior of a soil chemically stabilized and containing carbon nanotubes, the following experimental testing plan was designed: (i) Unconfined compressive strength (UCS) tests were performed on samples stabilized with four different surfactant types, varying in charge (nonionic and cationic), molecular weight and concentration (ranging from 0.1 to 3%). (ii) UCS tests were performed on samples stabilized with a CNT concentration of 0.001% and 0.01%. (iii) UCS tests were made on stabilized samples with different curing times (7 and 28 days). Table 1 summarizes the testing plan. The experimental work was complemented with particle size distribution tests to characterize the quality of CNT dispersions and with leaching tests to assure that CNTs are not released from the chemically stabilized soil matrix.

Table 1. Experimental testing plan.

Surfactant		CNTs	Time
Name	**Conc. (%)**	**Conc. (%)**	**(Days)**
-	-	-	7/28
		0.001	
		0.01	
Viscocrete	3	-	7/28
		0.001	
		0.01	
Glycerox	0.5/1/2	-	7
		0.001	
		0.01	
Amber 2001	0.1 */0.5/1 */2	-	7 *
		0.001	
		0.01	
Amber 4001	0.5/1/2/3	-	7
		0.001	
		0.01	

conc. = concentration; * there is a test for a time of 28 days but without CNTs.

2.2. Materials

A soil collected in central Portugal, near the city of Coimbra, was used in the experimental study. The soil is mainly composed of silt (~66%) with some clay (8–12%) and sand (17–22%) particles, having in its composition a high organic matter content (9.3%), which is mainly responsible for the plasticity characteristics of the soil (liquid limit $w_L \approx 71\%$ and plastic limit $w_P \approx 43\%$). The natural soil exhibits a high water content (80.9%), high void ratio (2.1) and low unit weight (14.6 kN/m^3), and is classified by the Unified Soil Classification System [35] as OH, organic silt with high plasticity [36–38]. These characteristics give the soil a poor geomechanical behavior (low strength and high deformability); thus, a ground improvement technique should be adopted to allow for any construction on it [39]. In the present work, chemical stabilization was selected to improve the soil properties by mixing the soil with a binder and a suspension of "properly" dispersed carbon nanotubes. Although the soil is slightly acid (pH ≈ 4.5–5.3), which may restrain some binder reactions, it exhibits high silica (~62%) and alumina (~16%) content, allowing a long-term strength improvement [7,8].

The soil was chemically stabilized with Portland cement type I 42.5 R, applied in a quantity of 175 kg/m^3 (kilos per cubic meter of soil). Table 2 presents the main characteristic of the cement particles. The high specific surface of the cement particles and the fact that they are slightly negatively charged should be highlighted.

Table 2. Characteristics of cement particles (Portland cement type I 42.5 R).

CaO (%)	SiO_2 (%)	Al_2O_3 (%)	Fe_2O_3 (%)	MgO (%)	SO_3 (%)	Cl^- (%)	S (m^2/kg)	Z (mV)
62.8	19.2	4.9	3.2	2.5	3.4	0.01	349	−2.14

S = specific surface; Z = zeta potential (evaluated by electrophoretic light scattering).

Carbon nanotubes were selected as an additive for the chemical stabilization of the soil, aiming to improve the geomechanical behavior of the composite material while promoting a reduction of the quantity of Portland cement added. Multiwall carbon nanotubes (MWCNTs) were chosen for the present work mainly due to economic factors; MWCNTs are significantly less expensive than singlewall carbon nanotubes and, so far, only MWCNTs are produced at an industrial level. According to data provided by the manufacturer, MWCNTs CN7000 have a mean diameter of 9.5 nm, a mean length of 1500 nm, a mean specific surface of 275,000 m^2/kg (1000 times greater than cement particles) and are composed of 90% pure carbon with some metal oxides (10%). MWCNT characterization was complemented with the evaluation of its density (1.7 g/cm^3) and charge (−25.2 mV) [8,21]. MWCNTs were applied in two small concentrations (0.001 and 0.01% w/w referred to the dry binder mass) to keep costs under control.

For the present work, four different surfactant types were selected: two commercial ones (Viscocrete and Glycerox supplied by Sika and Lubrizol, respectively) and two other noncommercial types (Amber 2001 and Amber 4001 developed by Aquatech), varying in charge (nonionic and cationic), molecular weight and concentration (ranging from 0.1 to 3%), as presented in Table 3. The choice of surfactants was determined by the charge of the additive (−25.2 and −2.14 mV for MWCNTs and cement particles, respectively), so cationic or nonionic surfactant types were selected, differing in molecular weight and size. The surfactants were added with the aim to disperse "properly" the MWCNTs particles, avoiding the loss of their beneficial properties. It is important to notice that the surfactants also have the potential to disperse the cement particles, justifying the tests without MWCNTs, included in the experimental testing program (Table 1).

Table 3. Characteristics of surfactants [36–38].

Surfactant	Market Condition	Charge (-)	Z (mV)	D_z (nm)	MW (kDa)
Viscocrete	commercial	nonionic	−2.8	4.65	242
Glycerox	commercial	nonionic	~0	41.93	4265
Amber 2001	noncommercial	cationic	66.7	170.84	1155
Amber 4001	noncommercial	cationic	-	5.65	54

Z = zeta potential (evaluated by electrophoretic light scattering); D_z = hydrodynamic diameter (evaluated by dynamic light scattering); MW = molecular weight (evaluated by static light scattering).

2.3. Sample Preparation and Tests

The samples of the soil chemically stabilized with Portland cement additivated or not with MWCNTs were prepared following the laboratory procedure presented in Figure 1.

After the soil homogenization, a representative sample was collected (phase I). A slurry composed of Portland cement (phase II) plus MWCNTs dispersed with the aid of surfactant (phase II.C) or not (phase II.A) was prepared. The MWCNTs were dispersed with ultrasonic energy, using an ice bath with water flux to control the temperature, applied to an aqueous suspension of MWCNTs (phase II.A) or to an aqueous solution of surfactant with MWCNTs (phase II.C). The ultrasounds were applied using a probe-sonicator (Vibracell 501 from Sonics), during 5 min with a frequency of 20 Hz and power of 500 W. The quality of the MWCNTs dispersion was evaluated by the particle size distribution analysis obtained using DLS (the smaller the particle size, the better the quality of dispersion). As surfactants can also promote the dispersion of binder and/or soil particles, tests with only a surfactant solution, cement and soil (phase II.B) were prepared. The water present in the slurry/aqueous suspension or solution increases the water content from the natural value (80.9%) to 113%. Afterward, the materials were mechanically mixed (phase III) and the paste produced was introduced in PVC molds in six layers (phase IV). Each layer was slightly compressed with a circular plate and vibrated with the help of a hand drill to remove possible air bubbles that may exist in the paste. Two geotextile filters were applied at the bottom and top of the sample. After curing under water for a period of 7 or 28 days (phase V) the samples were demolded and carefully cut to the final height of 76 mm (phase VI). Finally, the sample was placed on the compression load frame (Tristar 5000 from Wykeham Farrance) and the UCS test was performed at a constant strain rate of 0.76 mm/min (in agreement with BSI 1377-7 [40] and ASTM D2166 [41]). During the test, automatic readings were taken from the load cell and the vertical displacement transducer, allowing the definition of the stress–strain curve. All the tests were repeated twice to guarantee the reliability of the results. In order to assure that MWCNTs were not released from the chemically stabilized soil matrices, leaching tests were performed. More details can be found in Casaleiro [36], Figueiredo [37] and Moura [38].

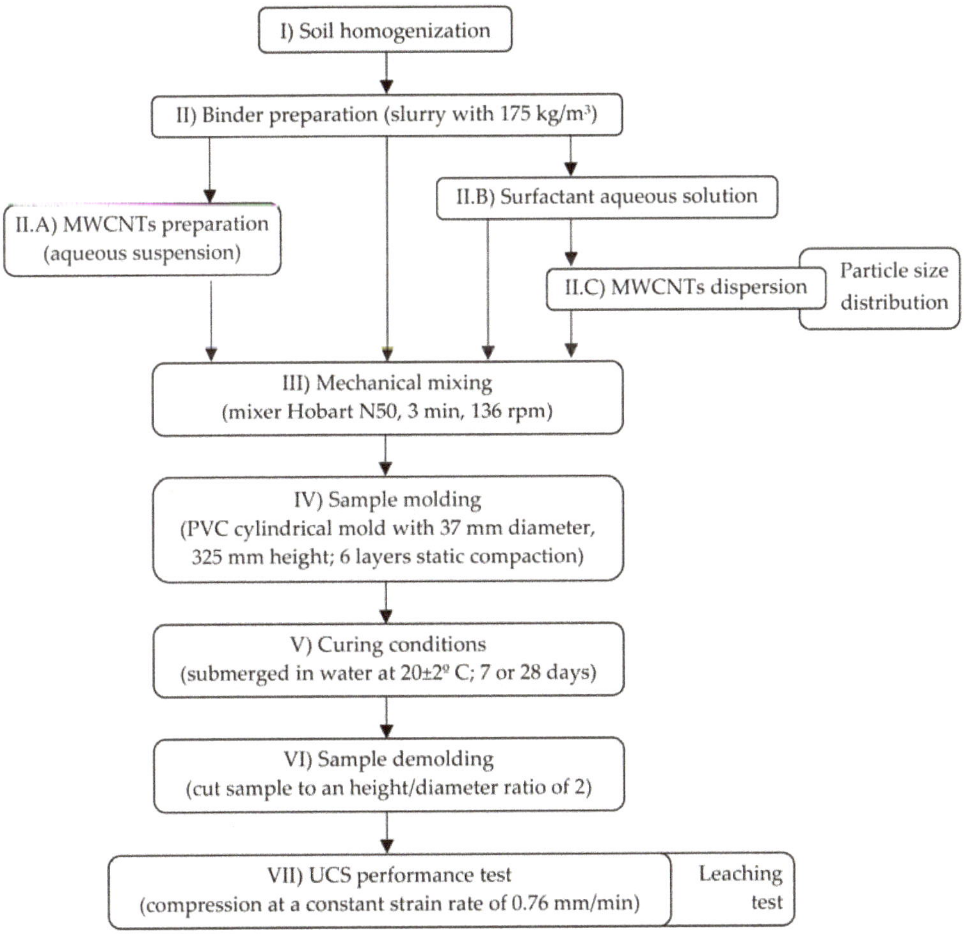

Figure 1. Laboratory procedure for preparation of soil samples chemically stabilized with Portland cement additivated or not with MWCNTs.

3. Results and Discussion

The results present some scattering between the samples tested for each different test condition due to the experimental nature of the study. Nevertheless, only the tests that comply with the conformity criterion (±15% of the variation of the unconfined compressive strength compared to the average value) were accepted.

The impact of the different parameters (surfactants, MWCNTs concentration, time) on the chemically stabilized soil behavior are expressed by the unconfined compressive strength improvement factor (*IF*), defined as the ratio between the unconfined compressive stress (q_u) of a specific test condition and the unconfined compressive strength of the reference test ($q_{u\,max}^{ref}$, for the test condition without surfactants or MWCNTs):

$$IF = \frac{q_u}{q_{u\,max}^{ref}} \tag{1}$$

The results are presented, preferably, as a function of the strength improvement factor, allowing in this way a direct reading of the impact of the parameters under study on the chemically stabilized soil behavior.

Figure 2 presents the stress–strain curves for the reference tests at 7 and 28 curing days. It should be noted that independent of the curing time, 7 and 28 days, both tests, T1 and T2, show very similar stress–strain curves, demonstrating good reproducibility of the laboratory procedure. As expected, the strength and stiffness increase with time as a result of the development of the physicochemical reactions of the Portland cement responsible for producing a stronger stabilized matrix [24,39,42–45].

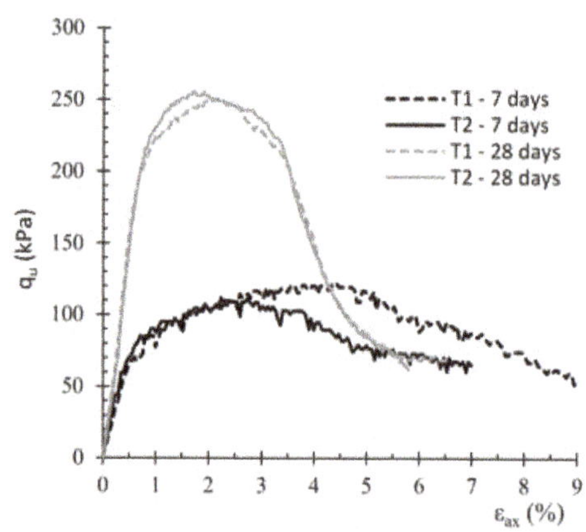

Figure 2. Stress–strain curves for the reference test conditions (soil chemically stabilized with Portland cement).

3.1. Effect of Surfactant Type

Figures 3 and 4 summarize the results of the effect of the surfactant type added to the chemically stabilized soil samples with a curing time of 7 days. With the exception of the higher concentrations of the surfactant Amber 4001, the addition of the surfactants has a positive impact on the geomechanical behavior of the stabilized soil, proving that the surfactant has the potential to better disperse the binder and/or the soil particles, thus, allowing the construction of a denser and stronger solid skeleton matrix. The best result was obtained with the surfactant Viscocrete (*IF* = 1.55), a nonionic surfactant type applied in a concentration of 3%. The results obtained for the other nonionic surfactant (Glycerox) are also positive but with lower strength improvement factors (*IF* ranging from 1.07 to 1.21). However, the IF increases with the Glycerox concentration, suggesting that for a higher Glycerox concentration better results could be obtained. This result may be explained by the fact that Viscocrete presents a smaller hydrodynamic diameter and molecular weight than Glycerox; thus, a smaller size of the surfactant molecules allows better adsorption on the surface of the solid particles (soil and binder), ensuring better dispersion. This is also valid when comparing the two cationic surfactants, justifying the better results with the Amber 4001, which has a smaller hydrodynamic diameter and molecular weight than Amber 2001. Moreover, the fact that the particle surface is not so much covered by surfactant when lower molecular weight and hydrodynamic diameter surfactants are used, may favor the cementitious reactions.

When comparing the two cationic surfactants (Amber 2001 and Amber 4001) with the nonionic surfactants for equal concentrations, it may be seen that for lower concentrations (0.5% and 1%) the best results are obtained with the cationic surfactants. However, for higher concentrations, the results of the Amber 2001 (2% conc.) are of the same order as those for Glycerox while for Amber 4001 (2% and 3% conc.) the results are negative ($IF < 1.0$) and lower than those for Glycerox and Viscocrete, respectively, for a concentration of 2% and 3%. Thus, the surfactant choice is not so dependent on the surfactant charge but rather on the balance between the concentration and the hydrodynamic diameter/molecular weight.

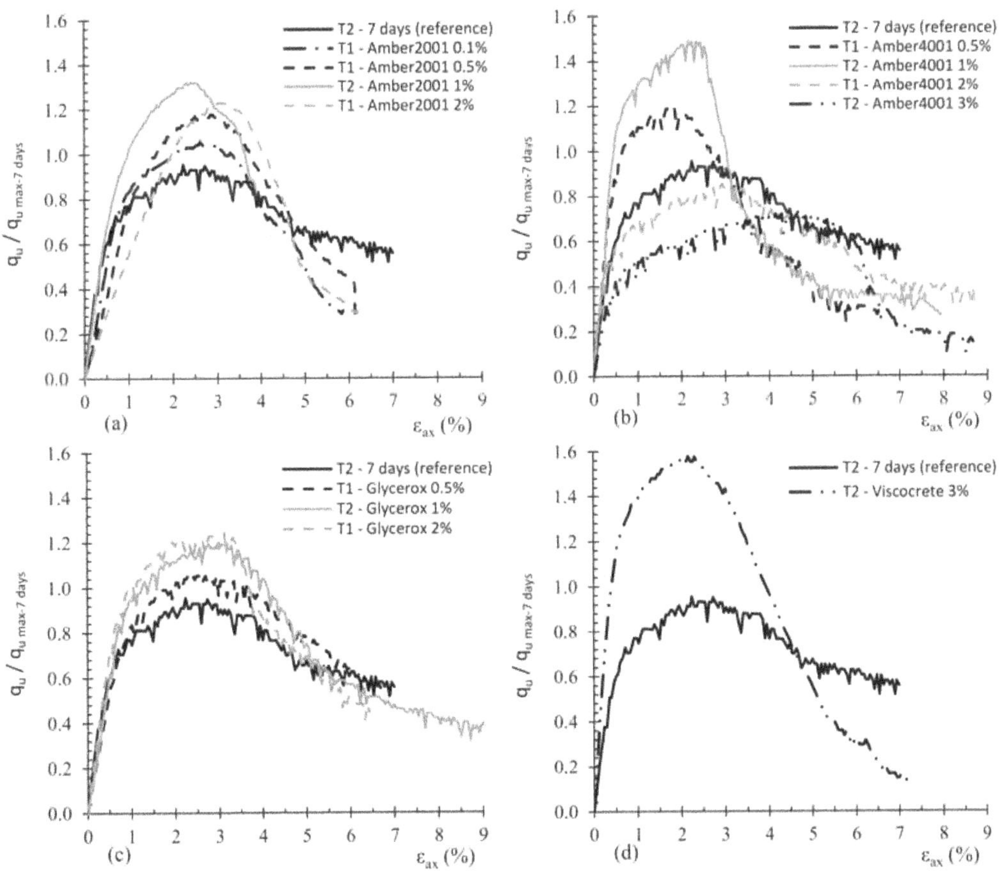

Figure 3. Stress–strain normalized curves (for samples with 7 curing days) for the stabilized soil with only a surfactant solution of (**a**) Amber 2001, (**b**) Amber 4001, (**c**) Glycerox and (**d**) Viscocrete.

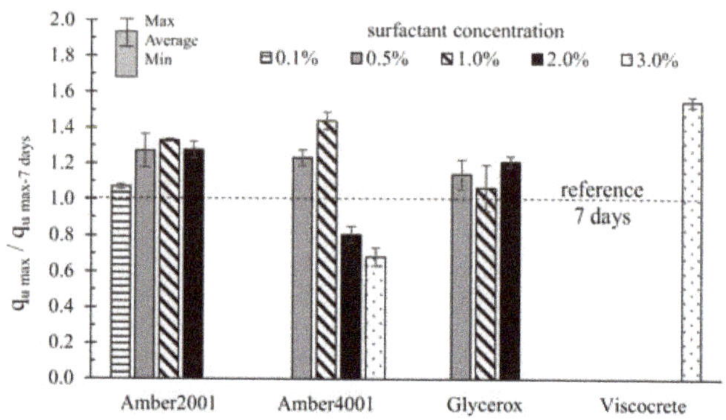

Figure 4. Strength improvement factor (for samples with 7 curing days) for the stabilized soil with only surfactant solutions.

As time evolves, the impact of the surfactant on the geomechanical behavior of the chemically stabilized soil may change since the surfactant presence may have effect on the time development of the physicochemical reactions of the Portland cement. To study this effect, samples were prepared with nonionic and cationic surfactants for the best concentrations, i.e., Viscocrete and Amber 2001 applied in a concentration of 3% and 1%, respectively. From the results (Figure 5), it may be seen that the cationic surfactant now has a negative effect while the nonionic surfactant still presents a positive but smaller effect on the geomechanical behavior of the chemically stabilized soil. Indeed, as the cementitious products are produced over time the solid matrix becomes denser, justifying the decrease in the importance of the surfactant. The decrease of the strength improvement factor is higher for the surfactant with larger molecule size (Amber 2001, IF decreases from 1.33 to 0.81), which may be explained by two factors: (i) surfactants with larger molecules potentially form micelles for lower surfactants concentrations, making the occurrence of cementitious reactions more difficult; (ii) surfactants with larger molecules adsorbed on the surface of solid particles may prevent the establishment of some cementitious bonds, thus promoting a solid matrix with less strength. Nevertheless, more tests should be performed for other surfactants and concentrations in order to deeper understand the surfactant effect over time.

3.2. *Effect of MWCNTs*

Figure 6 presents the results of the chemically stabilized soil samples additivated with MWCNTs dispersed in aqueous solution (without surfactant) for curing times of 7 and 28 days. It is clear that independent of the curing time the addition of the MWCNTs has a negligible effect on the geomechanical behavior of the chemically stabilized soil. This result contradicts previous studies with carbon nanotubes where it was found that the addition of a small concentration of CNTs has a significant impact on the geomechanical properties of a soil [8,32–34]. These results make clear that more important than the introduction of MWCNTs in a soil matrix is the need to ensure that they are properly dispersed in order to avoid the loss of their beneficial properties. Thus, it is crucial to add to the MWCNTs aqueous suspension a surfactant that may help in the dispersion process. Figure 7 shows several aqueous suspensions containing MWCNTs with and without surfactants, all dispersed with ultrasonic energy. The MWCNTs suspension without surfactant (Figure 7a) has aggregates clearly visible to the naked eye, whereas in the case where surfactant was added at the lowest concentration (Figure 7b,c) homogeneous suspensions are present.

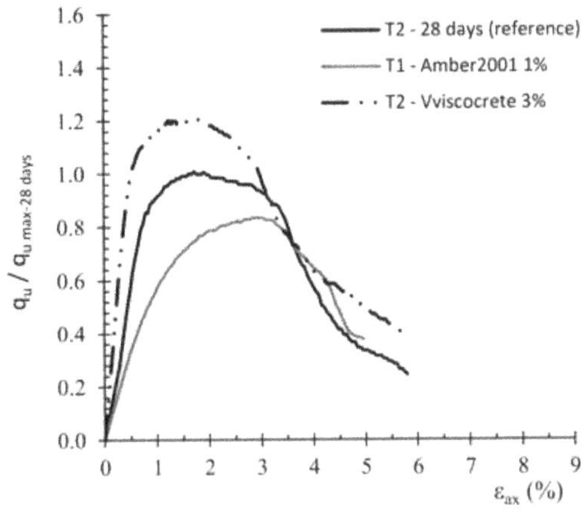

Figure 5. Stress–strain normalized curves (for samples with 28 curing days) for the stabilized soil with only a surfactant solution of Amber 2001 (conc. of 1%) and Viscocrete (conc. of 3%).

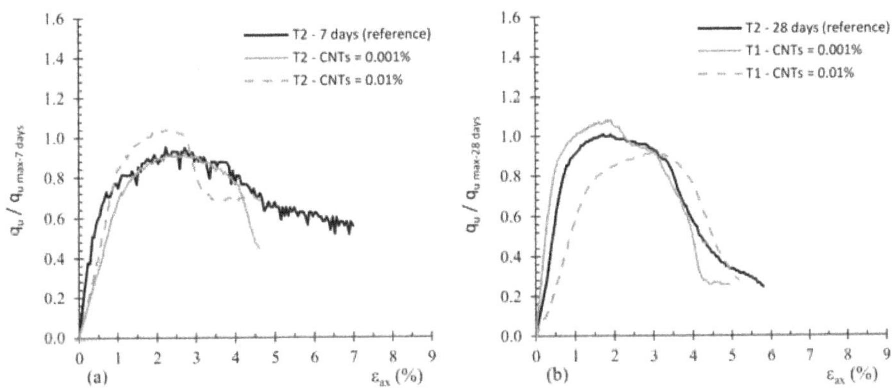

Figure 6. Stress–strain normalized curves for the stabilized soil additivated with MWCNTs and no surfactant for samples with (**a**) 7 curing days and (**b**) 28 curing days.

Figure 7. Photographs of MWCNT suspensions immediately after ultrasonication for a MWCNT concentration of 0.01%: (**a**) sample without surfactant, (**b**) sample with Glycerox at concentration of 0.5% and (**c**) sample with Amber 4001 at concentration of 0.5%.

Figures 8 and 9 summarize the results of the impact of adding MWCNTs (in concentrations of 0.001% and 0.01%) dispersed in a solution of surfactants to chemically stabilized soil samples for a curing time of 7 days. With the exception of the higher concentrations of the surfactant Amber 4001, the addition of MWCNTs for both concentrations (0.001% and 0.01%) dispersed in a solution of surfactants have a positive impact on the geomechanical behavior of the stabilized soil, proving that the addition of MWCNTs to a chemically stabilized soil can only be effective if the MWCNTs are "properly" dispersed.

The best results were obtained with the surfactants Amber 4001, Viscocrete and Glycerox, applied in concentrations of 1%, 3% and 2%, respectively, independent of the MWCNTs concentration (IF = 1.76, 1.62 and 1.63 for a MWCNTs conc. of 0.01%; IF = 1.66, 1.63 and 1.48 for a MWCNTs conc. of 0.001%). The results prove that the size of the surfactant molecules (Table 3) have a major role on the MWCNTs dispersion process (Table 4) and, ultimately, on the geomechanical behavior of the composite materials. Indeed, the smaller the size of the surfactant molecules, the better the quality of the dispersion (Table 4), i.e., smaller surfactant molecules adsorb more easily on the surface of the solid particles (MWCNTs, binder and soil), ensuring better dispersion and geomechanical behavior. These results agree with previous findings regarding the surfactant effect when no MWCNTs were added, but now the presence of MWCNTs enhances even more the strength improvement factor. However, it should be emphasized that better geomechanical behavior is not always associated with better MWCNTs dispersion because the medium where dispersion occurs is different; the characterization of MWCNT dispersion occurs in an aqueous medium enriched with surfactant while for the UCS tests in the medium there are chemically reacting cement particles and soil particles.

It should be noted that as the surfactant concentration increases, there may be formation of micelles when the critical micelle concentration (CMC) is exceeded, an effect that potentially may happen more easily given a larger size of the surfactant molecules. As seen from Table 4, for the surfactants with largest molecule sizes (Glycerox and Amber 2001), the hydrodynamic diameter of the MWCNTs dispersion increases slightly for a Glycerox concentration above 1% (suggesting the CMC is somewhat between 1% and 3%), while for the surfactant Amber 2001 the hydrodynamic diameter always increases with surfactant concentration (suggesting the CMC can be less than 0.5%). However, this is compatible with good MWCNTs dispersion since the formation of micelles is not necessarily detrimental for particle dispersion [8]. Thus, as stated before, the surfactant choice should be based on the balance between the concentration and the hydrodynamic diameter/molecular weight.

The surfactant Amber 2001 exhibits the worst performance since it has the largest hydrodynamic diameter, leading to MWCNTs dispersions of bad quality, as it may be seen from Table 4. The other cationic surfactant (Amber 4001) has an effect on the geomechanical behavior of the stabilized soil that depends on the surfactant concentration. For concentrations up to 1% (probably below the CMC), there is a significant beneficial impact that may be attributed to the cationic charge of the surfactant Amber 4001, which favors adsorption to MWCNTs and binder particles, thus promoting better dispersion and geomechanical behavior. On the contrary, for higher surfactant concentrations (2% and 3%, probably above the CMC) the MWCNT and binder particle dispersion is of poor quality, producing a negative effect in terms of geomechanical behavior of the chemically stabilized soil additivated with MWCNTs. Generally, the Glycerox surfactant presented better results regarding the geomechanical behavior of stabilized soils additivated with MWCNTs, especially for concentrations of 2% or higher. This can be associated with better dispersion of MWCNTs obtained with this surfactant (see Table 4). Indeed, it was observed that for higher surfactant concentrations (2% and 3%) a nonionic surfactant type (Glycerox and Viscocrete) assures better geomechanical behavior as long as it has a good MWCNTs dispersion.

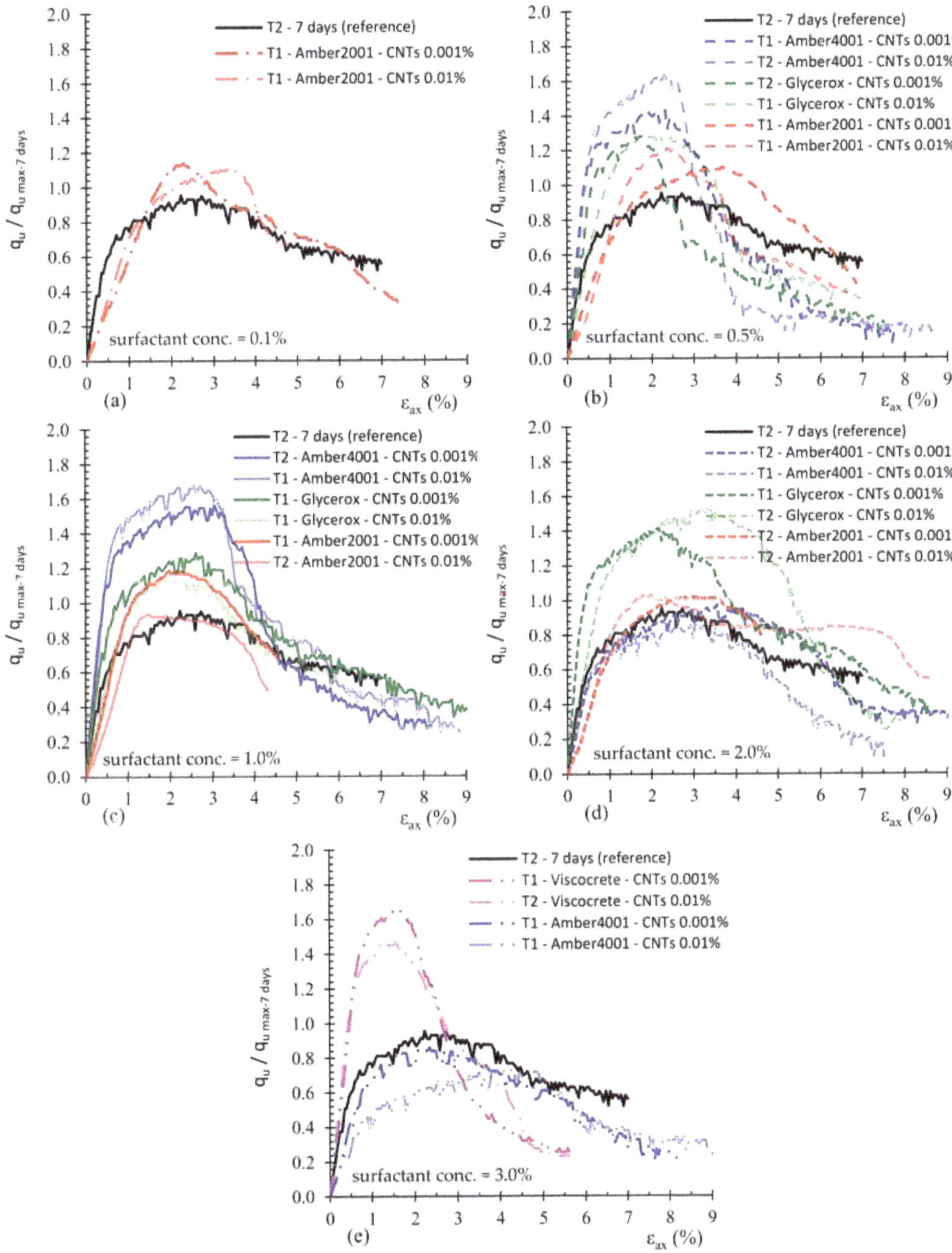

Figure 8. Stress–strain normalized curves (for samples with 7 curing days) for the stabilized soil with MWCNTs suspensions for different surfactants applied in concentrations of (**a**) 0.1%, (**b**) 0.5%, (**c**) 1.0%, (**d**) 2.0% and (**e**) 3.0%.

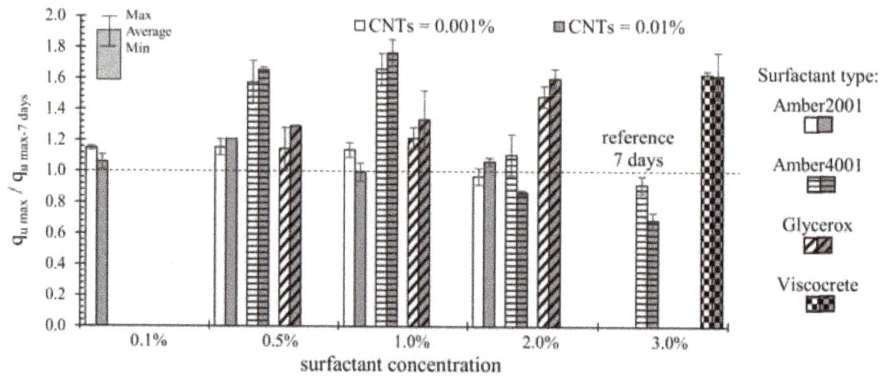

Figure 9. Strength improvement factor (for samples with 7 curing days) for the stabilized soil with MWCNT suspensions for the surfactants studied, applied in different concentrations.

Table 4. Dispersion characterization of MWCNTs suspensions.

Surfactant		D_z (for MWCNTs = 0.001%/0.01%)
Name	**Conc. (%)**	**(nm)**
Viscocrete	3	155.1/119.5
Glycerox	0.5	-/197.2
	1	-/167.6
	3	-/175.2
Amber 2001	0.1	-/-
	0.5	548.0/684.5
	1	-/718.2
	2	-/954.2
Amber 4001	0.5	-/521.5
	1	-/322.9
	3	-/316.8

conc. = concentration; D_z = hydrodynamic diameter (evaluated by dynamic light scattering).

Generally, better results were obtained for higher MWCNT concentrations, which may be explained by the fact that the presence of a higher number of nanoparticles will nanofill the matrix voids, allowing the development of a denser stabilized matrix. At the same time, due to the extraordinary mechanical properties of the MWCNTs (very high strength and stiffness), their presence in a higher quantity in a stabilized matrix will promote the enhancement of the geomechanical behavior since MWCNTs are bonded to the cementitious products produced by the Portland cement. In summary, when the MWCNTs are dispersed in a soil-binder matrix they may act as a nanofiller and nanoreinforcement, promoting a denser and stronger stabilized matrix.

As observed for the effect of surfactant, the impact of the MWCNTs on the geomechanical behavior of the chemically stabilized soil decreases as time evolves for all the surfactants and concentrations studied, independent of the quantity of MWCNTs (Figure 10). As the curing time evolves from 7 to 28 days, the cationic surfactant no longer has a positive effect on the geomechanical behavior, presenting a detrimental effect: for a MWCNTs concentration of 0.001%, the strength improvement factor decreases from 1.15/1.14 to 0.97/0.89 for surfactant concentration of 0.1%/1%, respectively; while for the case of a MWCNTs concentration of 0.01% the IF decreases from 1.06/0.99 to 0.93/0.78 for the same surfactant concentrations. A similar observation can be made for the nonionic surfactant

Viscocrete, where the IF decreases from 1.63 at 7 days to 1.36 at 28 days. Even though there is a decrease, in the case of the surfactant Viscocrete, the effect at 28 days is still positive contrary to what is observed for the Amber 2001, which may be related to the size of surfactant molecules. Indeed, Viscocrete is characterized by molecules of smaller hydrodynamic diameter than Amber 2001 (Table 3); thus, Viscocrete is less likely to form micelles that may hinder the occurrence of cementitious reactions and at the same time the smaller size of the Viscocrete molecules allow the bonding of the cementitious products to the surface of the solid particles (MWCNTs and soil), thus promoting a denser and stronger solid matrix. The relative decrease of the effect of MWCNTs with time is explained by the development over time of the physicochemical reactions of the Portland cement producing a greater quantity of cementitious products. Thus, with time the matrix becomes denser and stronger and, as a consequence, the relative impact of MWCNTs presence decreases.

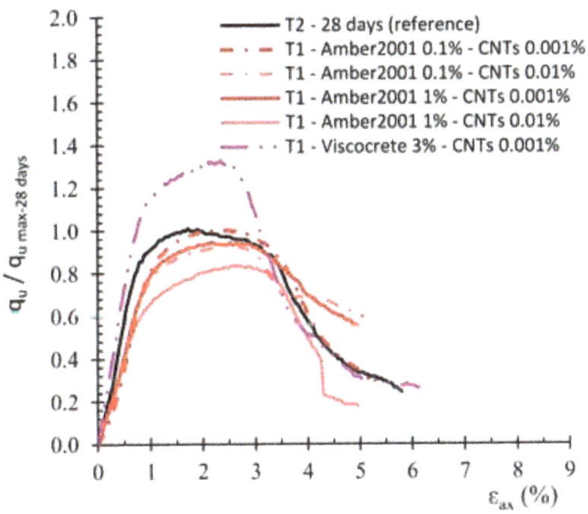

Figure 10. Stress–strain normalized curves (for samples with 28 curing days) for the stabilized soil with MWCNTs suspensions of the surfactant Amber 2001 (conc. of 0.1% and 1%) and Viscocrete (conc. of 3%).

One important issue related with the application of CNTs is their impact on environment if CNTs are released, namely, on human life, animals and plants. In order to investigate a possible release of MWCNTs from the stabilized soil matrices, leaching tests on chemically stabilized soil samples additivated with MWCNTs dispersed in a surfactant solution were done at 1, 4, 7 and 14 curing days. These tests were complemented with optical microscope images and SEM images of the leachate. Tests were performed for the "worst scenario", i.e., adding the highest quantity of MWCNTs (0.01%) dispersed in an Amber 2001 solution (conc. = 0.1%). As seen in Figure 11, two kinds of materials were identified in the SEM images: needle-shaped materials are calcium silicate associated with the Portland cement reactions with water, and materials in the form of irregular polyhedrons are soil particles. By analysis of Figure 11, it can be concluded that there are no traces of MWCNTs in the leachate, proving that the MWCNTs are entrapped in the stabilized soil matrix. In summary, the analyses from the leachate resulting from leaching tests, it is concluded that the amount of material released is not significant and does not contain traces of MWCNTs.

Figure 11. Optical and SEM images of the leachate at 7 days from a chemically stabilized soil samples additivated with MWCNTs (conc. = 0.01%) dispersed in a solution of Amber 2001 (conc. = 0.1%) with 7 curing days: (**a**,**b**) optical images and (**c**,**d**) SEM images.

4. Conclusions

The present work is a contribution to identify the most important or fundamental parameters that control the geomechanical behavior of a chemically stabilized soil additivated with MWCNTs dispersed or not in a surfactant solution. Based on the laboratory tests performed, it was found that a simple addition of MWCNTs to the stabilized soil does not produce any improvement of its geomechanical behavior since the nanoparticles are in an aggregate condition, which inhibits its ability to take advantage of the extraordinary properties of MWCNTs. The introduction of a surfactant to the chemically stabilized soil promotes the dispersion of the cement and soil particles, allowing the development of a denser and stronger solid skeleton matrix, which is reflected in an improvement of the geomechanical behavior up to 155% compared with the reference test. When the MWCNTs are combined with a surfactant solution and good MWCNTs dispersion is achieved, there is an enhancement of the geomechanical behavior of the stabilized soil up to 176% or 185% (for a MWCNTs concentration of 0.001% or 0.01%, respectively) compared with the reference test. Thus, for the concentrations examined, the MWCNTs concentration seems not to be a fundamental parameter since similar improvements can be achieved for a concentration that is 10 times lower. On the contrary, the characteristics of the surfactant seem to be a fundamental parameter affecting the geomechanical behavior of the stabilized soil enriched with MWCNTs. Indeed, the surfactant choice should depend on the balance between the concentration and the hydrodynamic diameter/molecular weight. A smaller size of the surfactant molecules allows better adsorption on the surface of the solid particles (MWCNTs, binder and soil), ensuring better dispersion without interfering with the cementitious reactions.

As time evolves from 7 to 28 curing days, a relative decrease of the effect of surfactants and MWCNTs on the geomechanical behavior of stabilized soil samples was observed, explained by the fact that the physicochemical reactions of the Portland cement develop with time, producing a greater quantity of cementitious products responsible for making the stabilized matrix denser and stronger and, as a consequence, the additives importance diminishes. Thus, time seems to be a fundamental parameter since better results are achieved for shorter times.

In summary, the addition of MWCNTs, "properly" dispersed, with the objective to contribute to the chemical stabilization of a soil, has potential to improve its geomechanical behavior. Thus, for the same level of unconfined compressive strength, MWCNTs applied in a proper concentration and enriched with a specific surfactant may provide a quick and valid alternative to the partial replacement of Portland cement. Moreover, it was demonstrated from the leaching tests that no traces of MWCNTs were observed in the leachate, proving that the MWCNTs are entrapped in the stabilized soil matrices, thereby not presenting a risk for people, animals or plants.

Author Contributions: P.D.F.C., D.T.R.F. and M.S.M.R.M. conducted the experimental investigation; A.A.S.C. prepared the first draft of the manuscript and was responsible for the conceptualization, validation and supervision of the research, and review and editing of the manuscript; M.G.R. was co-responsible for the conceptualization, validation and supervision of the research and participated in the review and editing of the manuscript. All authors have read and agreed to the published version of the manuscript.

Funding: This research received financial support from the Strategic Research Centre (CIEPQPF) Project UIDB00102/2020, funded by FCT (Fundação para a Ciência e Tecnologia), Portugal.

Institutional Review Board Statement: Not applicable.

Informed Consent Statement: Not applicable.

Acknowledgments: The authors would like to express their thanks to CIMPOR and AQUATECH for supplying the binder and the cationic surfactants, respectively.

Conflicts of Interest: The authors declare no conflict of interest.

References

1. Bolt, G.H. Composition and Physical Properties of Soils. In *Developments in Soil Science*; Elsevier: Amsterdam, The Netherlands, 1983; pp. 1–36.
2. Conklin, A.R., Jr. *Introduction to Soil Chemistry: Analysis and Instrumentation*; Wiley: Hoboken, NJ, USA, 2005.
3. Chu, J.; Varaksin, S.; Menge, U.K.P. Construction Processes—State of the Art Report. In Proceedings of the 17th International Conference on Soil Mechanics and Geotechnical Engineering, Alexandria, Egypt, 5–9 October 2009; pp. 3006–3135.
4. Venda Oliveira, P.J.; Correia, A.A.S.; Lopes, T.J.S. Effect of organic matter content and binder quantity on the uniaxial creep behavior of an artificially stabilized soil. *J. Geotech. Geoenviron. Eng.* **2014**, *140*, 04014053. [CrossRef]
5. Venda Oliveira, P.J.; Correia, A.A.S.; Garcia, M.R. Effect of Stress Level and Binder Composition on Secondary Compression of an Artificially Stabilized Soil. *J. Geotech. Geoenviron. Eng.* **2013**, *139*, 810–820. [CrossRef]
6. Mithcell, J.K. Soil Improvement—State-of-the-Art Report. In Proceedings of the 10th International Conference on Soil Mechanics and Foundation Engineering, Stockholm, Sweden, 15–19 June 1981; pp. 509–565.
7. Correia, A.A.S.; Oliveira, P.J.V.; Lemos, L.J.L. Strength assessment of chemically stabilised soft soils. *Proc. Inst. Civil Eng. Geotech. Eng.* **2019**, *172*, 218–227. [CrossRef]
8. Figueiredo, D.T.R.; Correia, A.A.S.; Hunkeler, D.; Rasteiro, M.G.B.V. Surfactants for dispersion of carbon nanotubes applied in soil stabilization. *Colloids Surf. A Physicochem. Eng. Asp.* **2015**, *480*, 405–412. [CrossRef]
9. Lorenzo, G.A.; Bergado, D.T. Fundamental Parameters of Cement-Admixed Clay—New Approach. *J. Geotech. Geoenviron. Eng.* **2004**, *130*, 1042–1050. [CrossRef]
10. Correia, A.A.S.; Oliveira, P.J.V.; Custódio, D.G. Effect of polypropylene fibres on the compressive and tensile strength of a soft soil, artificially stabilised with binders. *Geotext. Geomembr.* **2015**, *43*, 97–106. [CrossRef]
11. Edil, T.; Staab, D. Practitioner's Guide for Deep-Mixed Stabilization of Organic Soils and Peat. Final Report, The National Deep Mixing Research Program, Project Number NDM302. 2005; p. 60.
12. Porbaha, A. State of the art in deep mixing technology: Part I. Basic concepts and overview. *Proc. Inst. Civil Eng. Ground Improv.* **1998**, *2*, 81–92. [CrossRef]
13. Kitazume, M.; Terashi, M. *The Deep Mixing Method*; CRC Press: Leiden, The Netherlands, 2013.

14. Janz, M.; Johansson, S.-E. *The Function of Different Binding Agents in Deep Satbilization*; Swedish Deep Stabilization Research Centre: Linköping, Sweden, 1992; p. 47.
15. Ghasabkolaei, N.; Choobbasti, A.J.; Roshan, N.; Ghasemi, S.E. Geotechnical properties of the soils modified with nanomaterials: A comprehensive review. *Arch. Civ. Mech. Eng.* **2017**, *17*, 639–650. [CrossRef]
16. Huang, Y.; Wang, L. Experimental studies on nanomaterials for soil improvement: A review. *Environ. Earth Sci.* **2016**, *75*, 10. [CrossRef]
17. Venda Oliveira, P.J.; Neves, J.P.G. Effect of Organic Matter Content on Enzymatic Biocementation Process Applied to Coarse-Grained Soils. *J. Mater. Civ. Eng.* **2019**, *31*, 04019121. [CrossRef]
18. DeJong, J.T.; Soga, K.; Kavazanjian, E.; Burns, S.; Paassen, L.A.V.; Qabany, A.A.; Aydilek, A.; Bang, S.S.; Burbank, M.; Caslake, L.F.; et al. Biogeochemical processes and geotechnical applications: Progress, opportunities and challenges. *Géotechnique* **2013**, *63*, 287–301. [CrossRef]
19. Ramdas, V.M.; Mandree, P.; Mgangira, M.; Mukaratirwa, S.; Lalloo, R.; Ramchuran, S. Review of current and future bio-based stabilisation products (enzymatic and polymeric) for road construction materials. *Transp. Geotech.* **2021**, *27*, 100458. [CrossRef]
20. Axelsson, K.; Johansson, S.-E.; Anderson, R. *Stabilization of Organic Soils by Cement and Puzzolanic Reactions—Feasibility Study*; Swedish Deep Stabilization Research Centre: Linköping, Sweden, 2002.
21. Correia, A.A.S.; Casaleiro, P.D.F.; Rasteiro, M.G.B.V. Applying multiwall carbon nanotubes for soil stabilization. *Procedia Eng.* **2015**, *102*, 1766–1775. [CrossRef]
22. Licht, S. Co-production of cement and carbon nanotubes with a carbon negative footprint. *J. CO2 Util.* **2017**, *18*, 378–389. [CrossRef]
23. Manzur, T.; Yazdani, N. Optimum mix ratio for carbon nanotubes in cement mortar. *KSCE J. Civ. Eng.* **2015**, *19*, 1405–1412. [CrossRef]
24. Taylor, H. *Cement Chemistry*; Thomas Telford: London, UK, 1997.
25. Makar, J. The Effect of SWCNT and Other Nanomaterials on Cement Hydration and Reinforcement. In *Nanotechnology in Civil Infrastructure*; Gopalakrishnan, K., Birgisson, B., Taylor, P., Attoh-Okine, N.O., Eds.; Springer: Berlin/Heidelberg, Germany, 2011.
26. Luo, J.; Duan, Z.; Xian, G.; Li, Q.; Zhao, T. Damping Performances of Carbon Nanotube Reinforced Cement Composite. *Mech. Adv. Mater. Struct.* **2015**, *22*, 224–232. [CrossRef]
27. Šmilauer, V.; Hlavácek, P.; Padevet, P. Micromechanical Analysis of Cement Paste with Carbon Nanotubes. *Acta Polytech.* **2012**, *52*, 22–28. [CrossRef]
28. Wang, B.; Pang, B. Properties improvement of multiwall carbon nanotubes-reinforced cement-based composites. *J. Compos. Mater.* **2019**, *54*, 2379–2387. [CrossRef]
29. Sikora, P.; Elrahman, M.A.; Chung, S.Y.; Cendrowski, K.; Mijowska, E.; Stephan, D. Mechanical and microstructural properties of cement pastes containing carbon nanotubes and carbon nanotube-silica core-shell structures, exposed to elevated temperature. *Cem. Concr. Compos.* **2019**, *95*, 193–204. [CrossRef]
30. Zhou, C.; Li, F.; Hu, J.; Ren, M.; Wei, J.; Yu, Q. Enhanced mechanical properties of cement paste by hybrid graphene oxide/carbon nanotubes. *Constr. Build. Mater.* **2017**, *134*, 336–345. [CrossRef]
31. Taha, M.R.; Alsharef, J.M.A. Performance of Soil Stabilized with Carbon Nanomaterials. *Chem. Eng. Trans.* **2018**, *63*, 757–762.
32. Taha, M.R.; Ying, T. Effects of Carbon Nanotube on Kaolinite: Basic Geotechnical Behavior. In Proceedings of the ICCE 18—18th International Conference on Composites/Nano-engineering, Anchorage, AK, USA, 4–10 July 2010.
33. Arabani, M.; Haghi, A.; Moradi, Y. Evaluation of Mechanical Properties Improvement of Clayey Sand by Using Carbon Nanotubes. 2012.
34. Correia, A.A.S.; Rasteiro, M.G. Nanotechnology Applied to Chemical Soil Stabilization. *Procedia Eng.* **2016**, *143*, 1252–1259. [CrossRef]
35. ASTM. *ASTM D2487, Standard Practice for Classification of Soils for Engineering Purposes (Unified Soil Classification System)*; ASTM International: West Conshohocken, PA, USA, 1998.
36. Casaleiro, P.D.F. Chemical Stabilization of the Soft Soil of Baixo Mondego by Nanomaterials. Master's Thesis, University of Coimbra, Coimbra, Portugal, 2014.
37. Figueiredo, D.T.R. Characterization of Carbon Nano-Tubes Dispersions for Application in Soil Stabilization. Master's Thesis, University of Coimbra, Coimbra, Portugal, 2014.
38. Moura, M.S.M.R. Improvement of Carbon Nanotube Dispersions for Application in Soil Chemical Stabilization. Master's Thesis, University of Coimbra, Coimbra, Portugal, 2015.
39. Correia, A.A.S. Applicability of Deep Mixing Technique to the Soft Soil of Baixo Mondego. Ph.D. Thesis, University of Coimbra, Coimbra, Portugal, 2011.
40. BSI. *BSI 1377-7: Methods of Test for Soils for Civil Engineering Purposes. Shear Strength Tests (Total Stress)*; British Standards Institution: London, UK, 1990.
41. ASTM. *ASTM D2166: Standard Test Method for Unconfined Compressive Strength of Cohesive Soil*; ASTM International: West Conshohocken, PA, USA, 2000.
42. Venda Oliveira, P.J.; Correia, A.A.S.; Garcia, M.R. Effect of Organic Matter Content and Curing Conditions on the Creep Behavior of an Artificially Stabilized Soil. *J. Mater. Civ. Eng.* **2012**, *24*, 868–875. [CrossRef]

43. Horpibulsuk, S. Analysis and Assessment of Engineering Behavior of Cement Stabilized Clays. Ph.D. Thesis, Saga University, Saga, Japan, 2001.
44. Correia, A.A.S.; Lopes, L.; Reis, M.S. Advanced predictive modelling applied to the chemical stabilisation of soft soils (Ahead of print). *Proc. Inst. Civil Eng. Geotech. Eng.* **2021**, 1–11. [CrossRef]
45. Venda Oliveira, P.J.; Correia, A.A.S.; Lemos, L.J.L. Numerical modelling of the effect of curing time on the creep behaviour of a chemically stabilised soft soil. *Comput. Geotech.* **2017**, *91*, 117–130. [CrossRef]

applied sciences

MDPI

Article

Recent Development and Future Perspectives of Quality Control and Assurance for the Deep Mixing Method

Masaki Kitazume

Department of Civil and Environmental Engineering, Tokyo Institute of Technology, Tokyo 152-8550, Japan; kitazume@cv.titech.ac.jp

Abstract: The deep mixing method (DMM), an in situ soil stabilization technique, was developed in Japan and Nordic countries in the 1970s and has gained increased popularity in many countries. The quality of stabilized soil depends upon many factors, including its type and condition, the type and amount of binder, and the production process. Quality control and quality assurance (QC/QA) practices focus on stabilized soil, and comprises laboratory mix tests, field trial tests, monitoring and controlling construction parameters, and verification. QC/QA is one of the major concerns for clients and engineers who have less experience with the relevant technologies. In this manuscript, the importance of QC/QA-related activities along the workflow of deep mixing projects is emphasized based on the Japanese experience/results with mechanical mixing technology by vertical shaft mixing tools with horizontal rotating circular mixing blade. The current and recent developments of QC/QA are also presented.

Keywords: quality control; quality assurance; deep mixing method

Citation: Kitazume, M. Recent Development and Future Perspectives of Quality Control and Assurance for the Deep Mixing Method. *Appl. Sci.* **2021**, *11*, 9155. https://doi.org/10.3390/app11199155

Academic Editors: António Alberto Santos Correia and Paulo José da Venda Oliveira

Received: 30 July 2021
Accepted: 25 September 2021
Published: 1 October 2021

Publisher's Note: MDPI stays neutral with regard to jurisdictional claims in published maps and institutional affiliations.

Copyright: © 2021 by the author. Licensee MDPI, Basel, Switzerland. This article is an open access article distributed under the terms and conditions of the Creative Commons Attribution (CC BY) license (https://creativecommons.org/licenses/by/4.0/).

1. Introduction

The deep mixing method, an in situ soil admixture stabilization technique using cement and/or lime, was developed and put into practice in Japan and Nordic countries in the 1970s. Due to its wide applicability and high improvement effect, the method spread worldwide. More than two decades of practice have seen the equipment improved, stabilizing agents changed, and the applications diversified [1]. The quality of stabilized soil depends upon the type and condition of the soil, the type and amount of binder, and the production process. Quality control and quality assurance (QC/QA) practices have been accepted worldwide for more than five decades [2]. They are comprised of laboratory mix tests, field trial tests, monitoring and control of construction parameters during production, and verification by the engineering characteristics of stabilized soil.

The purpose of deep mixing is not only to produce good-quality stabilized soil, but also to construct improved ground that guarantees the performance of the superstructure. The ground improved by the method is a composite system comprising stabilized soil columns and original soils [1]. Construction control parameters during production include the penetration and withdrawal speeds of the mixing shaft, rotation speed of the mixing blade, flow rate of the binder, and water/binder ratio (for the wet method). However, the field-stabilized soil strength has relatively large variability, even if the production adheres to the established mixing machine and procedure. One of the reasons behind the strength variability is the type of soil. Some cohesive soils are so "sticky" that the soil and binder mixture adheres to and rotates with the mixing blade without efficient mixing (entrained rotation phenomenon). Several approaches can be adopted for such soils: new type special cement, injecting chemical additives, and injecting air.

In this manuscript, the recent development and future perspective of the QC/QA procedures are briefly discussed based on the Japanese experience/results with mechanical mixing technology by vertical shaft mixing tools with horizontal rotating circular mixing blade, as well as the countermeasures for problematic soils.

2. Deep Mixing Method

2.1. Classification of Techniques

There are a wide variety of technologies classified into the deep mixing methods: the mixing by vertical shaft mixing tools with horizontal rotating circular mixing blade, and the mixing by horizontal shaft mixing tools with vertical rotating circular mixing blade. The quality control and assurance is different in each technology. The former technology is focused on in this manuscript. The Japanese deep mixing method can be classified into mechanical mixing, high-pressure injection, and the combination of mechanical and high-pressure injection, as shown in Figure 1. In mechanical mixing, a binder is injected into the original ground and mixed by the mixing blades. The binder is used either with a slurry [3] or dry form [4]. In the high-pressure injection technique, the original ground is disturbed and softened by high-pressure injection of water and/or air at a pressure of 10 to 60 MN/m^2, while the binder is injected and mixed with soil. The third technique exploits the features of both mechanical mixing and high-pressure injection [5]. In this manuscript, the quality control and assurance of the mechanical mixing technique is introduced.

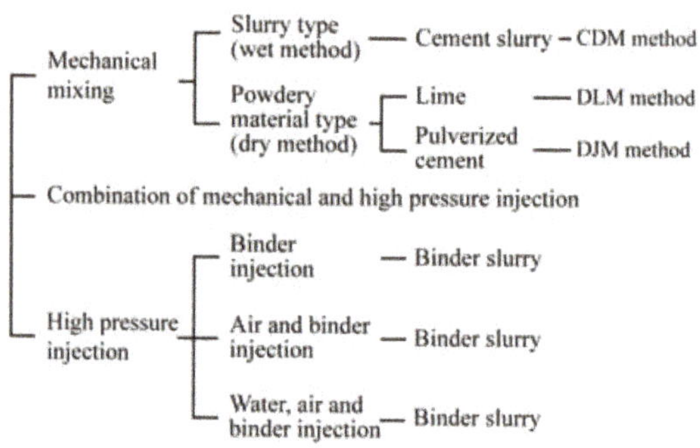

Figure 1. Classification of the Deep Mixing Method.

2.2. Production Procedure

There are two construction procedures which primarily depend on the injection sequence of binder (Figure 2): (a) injecting binder during the penetration of a mixing tool (penetration injection method) and (b) injecting binder during the withdrawal of a mixing tool (withdrawal injection method). Many Japanese deep mixing equipment install the two injection nozzles for the two methods. The injection nozzle, installed close to the lowermost mixing blades, is used for the penetration injection, but one installed close to the uppermost mixing blade is used for the withdrawal injection. In the case of the penetration injection, the mixing blades cut and disturb the soft soil, and the binder was injected at a constant flow rate during the penetration of the mixing tools into the ground. In the withdrawal stage, the mixing blades rotate reversibly, and mix the soft soil and binder again. In the case of the withdrawal injection, the mixing blades cut and disturb the soft soil during the penetration of the mixing tools into the ground, and the binder was injected at a constant flow rate and mixed the soft soil in the withdrawal stage.

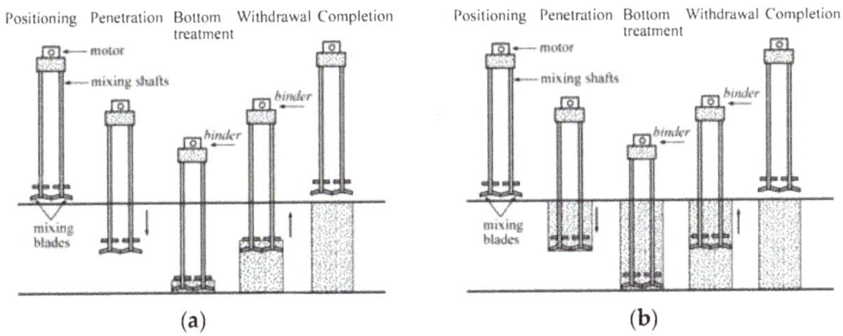

Figure 2. Production of the DM-stabilized soil column. (**a**) Penetration injection method; (**b**) withdrawal injection method.

3. Laboratory, Field, and Design Strengths

As the field mixing and curing conditions are quite different from the laboratory mix test, the field-stabilized soil strength is usually different from the laboratory-stabilized soil strength. The field-stabilized soil has a relatively large strength variability. The average strength, $\overline{q_{ul}}$, and deviation of the laboratory stabilized soil and the average strength, $\overline{q_{uf}}$, and deviation of field stabilized soil are schematically shown in Figure 3. The design standard strength, q_{uck}, is derived from $\overline{q_{uf}}$ by incorporating the strength deviation as Equations (1) and (2). The magnitude of the parameter K can be determined by the probability, which is defined as the frequency of the strength lower than the q_{uck}, as a portion drawn with a hatch in Figure 3.

$$q_{uck} \leq \overline{q_{uf}} - K \cdot \sigma \tag{1}$$

$$\overline{q_{uf}} = \lambda \cdot \overline{q_{ul}} \tag{2}$$

where K = coefficient, q_{uck} = design standard strength (kPa), $\overline{q_{uf}}$ = average unconfined compressive strength of field stabilized soil (kPa), $\overline{q_{ul}}$ = average unconfined compressive strength of laboratory stabilized soil (kPa), σ = standard deviation of field strength (kPa), and λ = ratio of $\overline{q_{uf}}/\overline{q_{ul}}$.

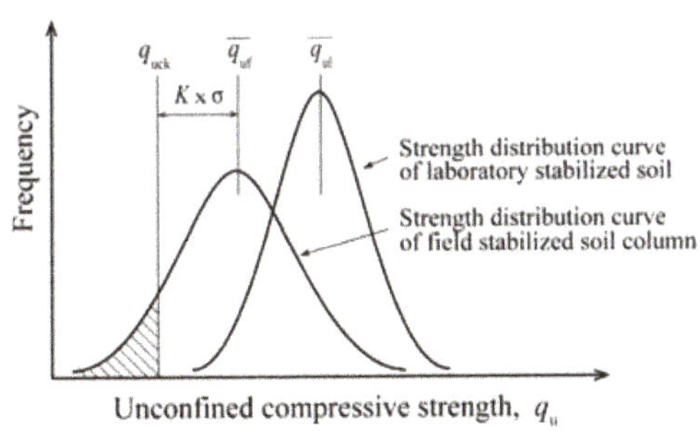

Figure 3. Field and laboratory strengths of stabilized soils.

4. Flow of the Deep Mixing Project and QC/QA

4.1. Dep Mixing Project Work Flow

Quality assurance to fulfill the requirements of geotechnical design cannot only be achieved by process control during production, but should involve relevant activities that are carried out prior to, during, and after construction.

QC/QA is one of the major concerns for clients and engineers who have less experience with the deep mixing technique. The procedure of the geotechnical design is different depending on the application, and the construction control parameters and values are also different for different construction equipment. However, the overall workflow shown in Figure 4 (after [6]) is common to deep mixing projects. In the figure, the project owner's functions are shown in white frame, activities related to the geotechnical design are in slight gray frame, activities related to the process design and actual production with QC are in light gray frame, and accumulated experience and database on locally available production processes is shown in gray frame.

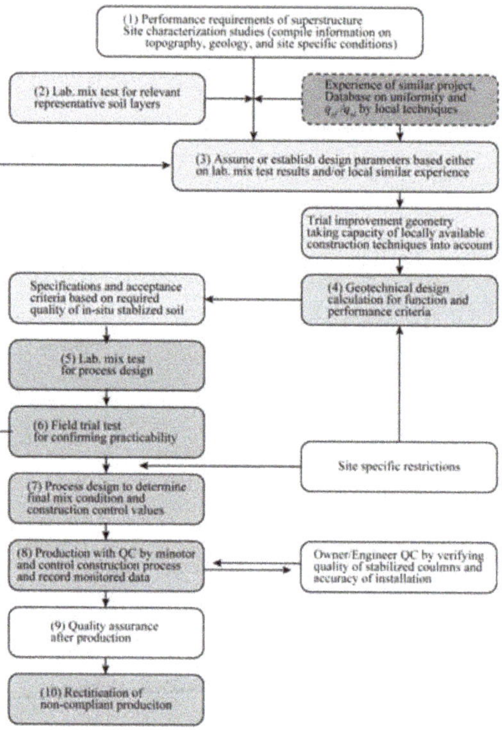

Figure 4. Flow of the deep mixing project and QC/QA (after [6]).

4.2. *QC/QA for Stabilized Soil-Current Practice*

To ensure sufficient quality of the stabilized soil column, quality control and quality assurance are required prior to, during, and after production. For this purpose, QC consists of (i) laboratory mix tests, (ii) field trial tests, (iii) quality control during construction, and (iv) quality assurance after construction through laboratory tests on core samples.

4.2.1. Laboratory Mix Test

Laboratory mix tests should be conducted on soil samples retrieved from all soil layers to be stabilized to determine a suitable type and quantity of binder to ensure the design strength.

4.2.2. Field Trial Test

Field trial tests are an important preproduction QA, especially when no comparable experience is available. It is recommended to conduct field trial tests in advance in or adjacent to the construction site to confirm the actual field strength and uniformity and determine the construction control parameters and final mix design for production. The change in electric or hydraulic power consumption, change in torque, and/or change in penetration speed are measured to establish the construction control criteria for the bottom treatment.

4.2.3. Quality Control during Production

Stabilized soil columns must be produced to satisfy the specified geometric layout and quality of stabilized soil. The rig operator should control, monitor, and record the geometric layout of each column.

As shown in Figure 5, the operation monitoring covers the quality and quantity control monitoring. The gauges and meters normally used are marked 1 to 8 in Figure 5.

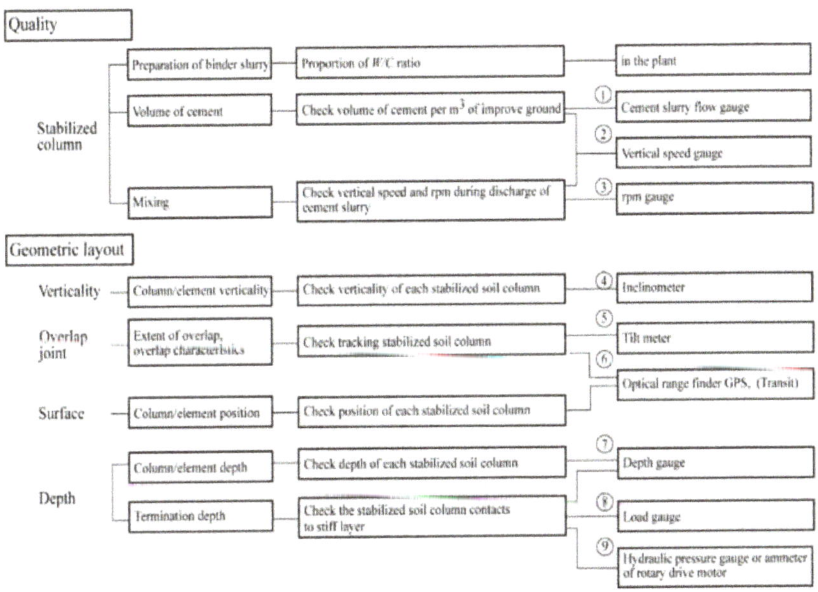

Figure 5. Operation monitoring for the CDM method on land works (after [3]).

Quality control also includes binder storage, binder slurry preparation, and control of the mixing process. Storage and proportioning of binder, chemical additives, and mixing water are normally controlled, monitored, and recorded at the plant. Construction control parameters during column production include the penetration and withdrawal speeds, rotation speed of the mixing blade, flow rate of the binder, and water/binder ratio (for the wet method). The construction control values are predetermined by the process design considering the laboratory mix test result, field trial test, and contractors' experience. During column production, construction control values are controlled, monitored, and displayed in the control room for the plant and rig operators. The mixing shaft and blade should be frequently observed for any possible defects during production.

The mixing degree mostly depends on the rotation speed and the penetration and withdrawal speeds of the mixing blade. An index named "blade rotation number", T, has been introduced to evaluate the mixing degree, which means the total number of mixing blade passes during 1 m of mixing shaft movement, and is defined by Equations (3) and (4) for penetration injection and withdrawal injection. According to accumulated research, the "blade rotation number" should be approximately 270 or larger to ensure sufficient mixing for Japanese wet and dry methods [3,7,8].

$$T = \sum M \cdot \left(\frac{N_d}{V_d} + \frac{N_u}{V_u} \right) \text{ for penetration injection} \quad (3)$$

$$T = \sum M \cdot \left(\frac{N_u}{V_u} \right) \text{ for withdrawal injection} \tag{4}$$

where N_d = number of rotations of the mixing blade during penetration (N/min), N_u = number of rotations of the mixing blade during withdrawal (N/min), T = blade rotation number (N/m), V_d = penetration speed of the mixing blade (m/min), V_u = withdrawal speed of the mixing blade (m/min), and ΣM = total number of mixing blades.

4.2.4. Quality Verification

After the improvement, the quality of field-stabilized soil columns should be verified in advance of the construction of superstructures to confirm the design quality, such as continuity, uniformity, strength, permeability, or dimension. Core samples were taken throughout the depth to verify the uniformity and continuity of the stabilized soil column (Figure 6). The engineering properties of the stabilized soil are evaluated by the unconfined compressive strength of the core samples.

Figure 6. Core boring machine.

The reliability and accuracy of the unconfined compressive strength of the core sample depends upon the quality of the core sample, and the quality of the sample depends upon the drilling and coring method and drillers' skill. The Denison-type sampler, double tube core sampler, or triple tube core sampler can be used for stabilized soil columns whose unconfined compressive strength ranges from 100 to 6000 kPa. It is advisable to use samplers of a relatively large diameter to take high-quality samples. The quality of the retrieved core is evaluated by visual observation and the rock quality designation (*RQD*) iindex. The *RQD* index is defined by Equation (5). The *RQD* index measures the percentage of "good rock" within a borehole and provides the rock quality. The *RQD* index was originally applied for evaluating the quality of rock sample, as shown in Table 1.

$$RQD = \frac{\sum \text{length of core pieces} > 10 \text{ cm}}{\text{Total core run length}} \cdot 100\ (\%) \tag{5}$$

Table 1. *RQD* index and rock quality (Deere, 1989).

RQD	Description of Rock Quality
0–25%	Very poor
25–50%	Poor
50–75%	Fair
75–90%	Good
90–100%	Excellent

A variety of verification test procedures to examine the engineering characteristics of stabilized soil have been proposed [9–11]. However, actual practices rely upon the unconfined compression test on core samples. Most of the other procedures seem to be used only for research purposes or settling noncompliance.

5. Problems Associated with Difficult Soil Conditions

5.1. Stabilized Soil Strength Variability

The field-stabilized soil columns have relatively large strength variability. One of the reasons behind the strength variability is the type of soil. Some cohesive soils are so sticky that the soil and binder mixture adheres to and rotates with the mixing blade without an efficient mixing "entrained mixing phenomenon", a condition in which disturbed soil adheres to and rotates with the mixing blades, as shown in Figure 7.

Figure 7. Example of "entrained mixing phenomenon".

When encountering stiff, cohesive, and sticky soil, water is often injected from the bottom outlet of the mixing shaft to increase the fluidity of the original soil. Water injection may be effective in increasing the fluidity of the original soil, but its effect is limited for some types of soil. However, water injection causes a decrease in the long-term stabilized soil strength and increases the ground heaving and horizontal movement during column installation. Several alternative approaches will be introduced in the following section: using a new type of special cement, and injecting chemical additives and air.

5.2. Using the New Type of Special Cement

Aoyama et al. [12] produced two special cements to stabilize cohesive soils that suppress short-term strength increases while guaranteeing long-term strength: Cement-A prevents electrostatic repulsion by negative charges to maintain fluidity, and Cement-B prevents electrostatic repulsion and polyvalent metal ions. Cement-A and Cement-B are mixed with soils T and V, respectively. The soil T (fines content less than 5 μm, *Fc* of 49%, and fines content less than 75 μm, *Fs* of 66%) contains a large amount of clay minerals and has a large cation exchange capacity, and the soil V (*Fc* of 53% and *Fs* of 99%) contains a large amount of polyvalent metal ions.

Figure 8a shows the shear strength of fresh stabilized soil measured by a hand vane apparatus. The strength of soil stabilized with ordinary Portland cement increases with time soon after mixing, which corresponds to the decrease in the fluidity of the soil and binder mixture. The stabilized soil with Cement-A shows almost no strength gains, even 60 min after mixing, which indicates that the fluidity of the soil and binder mixture is kept high. A similar phenomenon can be seen in Figure 8b, where the stabilized soil with ordinary Portland cement increases in strength, but one with Cement-B shows a slight increase in strength. It should be noted in the figure that soil V stabilized with Cement-A also shows a large strength gain, while slight strength increases were observed in the case of soil T. These results indicate that the strength increase in fresh stabilized soil can be suppressed for about one hour by adding chemical substances appropriate for the soil type.

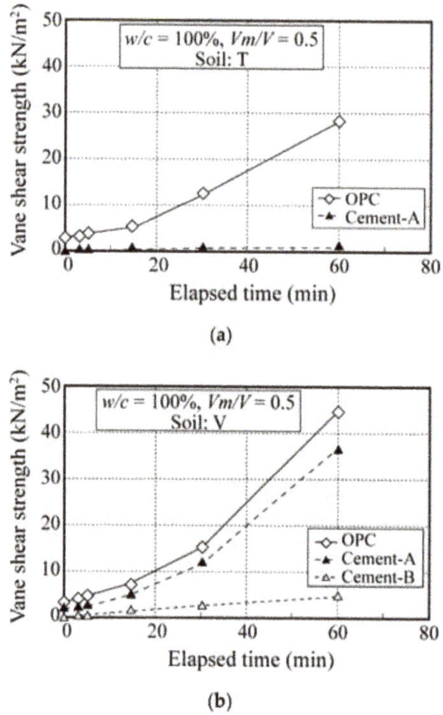

Figure 8. Increase in strength of the soil and cement mixture [12]. (**a**) Fresh stabilized soil, T; (**b**) fresh stabilized soil, V.

5.3. Adding Dispersant

Mizutani et al. [13] and Hirano et al. [14] showed the effect of two dispersants on the fluidity of soil and cement mixtures, where three types of soil, (1) artificial soil composed of kaolin clay (70%) and silica sand (30%), (2) artificial soil composed of kaolin clay (40%) and silica sand (60%), and (3) volcanic soil, were stabilized with cement-based special binder [15] mixed with dispersant. The shear strength of the fresh soil and cement mixture measured by a hand vane apparatus is shown in Figure 9. In the case of the glycolic acid dispersant (Figure 9a), the shear strength of the fresh soil cement mixture remained quite small by a very small amount of addition, and was almost constant within 60 min, remaining almost unaffected by the amount of addition. However, the shear strength increased rapidly with the elapsed time when nothing was added. In the case of the polycarboxylic acid dispersant (Figure 9b), the strength of the mixture was smaller than that of the mixture without dispersant and increased with elapsed time. According to these figures, the glycolic acid and polycarboxylic acid dispersant function to keep the fluidity of the soil and binder mixture high soon after mixing. It can thus be concluded that the addition of suitable dispersant functions maintains the fluidity of the mixture.

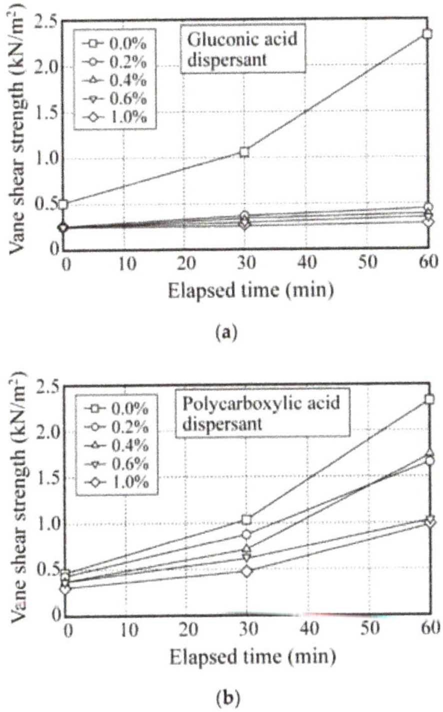

Figure 9. Effect of dispersant on increases in strength of the soil and binder mixture [14]. (**a**) Glycolic acid dispersant; (**b**) polycarboxylic acid dispersant.

5.4. Injecting Atomized Cement Slurry

A new type of construction machine is introduced, where cement slurry is accompanied by compressed air at the injection nozzle and atomized cement slurry is injected to mitigate the entrained rotation phenomenon, as shown in Figure 10 [16]. The injected air can function as a ball bearing to increase the fluidity of the soil and binder mixture. Figure 11 shows the stabilized soil strength distribution within the column obtained in field tests to compare the effect of atomized cement slurry injection (CI-CMC technique) and ordinary machines. The 28-day strength of the column stabilized by the ordinary machine shows a comparatively small strength on the order of 1000 to 1700 kPa, while that stabilized by the CI-CMC technique shows a large strength on the order of 3100 to 3600 kPa. The figure also shows that the strength variation of the CI-CMC technique is smaller than that of the ordinary machine. As a result, atomized cement slurry injection can function to keep the fluidity of fresh stabilized soil high, increase the strength, and decrease the strength variation.

Atomized cement slurry can disturb the soil structure and increase the fluidity of the soil and cement mixture, which can contribute to the decrease in the penetration resistance and required mixing energy and, in turn, the mixing machine can penetrate and mix a hard soil layer. Figure 12 shows the required auger torque to compare the CI-CMC and ordinary techniques [16]. In the case of the ordinary mixing machine, the mixing machine is stopped at the hard layer at a depth of approximately −8 m; however, the CI-CMC machine can penetrate through the hard layer. According to the case history, the CI-CMC technique can penetrate approximately 2.5 m into the mudstone layer with SPT *N*-values of approximately 25 to 50 (q_u = 4 to 5 MPa).

Figure 10. Mixing machine injecting atomized cement slurry [16].

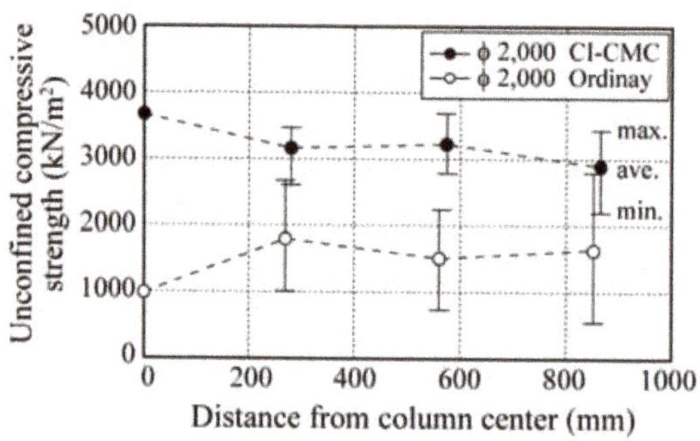

Figure 11. Stabilized soil strength distribution within the column [16].

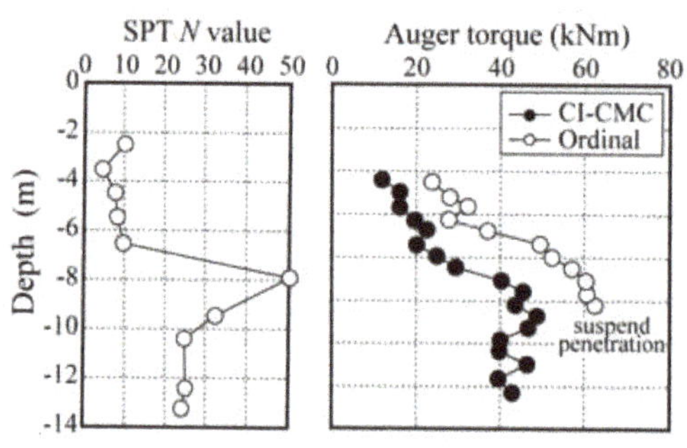

Figure 12. Effect of atomized cement slurry injection on penetration in the hard layer [16].

6. Prediction of Field Strength by Construction Log Data and Future Perspectives

6.1. Relationship between the W/C Ratio and Stabilized Soil Strength

There are many proposed formulas to predict the laboratory and field strengths [17–22]. Among them, one of the key parameters, the W/C ratio, is introduced, which is defined as the total weight of water contained in the soil and binder slurry against the weight of the binder. The ratio is also expressed by the water content of soil, binder content, and water to binder ratio.

$$\begin{aligned} W/C &= (W_{ws} + W_{wc})/W_c \\ &= (w/aw + w/c) \end{aligned} \tag{6}$$

where aw = binder content, w = water content of soil, w/c = water to binder ratio of binder slurry, W/C = total water to binder ratio, W_c = dry weight of binder (kg), W_{wc} = weight of water contained in binder slurry (kg), and W_{ws} = weight of water contained in soil (kg).

Figure 13 shows an example of the relationship between the *W/C* ratio and laboratory stabilized soil strength at 28 days of curing, q_{u28}, for various soils [23]. Furnace slag cement type B was used as a binder in the laboratory mix tests, where the cement dosage and the *w/c* ratio of cement slurry were changed to cover a wide range of *W/C* ratios. In the figure, test cases of the *w/c* ratio of 100% are shown, where the original soils are classified into three groups depending on their fines content, *Fc*: 0 to 30%, 30 to 70%, and 70 to 100%. The strength rapidly decreased with increasing *W/C* ratio irrespective of the fine content. It was also found that the strength became larger when the fine content became large. A similar relationship has been found in various soils and binders stabilized by the deep mixing method [24] and pneumatic flow mixing method [25]. The stabilized soil strength can be formulated as Equation (7):

$$q_u = a \times (W/C)^b \tag{7}$$

where a = parameter (kPa), b = parameter, and q_u = unconfined compressive strength (kPa).

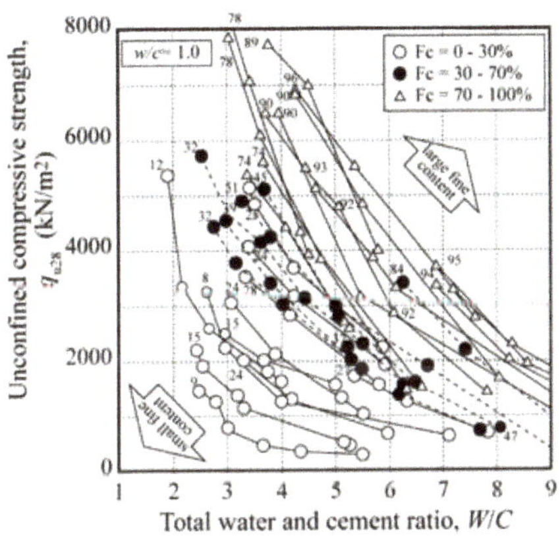

Figure 13. Relationship between the W/C ratio and unconfined compressive strength [23].

The parameters *a* and *b* are influenced by the fines content, *Fc*, where the value of *a* increases as *Fc* increases, and the value of *b* is negative and its absolute value decreases as *Fc* increases.

6.2. Construction Log Data

Figure 14 shows an example of deep mixing log data of a production column in a certain project. The mixing tool is equipped with two injection nozzles at the elevation of the bottom and top mixing blades for penetration injection and withdrawal injection, respectively. In Figure 14, the positions of the bottom and top mixing blades and the rotation speed of the mixing blade are plotted along the time of day. The ground at the construction site consists of four layers: the clay-1 and clay-2 layers are soft layers with natural water contents of approximately 65–80%, while the alluvial clay and alluvial sand layers are relatively hard layers with low natural water contents. Laboratory mix tests were carried out on the soils, and the relationships between the *W/C* ratio by changing the cement dosage and the laboratory strength, q_{ul}, were examined to obtain parameters a = 37.8 MPa and b = −1.62 for the clay and a = 16.7 MPa and b = −0.98 for the sand in Equation (5).

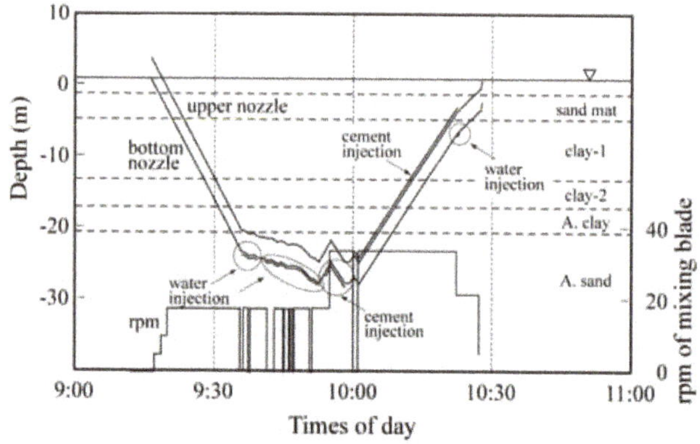

Figure 14. Example of DM construction log data.

In the production, as shown in Figure 14, the mixing tool penetrated clay-1, clay-2, and alluvial clay at a constant speed of 1 m/min, while the rotation speed of the mixing blade was kept almost constant at 18 rpm. When the bottom end of the mixing blade reached the alluvial sand layer at −23.2 m, the penetration speed was reduced to 0.13 m/min due to the hard alluvial sand layer where the required torque of the mixing blade was increased. The mixing shaft continued to penetrate into the alluvial sand layer to a depth of −27.5 m with a low penetration speed, while the mixing blade was rotated at a constant speed with several intentional stops. During penetration, water was injected approximately 1.1 m^3 at depths of −23.2 m to −23.8 m and approximately 5.09 m^3 at depths of −24.7 to −27.5 m from the bottom nozzle for the purpose of softening the alluvial sand layer for ease of penetration. After reaching the design depth of −27.5 m, the mixing tool was moved up and down for the bottom treatment, while cement slurry of 80% *w/c* was injected from the bottom nozzle (penetration injection).

6.3. Analysis of Construction Log Data

The construction log data shown in Figure 14 were analyzed to evaluate the mixing condition every meter along the depth, and the analyzed result is shown in Figures 15–17. Figure 15 shows the distribution of the cement factor along the depth, which is defined as the weight of cement (kg) over the soil volume (m^3). The cement factor is almost constant at approximately 150 kg/m^3 as the design criteria in the clay-1, clay-2, and alluvial clay layers. In the alluvial sand layer, the cement factor is scattered, ranging from 118 to 289 kg/m^3, as the cement slurry and water were injected for the bottom treatment and flushing the delivery tube.

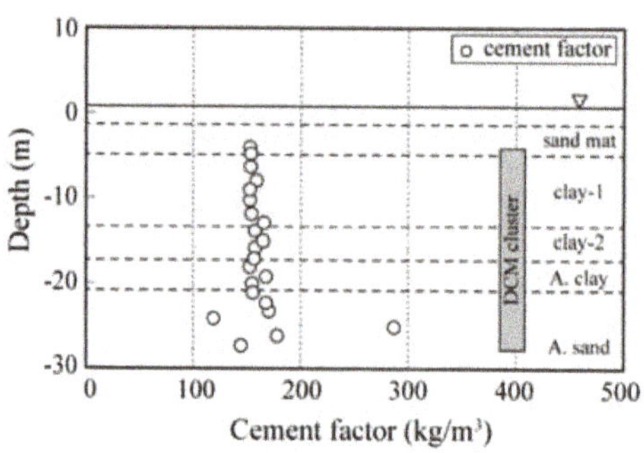

Figure 15. Distribution of binder factor along depth.

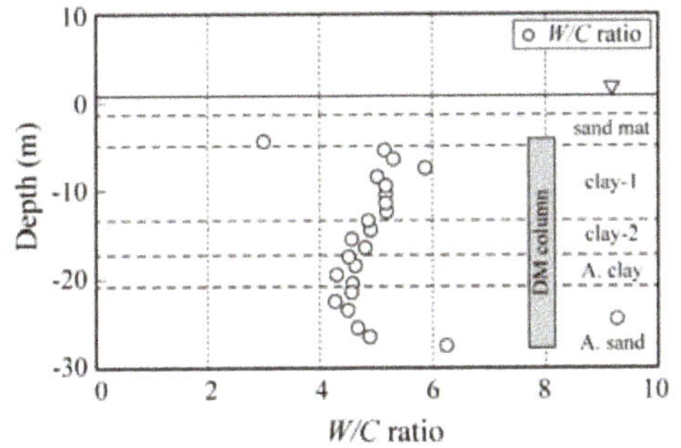

Figure 16. Distribution of W/C ratio along depth.

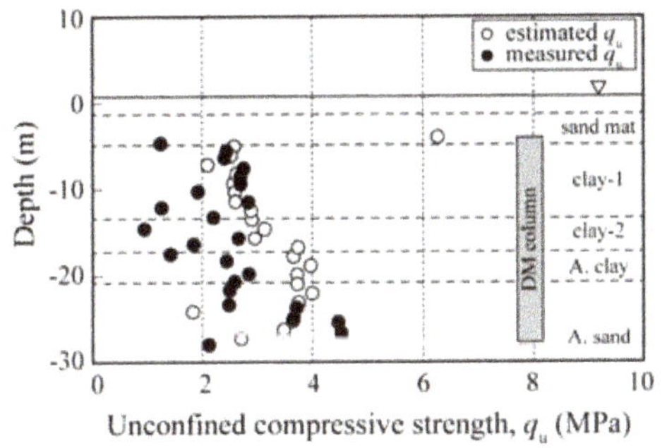

Figure 17. Distribution of predicted laboratory strength and measured field strength along depth.

The *W/C* ratio was calculated by Equation (4) using the estimate water content of soil at the depth and plotted along the depth in Figure 16. In the clay-1 layer, the *W/C* ratio is almost constant at approximately 4.8, except at approximately −7 m, where 0.48 m^3 of water was injected to flush the delivery tube after the withdrawal injection. In the clay-2 and alluvial clay layers, the *W/C* ratio is smaller than that of the clay-1 layer, even though the amount of cement is almost the same, which is due to the low natural water content of the layers rather than that of the clay-1 layer. The *W/C* ratio in the alluvial sand layer is scattered from 4.3 to 9.3, which is due to the scatter in the amount of injected cement slurry and water.

The estimated unconfined compressive strength of laboratory soil, q_{ul}, is shown along the depth in Figure 17. The predicted strength at a depth of approximately −7 m is approximately 2.1 MPa, which is approximately 20% smaller than that of the others and is due to the water injection there. In the alluvial clay layer, the predicted strength is almost constant at approximately 3.9 MPa. In the alluvial sand layer, the predicted strength decreases with depth. The estimated strength at a depth of approximately −26.0 m is quite small at approximately 1.9 MPa.

The core sample was taken from the column, and unconfined compression tests were carried out to measure the field strength, q_{uf}. These measured strengths are plotted in Figure 17 by solid circles. The measured strengths in clay-1 and clay-2 are smaller than the estimated strength. The ratio of the measured and estimated strength, q_{uf}/q_{ul}, is roughly obtained as approximately 0.63. It should be noted that the measured strength is scattered rather than the estimated strength, which might be due to the local soil condition, mixing condition (entrained rotation phenomenon), and the coring and unconfined compression testing machine and techniques.

6.4. Future Perspectives

As described above, the W/C ratio has a close relationship with the unconfined compressive strength and can be applied in QC/QA during construction to assure the design strength. Many construction parameters are continuously measured during the production in Japan, which includes plotting the position of mixing shaft, penetration and withdrawal speeds of mixing shaft, rotation speed of mixing blade, electric current for driving mixing blade, lifting load of mixing blades, and accumulated cement volume. It is possible to estimate the soil type and water content at each depth based on their measured values, and the field-stabilized soil strength can be estimated by Equations (4) and (5). Once the strength during production has been predicted, several countermeasures can be adopted and carried out during production to assure the design criteria.

In fact, many attempts have been made to estimate the types and properties of ground from the construction log data in several ground improvement methods. However, due to the errors and scatters in the measured data and large variations in ground properties, it was not possible to estimate them with acceptable accuracy. It is expected that the types and properties of ground can be estimated by referring to a vast database using AI system in near future. In recent years, information communication technology (ICT) has been applied in the construction industry to increase the productivity and reliability of construction. ICT construction machines feature advanced technology, such as machine guidance systems and production control systems, to assist the operator and site management systems in processing productivity and work progress data.

7. Concluding Remarks

In the manuscript, the current practice of quality control and assurance was briefly introduced based on the Japanese experience/results with mechanical mixing technology by vertical shaft mixing tools with horizontal rotating circular mixing blade. The QC/QA practice comprises laboratory mix tests, field trial tests, monitoring, and control of construction parameters during production and verification by the engineering characteristics of stabilized soil.

It is well known that field-stabilized soil columns/elements have relatively large strength variability. One of the reasons behind the strength variability is the type of soil. Some cohesive soils are so sticky that the soil and binder mixture adheres to and rotates with the mixing blade without efficient mixing. For such soils, several alternative approaches are introduced in the manuscript: using a new type of special cement, injecting chemical additives, and injecting air.

The W/C ratio had a close relationship with the unconfined compressive strength and can be applied in QC/QA during construction to assure the design strength. The field-stabilized soil strength could be estimated by the production log data during production. Once the strength during production is predicted, several countermeasures can be adopted and carried out during production to assure the design criteria.

As ICT and AI technologies are developed very rapidly, ICT deep mixing equipment will be working on site to produce stabilized soil columns/elements in the near future. In the machine, the type and properties of the ground profile are estimated by the measured construction control parameters (such as the hanging load, penetration speed of the mixing tool, and driving torque of the mixing blade in the penetration stage) and the appropriate mixing conditions (such as the amount of binder and blade rotation number, obtained with the help of the W/C concept and the accumulated database), together with AI technology to achieve the design strength along the depth. The stabilized soil column is produced automatically by ICT deep mixing equipment according to the obtained mixing conditions.

It is noted that the conclusions presented here are only valid for these specific conditions, therefore any extrapolation to other DM techniques should be made with caution requiring more studies.

Funding: This research received no external funding.

Institutional Review Board Statement: Not applicable.

Informed Consent Statement: Not applicable.

Data Availability Statement: Not applicable.

Conflicts of Interest: The author declares no conflict of interest.

References

1. Kitazume, M.; Terashi, M. *The Deep Mixing Method*; CRC Press, Taylor & Francis Group: Boca Raton, FL, USA, 2013; 410p.
2. Terashi, M.; Kitazume, M. Current Practice and Future Perspective of QA/QC for Deep-Mixed Ground. In Proceedings of the International Symposium on Deep Mixing and Admixture Stabilization, Okinawa, Japan, 19–21 May 2009; pp. 61–99.
3. Cement Deep Mixing Method Association. *Cement Deep Mixing Method (CDM) Design and Construction Manual*; Cement Deep Mixing Method Association: Tokyo, Japan, 1999. (In Japanese)
4. Deere, D.U. *Rock Quality Designation (RQD) after 20 Years*; U.S. Army Corps of Engineers: Washington, DC, USA, 1989.
5. Dry Jet Mixing Method Association. *Dry Jet Mixing (DJM) Method Technical Manual*; Dry Jet Mixing Association: Narita, Japan, 2006. (In Japanese)
6. JACSMAN Association. *Technical Data for JACSMAN, Ver. 6*; JACSMAN Association, KU Publisher: Tokyo, Japan, 2011; 21p. (In Japanese)
7. Terashi, M. The state of practice in deep mixing method. Grouting and ground treatment. ASCE Geotechnical Special Publication No. 120;. In Proceedings of the 3rd International Conference on Grouting and Ground Treatment, New Orleans, LA, USA, 10–12 February 2003; Volume 1, pp. 25–49.
8. Coastal Development Institute of Technology. *Technical Manual of Deep Mixing Method for Marine Works, Revised Version*; Coastal Development Institute of Technology: Tokyo, Japan, 2008; 289p. (In Japanese)
9. Public Works Research Center. *Deep Mixing Method Design and Execution Manual for Land Works*; Public Works Research Center: Narita, Japan, 2004; 234p. (In Japanese)
10. Hosoya, Y.; Nasu, T.; Hibi, Y.; Ogino, T.; Kohata, Y.; Makihara, Y. Japanese Geotechnical Society Technical Committee Reports: An evaluation of the strength of soils improved by DMM. In Proceedings of the Second International Conference on Ground Improvement Geosystems, Tokyo, Japan, 14–17 May 1996; Volume 2, pp. 919–924.
11. Halkola, H. Keynote lecture: Quality control for dry mix method. In *Dry Mix Methods for Deep Stabilization*; Routledge: London, UK; A.A. Balkema: Rotterdam, The Netherlands, 1999; pp. 285–294.
12. Larsson, S. State of Practice Report-Execution, monitoring and quality control. In Proceedings of the International Conference on Deep Mixing-Best Practice and Recent Advances, Stockholm, Sweden, 23–25 May 2005; Volume 2, pp. 732–785.
13. Aoyama, K.; Miyamori, T.; Wakiyama, T.; Kikuchi, D. The Influence of Physico-Chemical Property on Improved Soil Character. *J. Jpn. Soc. Civ. Eng.* **2002**, *2002*, 207–219. (In Japanese)
14. Mizutani, Y.; Makiuchi, K.; Minegishi, K. *Effects of Surfactant Additive on Strength Stability of Soil Cement Pile*; Japanese Geotechnical Society: Tokyo, Japan, 2003; pp. 1047–1048. (In Japanese)
15. Hirano, S.; Mizutani, Y.; Nakamura, H.; Shimomura, S.; Sasada, H. *Quality Improvement of Deep Mixing Method by Dispersant Additives*; Japanese Geotechnical Society: Tokyo, Japan, 2015; pp. 881–882. (In Japanese)
16. Japan Cement Association. *Soil Improvement Manual Using Cement Stabilizer*, 3rd ed.; Japan Cement Association: Tokyo, Japan, 2007; 387p. (In Japanese)
17. Murakami, S. Deep mixing method using ejector discharge injection and new construction management system-CI-CMC Method. *Kisoko* **2017**, *6*, 44–47. (In Japanese)
18. Tsuchida, T.; Tang, Y.X. A consideration on estimation of strength of cement-treated marine clays. *Jpn. Geotech. J.* **2012**, *7*, 435–447. (In Japanese) [CrossRef]
19. Horpibulsuk, H.; Miura, N.; Nagaraj, T.S. A new method for predicting strength of cement stabilized clays. In Proceedings of the International Conference on Coastal Geotechnical Engineering in Practice, Yokohama, Japan, 20–22 September 2000; Volume 1, pp. 605–610.
20. Tang, Y.X.; Miyazaki, Y.; Tsuchida, T. Advanced reuses of dredging by cement treatment in practical engineering. In Proceedings of the International Conference on Coastal Geotechnical Engineering in Practice, Yokohama, Japan, 20–22 September 2000; Volume 1, pp. 725–731.
21. Yanagihara, M.; Horiuchi, S.; Kawaguchi, M. Long-term stability of coal-fly-ash slurry manmade island. In Proceedings of the International Conference on Coastal Geotechnical Engineering in Practice, Yokohama, Japan, 20–22 September 2000; Volume 1, pp. 763–769.
22. Miyazaki, Y.; Tang, Y.X.; Ochiai, H.; Yasufuku, N.; Omine, K.; Tsuchida, T. Utilization of cement treated dredged soil with working ship. *J. Jpn. Soc. Civ. Eng.* **2003**, *2003*, 193–204. (In Japanese)
23. Yoneda, K. Examination of strength characteristics of cement stabilized soils. In *Technical Forum*; Japan Geotechnical Consultants Association: Tokyo, Japan, 2011. (In Japanese)

24. Federal Highway Administration. *Design Manual: Deep Mixing for Embankment and Foundation Support*; Federal Highway Administration: Washington, DC, USA, 2013; 228p.
25. Kitazume, M.; Satoh, T. Development of Pneumatic Flow Mixing Method and its Application to Central Japan International Airport Construction. *J. Ground Improv.* **2003**, *7*, 139–148. [CrossRef]

applied sciences

MDPI

Article

Behaviour of Compacted Filtered Iron Ore Tailings–Portland Cement Blends: New Brazilian Trend for Tailings Disposal by Stacking

Nilo Cesar Consoli [1,*], Jordanna Chamon Vogt [2], João Paulo Sousa Silva [3], Helder Mansur Chaves [1], Hugo Carlos Scheuermann Filho [1], Eclesielter Batista Moreira [1] and Andres Lotero [1]

1 Graduate Program in Civil Engineering, Universidade Federal do Rio Grande do Sul, Porto Alegre 90035-190, Brazil; heldermansurchaves@hotmail.com (H.M.C.); hugocsf@gmail.com (H.C.S.F.); eclesielter_ebm@hotmail.com (E.B.M.); andreslotero@hotmail.com (A.L.)
2 Coproducts Business Development, VALE S.A., Nova Lima 34000-000, Brazil; jordanna.vogt@vale.com
3 Exploration and Mineral Projects—Mineral Development Centre, VALE S.A., Santa Luzia 33040-900, Brazil; joao.paulo.silva@vale.com
* Correspondence: consoli@ufrgs.br

Abstract: Failures of tailings dams, primarily due to liquefaction, have occurred in Brazil in recent years. These events have prompted the Brazilian government to place restrictions on the construction of new dams, as iron ore tailings deposited behind upstream dams by spigotting have been shown to have low in situ densities and strengths and are prone to failure. This work proposes a new trend for tailings disposal: stacking compacted filtered ore tailings–Portland cement blends. As part of the proposal, it analyses the behaviour of compacted iron ore tailings–Portland cement blends, considering the use of small amounts of Portland cement under distinct compaction degrees. With the intention of evaluating the stress–strain–strength–durability behaviour of the blends, the following tests were carried out: unconfined compression tests; pulse velocity tests; wetting–drying tests; and standard drained triaxial compression tests with internal measurement of strains. This is the first study performed to determine the strength and initial shear stiffness evolution of iron ore tailings–Portland cement blends during their curing time, as well friction angle and cohesion intercept. This manuscript postulates an analysis of original experimental results centred on the porosity/cement index (η/Civ). This index can help select the cement quantity and density for important design parameters of compacted iron ore tailings–cement blends required in geotechnical engineering projects such as the proposed compacted filtered iron ore tailings–cement blends stacking.

Keywords: cemented iron ore tailings behaviour; filtered tailings stacking; Portland cement; compaction

Citation: Consoli, N.C.; Vogt, J.C.; Silva, J.P.S.; Chaves, H.M.; Scheuermann Filho, H.C.; Moreira, E.B.; Lotero, A. Behaviour of Compacted Filtered Iron Ore Tailings–Portland Cement Blends: New Brazilian Trend for Tailings Disposal by Stacking. *Appl. Sci.* **2022**, *12*, 836. https://doi.org/10.3390/app12020836

Academic Editors: Paulo José da Venda Oliveira and António Alberto Santos Correia

Received: 29 November 2021
Accepted: 9 January 2022
Published: 14 January 2022

Publisher's Note: MDPI stays neutral with regard to jurisdictional claims in published maps and institutional affiliations.

Copyright: © 2022 by the authors. Licensee MDPI, Basel, Switzerland. This article is an open access article distributed under the terms and conditions of the Creative Commons Attribution (CC BY) license (https://creativecommons.org/licenses/by/4.0/).

1. Introduction

Tailings are the residues derived from ore extraction and processing and are mainly constituted by crushed rock fines, chemicals and water [1,2]. This combination results in a material having an aqueous slurry consistency which facilitates the disposal in large impoundments designated as tailings dams. In this regard, the upstream method of construction (Figure 1a) is the cheapest manner to expand the dam once the initial embankment has been built. In brief, this methodology consists in founding the raising dam directly into the deposited tailings. Nonetheless, as the tailings are customarily found saturated at a loose condition, stability issues related to static and/or dynamic liquefaction may compromise the security of the dams assembled using the upstream method [3,4].

For this reason, since 2019, building upstream tailings dams has been prohibited in Brazil due to collapses that released massive mudslides that buried the surrounding areas, resulting in destruction, environment pollution and several deaths. According to the non-profit organization World Mine Tailings Failures [5], 45 tailings dam failures occurred between 2009 and 2019. A United Nations Environment Programme [6] report documented some of these significant failures, if not in terms of loss of life, then in terms of environmental damage. These are some of the incidents which occurred between 2015 and 2020: Fundão, 2015 (Brazil); New Walles, 2016 (USA); Tonglushan, 2017 (China); Mishor Rotem, 2017 (Israel); Brumadinho, 2019 (Brazil); and Hpakant, 2020 (Myanmar), among others [7,8].

Numerous other tailings failures have occurred worldwide but were not reported as they did not involve any fatalities. These catastrophic incidents may be caused, in many cases, by lack of control of the design, but to some extent they reflect a relatively poor understanding of the mechanics of tailings. Santamarina et al. [9] highlight how knowledge gaps and management shortcomings contribute to the catastrophic failures that claim thousands of lives around the world. Therefore, a deeper knowledge on the behaviour of these structures and materials, as well as the search for alternatives focusing risk mitigation, is crucial and of great concern to companies, government agencies and society.

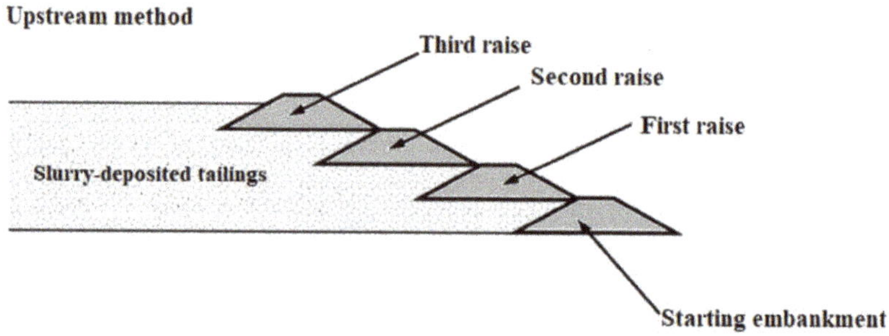

(a)

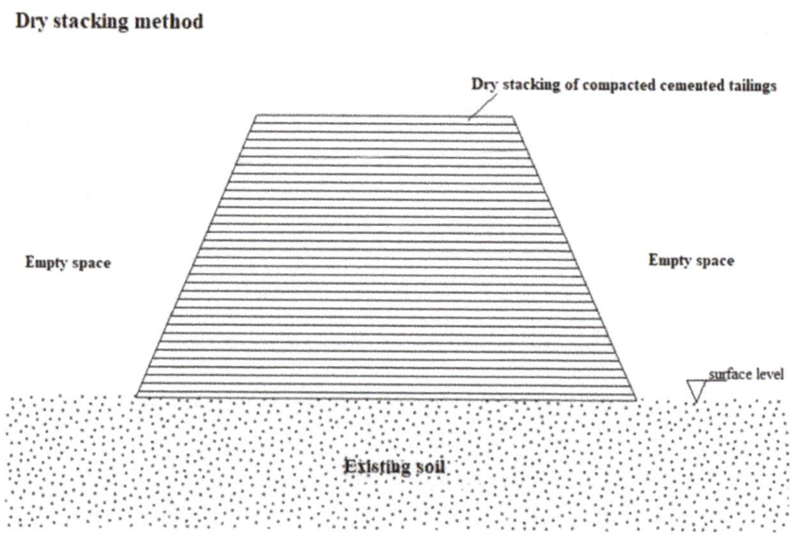

(b)

Figure 1. Schemes of tailings dam construction methods (a) upstream method (b) dry stacking method.

Dry stack tailings (Figure 1b) are being adopted in Brazil as a potential solution for reducing the risk of catastrophic dam failure and tailings runout. Essentially, they consist of the stacking of compacted dry tailings, forming piles of hundreds of meters. In this regard, the use of compacted filtered ore tailings–Portland cement blends stacking will allow tailings disposal sites, which currently do not use binders and thus have shallow slopes, to occupy smaller areas by creating steeper, more stable stacks which will consequently lead to less environmental and visual impact. The current research is studying the stress–strain–strength behaviour of artificially cemented (using Portland cement) compacted filtered iron mining tailings for stacking in order to drastically reduce the possibility of tailings liquefaction once cementitious bonds (cohesion) are built amongst tailings particles. The reason for studying especially iron mine tailings specifically is that Brazil is the second largest producer of iron ore (and consequently iron ore tailings) in the world, with approximately 388 million metric tons produced in 2020 [10].

Recognizing the topic's importance and based on concepts of ground improvement, this research aims to contribute to understanding the mechanical behaviour of compacted (considering distinct dry unit weights) iron ore tailings stabilized with early strength Portland cement (in distinct amounts), from unconfined compression, initial shear stiffness, performance under wetting–drying cycles and consolidated drained triaxial tests points of view.

2. Background

The characteristics of ore tailings are highly variable depending on the composition of the ores and the extraction processes used. In general, tailings can vary in size from colloidal to sand, with the degree of plasticity depending on the surface activity of the fines content [11]. The most common disposal method for tailings is hydraulic deposition, followed by sedimentation in an impoundment and consolidation under their own weight, which may take many years due to their relatively low hydraulic conductivity [12,13].

Frequently, the disposal conditions of relatively small size particles result in a saturated and low strength environment—often susceptible to liquefaction, caused by either static or seismic loading. In general, the large mudflows that follow dam failures imply the presence of loose, water-saturated sediments that want to contract upon shear. The water cannot drain fast enough, and grains become temporarily suspended, forming a dense fluid [9], which characterizes the liquefaction phenomenon. Conceptually, Jefferies and Been [14] define soil liquefaction as a phenomenon in which soil loses much of its strength or stiffness for a generally short time but nevertheless long enough to cause failures which result in large financial losses, environmental damage and, in the worst cases, loss of life. This is particularly important, since there are many incidents on tailings impoundments that are claimed to be related to liquefaction.

The stability performance of mine tailings is linked to their dry unit weight (γ_d) and consequently compaction could reduce liquefaction potential [15]. However, the existence of cementitious bonds amongst tailings particles (due to blends of tailings with Portland cement) prevents them of suffering liquefaction and enhances mechanical behaviour.

3. Experimental Program

3.1. Materials

The iron tailings used in the testing were taken from the Iron Quadrangle region, located in the province of Minas Gerais, Brazil (see Figure 2). The grain size distribution of the iron mine tailings is given in Figure 3. The iron mine tailings' physical properties are displayed in Table 1, being classified [16] as silty sand (SM). Mineralogical characterization of the iron tailings, acquired using an X-ray diffractometer, detected the presence of a few compounds: quartz [SiO_2], hematite [Fe_2O_3], goethite [$FeHO_2$], kaolinite [$Al_2H_4O_9Si_2$], and muscovite [$Al_3H_2KO_{12}Si_3$]. Regarding the chemical composition of the studied iron tailings, the following element concentrations were found after X-ray fluorescence: 69.7% of SiO_2, 24.0% of Fe_2O_3, 4.8% Al_2O_3, 0.40% of MnO, 0.25% of P_2O_5, 0.15% of K_2O, and 0.1% of SO_3, amongst others. The results of standard (600 kN.m/m^3) and modified (2700 kN.m/m^3) Proctor compaction tests are displayed in Figure 4.

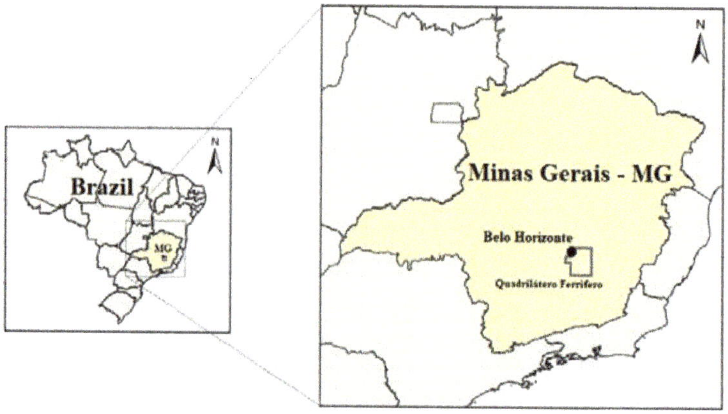

Figure 2. Location of the Iron Quadrilateral on the map of Brazil and Minas Gerais (MG) province.

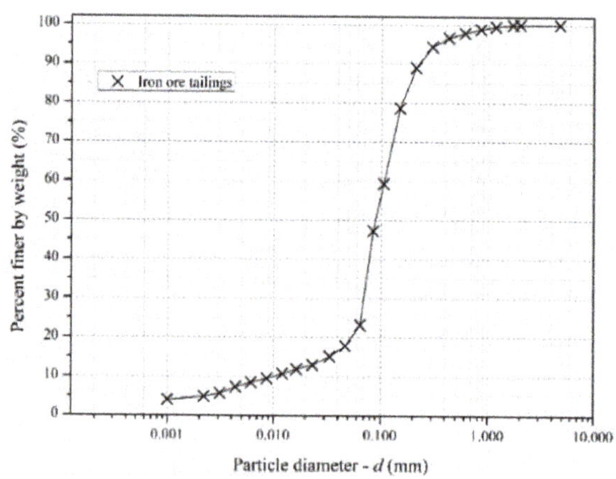

Figure 3. Iron ore tailings grain size distribution.

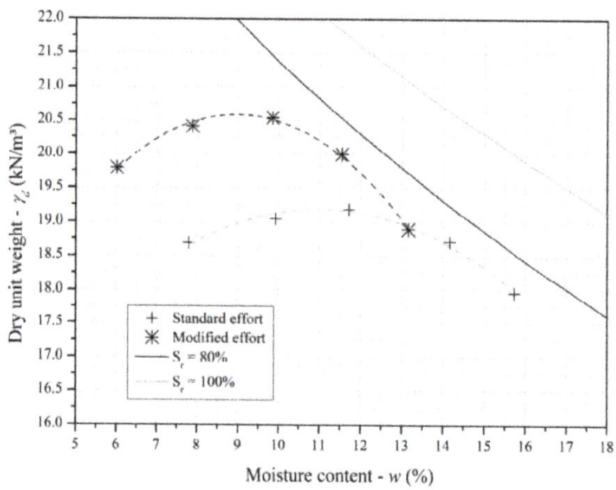

Figure 4. Compaction curves of iron ore tailings at standard and modified energies.

Table 1. Physical properties of studied iron ore tailings.

Physical Properties	Iron Ore Tailings (IOT)
Specific gravity of solids	2.916
Uniformity coefficient	10.7
Coefficient of curvature	3.9
Mean particle diameter-(mm)	0.085
Liquid limit (%)	-
Plastic limit (%)	-
Plasticity index (%)	Nonplastic
Medium sand (0.425 mm < *d* < 0.200 mm) (%)	4.0
Fine sand (0.075 mm < *d* < 0.425 mm) (%)	49.0
Silt (0.002 mm < *d* < 0.075 mm) (%)	42.0
Clay (*d* < 0.002 mm) (%)	5.0
USCS Classification (ASTM 2017)	SM
Maximum dry unit weight at standard energy compaction (kN/m^3)	19.2
Optimum moisture content at standard energy compaction (%)	11.6

High early strength (Type III) Portland cement [17] was used throughout this investigation. Its rapid strength gain allows blends to achieve important strength thresholds from short curing periods. Cement grains' specific gravity is 3.15.

Distilled water was utilized both for characterization tests and moulding specimens for the triaxial tests.

3.2. Methods

3.2.1. Moulding Portland Cement Stabilized Iron Ore Tailings (IOT) Specimens

Cylindrical specimens (50 mm in diameter and 100 mm in height) were moulded for the unconfined compression and initial shear stiffness tests, as well as for performance under wetting–drying and for consolidated isotropically drained triaxial tests using the undercompaction method [18]. A target dry unit weight (γ_d) for a particular specimen was then instituted as a result of the dry compacted iron tailings Portland cement mix divided by the total volume of the specimen [19]. As exhibited in Equation (1) [20], porosity (η) is a function of dry density (γ_d) of the mix and Portland cement content (PC). Each substance (iron mine tailings and Portland cement) has a unit weight of solids (γs_{IOT} and γs_{PC}), which also must be measured for computing porosity.

$$\eta = 100 - 100\left\{\left[\frac{\gamma_d}{1+\frac{PC}{100}}\right]\left[\frac{1}{\gamma s_{IOT}} + \frac{\frac{PC}{100}}{\gamma s_{PC}}\right]\right\} \quad (1)$$

After the weighing of the dry materials (i.e., iron mine tailings and cement), these were manually mixed with a spoon until a powder having a visual uniformity was obtained. Next, the correct amount of distilled water was supplemented to reach moisture content of 11.6% (optimum moisture content for standard Proctor compaction effort—see Table 1) for the iron tailings—Portland cement blend, and the mixture continued up to the formation of a homogeneous paste. Following, the specimen was statically compacted inside a cylindrical split mould to its target dry unit weight. Three layers were used in the compaction process, with the top of the first and second layers being slightly scarified in order to guarantee the adherence of the subsequent layer. Once the moulding was finished, the specimen was retrieved from the mould, measured, weighed and sealed inside a plastic bag (to maintain water content) and sent to be cured in a humid room at 23 ± 2 °C with relative moisture of about 95%. As acceptance criteria, the obtained dry unit weight (γ_d) should range within ± 1% of the target value, whereas the moisture content (w) should be around 0.5% of the previously assigned value. Within each tested dosage, the cement content was calculated over the mass of dry iron mine tailings and the dry unit weight (γ_d) was determined as the ratio between the mass of dry solids and the total volume of the test specimen.

3.2.2. Program of Unconfined Compression Tests

The unconfined compression tests followed the ASTM C39 standard [21]. Specimens were moulded with 11.6% of moisture content (optimum moisture content for standard Proctor compaction effort), dry densities of 17 kN/m^3, 18 kN/m^3 and 19 kN/m^3 (corresponding to 89%, 94% and 99% of degree of compaction of standard Proctor compaction effort, respectively), Portland cement contents of 1%, 2%, 3%, 4% and 5% (determined following international [22] and Brazilian [23,24] experience with soil–cement). Specimens were cured for 2, 4, 7, 28 and 90 days. One day prior to the test, the specimens were submerged in a water container for 24 h in order to reduce possible suction effects [20,23]. The temperature of the water tank was controlled according to the adopted curing temperature (i.e., 23 °C). Next, the unconfined compression test was performed using an automatic loading press with maximum capacity of 50 kN at a displacement rate of 1.14 mm/min; the maximum load measured using a load cell. A full factorial design setting was used to define the mix designs for the tests. For this reason, all possible combinations of amounts of cement and dry unit weight values were tested considering each curing period. Thus, 15 dosages were intended to be tested within each curing time; in triplicate moulded for each dosage.

3.2.3. Program of Pulse Velocity Tests and Ultrasonic Elastic Constants

Initial Shear Modulus (G_0) of artificially cemented soils can be determined using ultrasonic pulse velocity tests performed in accordance with ASTM D2845 [25]. For homogeneous and elastic media, G_0 may be calculated through the product between the bulk density and the square of the velocity of a shear wave passing through it [26]. Therefore, as this test is non-destructive, pulse velocity tests were performed on the same specimens moulded for an unconfined compression test, immediately before taking specimens to failure, using special transducers coupled on top and underneath the samples using a special coupler gel. An ultrasonic pulse device was used to emit compression (54 kHz) and shear waves (250 kHz) that are emitted and cross the cylindrical specimens, with the propagation times measured. Therefore, the shear modulus at very small deformations (G_0) can be obtained.

3.2.4. Program of Durability of Specimens Submitted to Wetting–drying Cycles

Durability tests consisting of wetting–drying cycles were carried out in accordance with ASTM D559 [27], but without brushing. Specimens were moulded with 11.6% of moisture content, dry densities of 17 kN/m^3, 18 kN/m^3 and 19 kN/m^3, and Portland cement contents of 1%, 2%, 3%, 4% and 5%. The same experimental design setting previously described for the strength tests was used herein, with the difference that only one specimen for each dosage was tested. This test method aims to simulate harsh on-field conditions over 12 cycles of such procedures [28]. After 2, 4, and 7-days of curing were completed, each specimen cycle started by immersing it in water for 5 h at 23 °C. Then, specimens were submitted to a drying process in an oven during 42h at 71 °C. Twelve cycles of these procedures are required to simulate harsh on-field conditions. After each one of the 12 cycles, the initial shear modulus (G_0) was measured in accordance with ASTM D2845 [25]. After the 12th cycle, specimens were taken to failure through unconfined compression tests in accordance with the ASTM C39 standard [21].

3.2.5. Program of Consolidated Isotropically Drained (CID) Triaxial Tests

A series of consolidated isotropically drained (CID) triaxial compression tests was conducted on artificially cemented compacted filtered iron mining tailings, with the aim of evaluating the deviatoric stress–axial strain–volumetric strain behaviour of the materials. The general procedures described by BS 1377 were followed [29]. In this regard, two representative dosages were chosen and tested under three effective confining pressures (σ'_3 = 50, 100 and 200 kPa). The first dosage contained 3% of cement and was moulded at a dry unit weight of 17 kN/m^3, and the second had the same amount of cement but was compacted to a dry density of 19 kN/m^3. The pressures throughout the tests were electronically monitored by pressure transducers, whereas the vertical load was assessed using a 20 kN high-resolution load cell. The axial displacements were globally measured using a linear variable differential transformer (LVDT) and locally assessed by Hall effect sensors positioned directly in contact with the test specimen [30]. The volumetric strain was measured by an Imperial College volume gauge [31] connected to the drainage outlet. To ensure the saturation of tailings specimens, a back pressure of approximately 500 kPa was applied to produce B parameters higher than 95%. All reported test specimens were isotropically consolidated to their desired consolidation pressure before shearing. Finally, shearing of specimens in triaxial tests occurred at a rate of 1 mm/h. For the calculation of the applied stresses, the area corrections proposed by La Rochelle et al. [32] were adopted.

4. Results and Analysis

4.1. Unconfined Compressive Strength (q_u)

Figure 5a portrays q_u as a function of porosity/cement index (η/C_{iv}) (stated as porosity (η) divided by the volumetric cement content (C_{iv}), the latter expressed as a percentage of cement volume to the total volume of the iron tailings–Portland cement mixes [33]) for the curing periods studied (2, 4, 7, 28 and 90 days). Diambra et al. [34] carried out the theoretical approach validating the shape of the equation. Figure 5a indicates that the η/C_{iv} index is useful in normalizing strength results for iron ore tailings–Portland cement blends. The results indicate that the behaviour of the studied blends presents the same trend, thus generating a single equation (Equation (2)).

$$q_u(\text{kPa}) = A \times 10^4 \times \left[\frac{\eta}{C_{iv}}\right]^{-D} \quad (2)$$

Scalar "D" has been found to be a constant (D = 1.3) to all curing times studied (from 2 to 90 days), while scalar "A" increases with curing time, as shown in Table 2. "A" changes from 1.63 (for 2 days of curing) to 4.89 (for 90 days of curing) and the coefficient of determination (R^2) varies in the range 0.92 to 0.97. From 2 days of curing to 4, 7, 28 and 90 days of curing, the strength increase percentages were of 63.2%, 82.8%, 147.9% and 200.0%, respectively.

Table 2. "A" and "C" scalars for Equations (2) and (3), respectively.

Curing Period	Strength Data—q_u		Stiffness Data—G_0	
	"A"	Coefficient of Determination (R^2)	"C"	Coefficient of Determination (R^2)
2 days	1.63	0.92	1.46	0.86
4 days	2.66	0.96	2.98	0.92
7 days	2.98	0.97	4.11	0.97
28 days	4.04	0.94	4.53	0.96
90 days	4.89	0.96	6.04	0.96

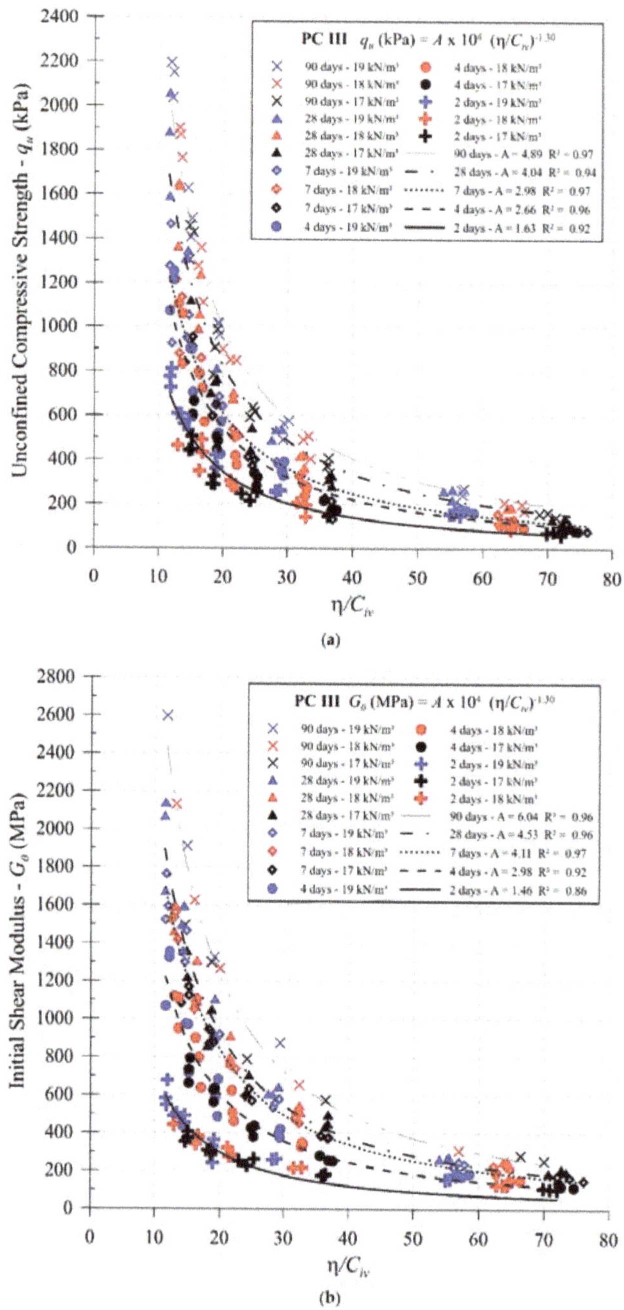

Figure 5. Compacted (γ_d = 17 kN/m^3, γ_d = 18 kN/m^3, γ_d = 19 kN/m^3) filtered iron mining tailings treated with early strength Portland cement (from 1% to 5%): (**a**) unconfined compressive strength (q_u) versus porosity/cement index (η/C_{iv}) considering distinct curing time periods and (**b**) initial shear stiffness (G_0) versus η/C_{iv} taking under consideration different curing time periods (2, 4, 7, 28 and 90 days of curing).

4.2. Initial Shear Modulus

Similarly, as presented for the unconfined compressive strength test results, the porosity/cement index was used for the initial shear modulus (G_0) results for the curing periods studied (2, 4, 7, 28 and 90 days), as presented in Figure 5b. Therefore, an adequate association between G_0 and the η/C_{iv} index (considering the same power shape as for strength) could be obtained as the coefficient of determination (R^2) varies in the range 0.86 to 0.97 for the studied curing times, in the format of a specific equation (Equation (3)).

$$G_0(\text{MPa}) = C \times 10^4 \times \left[\frac{\eta}{C_{iv}}\right]^{-E} \tag{3}$$

Scalar "E" has been found to be a constant (E = 1.3) for all curing times studied (from 2 to 90 days), while scalar "C" increases with curing time, as shown in Table 2. "C" changes from 1.46 (for 2 days of curing) to 6.04 (for 90 days of curing). From 2 days of curing to 4, 7, 28 and 90 days of curing, the initial shear modulus (G_0) increase percentage were of 104.1%, 181.5%, 210.3% and 313.7%, respectively. It is interesting to observe that the rate of increase of q_u and G_0 was not the same with curing time. The rate of increase of G_0 was higher up to 28 days of curing and the rate of increase of q_u was higher from 28 days to 90 days of curing.

4.3. Durability under Wetting–Drying Cycles

Figure 6 presents G_0 variation of iron ore tailings compacted at γ_d of 17, 18 and 19 kN/m^3 and treated with 1 to 5% of early strength Portland cement. Wetting–drying cycles were performed after 2, 4 and 7 days of curing. Such performance mimics the behaviour of the studied blends after being submitted to harsh on-field conditions over 12 cycles of such procedures. It is well established that increasing both the quantity of cement and γ_d improves the stiffness of the compacted iron ore tailings–Portland cement mixes considering wetting–drying cycles. Disregarding the initial curing time (2, 4 or 7 days), Figure 6 shows a comparable qualitative response regarding the impact of wetting–drying cycles: G_0 increased from zero to three wetting–drying cycles and then oscillated about an average, distinctive for each γ_d and quantity of cement employed, for additional cycles. The oven drying for 42 h at 71 ± 2 °C, during the drying part of the wetting–drying cycles, triggered the catalysis of the chemical reactions of the Portland cement, bringing about the increase of G_0 of iron tailings–Portland cement mixes in the initial cycles. Distinct results were achieved by Consoli et al. [28], who assessed the effect of wetting–drying cycles on G_0 of a nonplastic silt. Test results by Consoli et al. [28] indicated that G_0 of nonplastic silt–Portland cement (also early strength) blends mostly reduced with more wetting–drying cycles, reaching a steadiness at about six wetting–drying cycles.

Figure 7 presents the correlation of q_u and G_0 as a function of η/C_{iv} index after 12 wetting–drying cycles. Looking at q_u results Figure 7a, it can be noticed that after 12 wetting–drying cycles, q_u is related to η/C_{iv} index through Equation (4). This equation has the same form as Equation (2) and the scalar of present equation is found above results of 90 days of curing at 23 °C. The q_u, after 12 wetting–drying cycles, being above the results of 90 days of curing at 23 °C is an example of enhancement triggered by the catalysis of the chemical reactions of the Portland cement, due to oven drying for 42 h at 71 ± 2 °C.

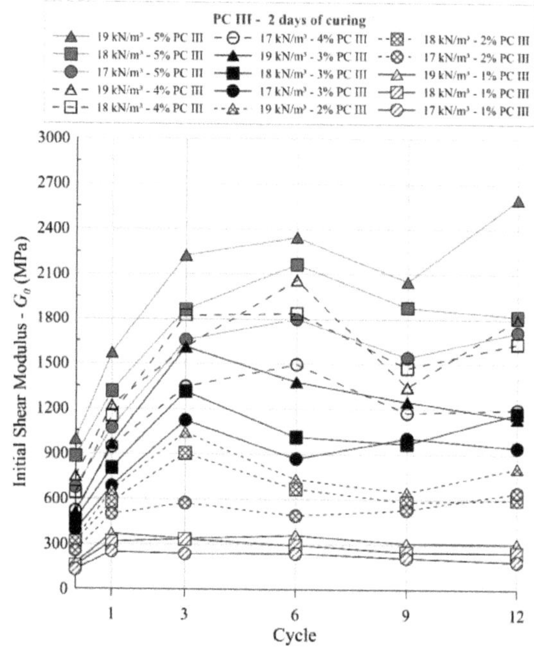

(a)

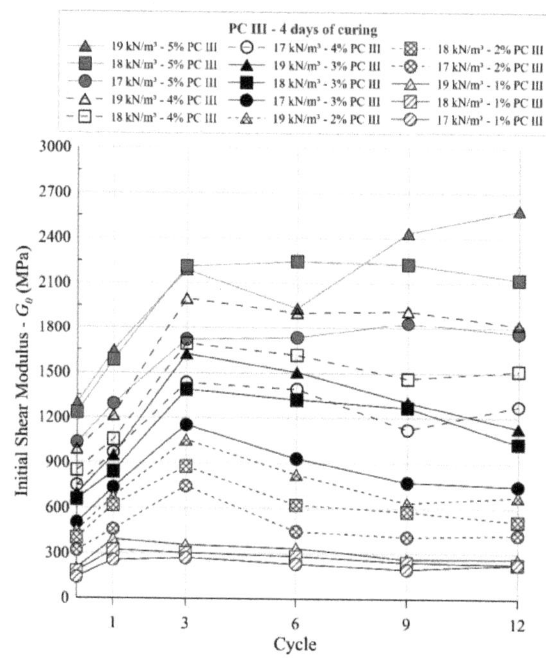

(b)

Figure 6. *Cont.*

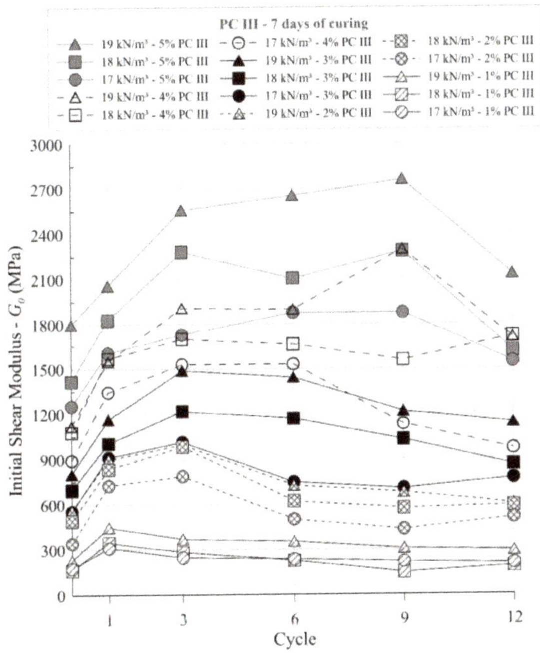

(c)

Figure 6. Performance (initial shear stiffness (G_0) variation) of compacted iron ore tailings treated early strength Portland cement after wet–dry cycles after: (**a**) curing for 2 days, (**b**) curing for 4 days and (**c**) curing for 7 days.

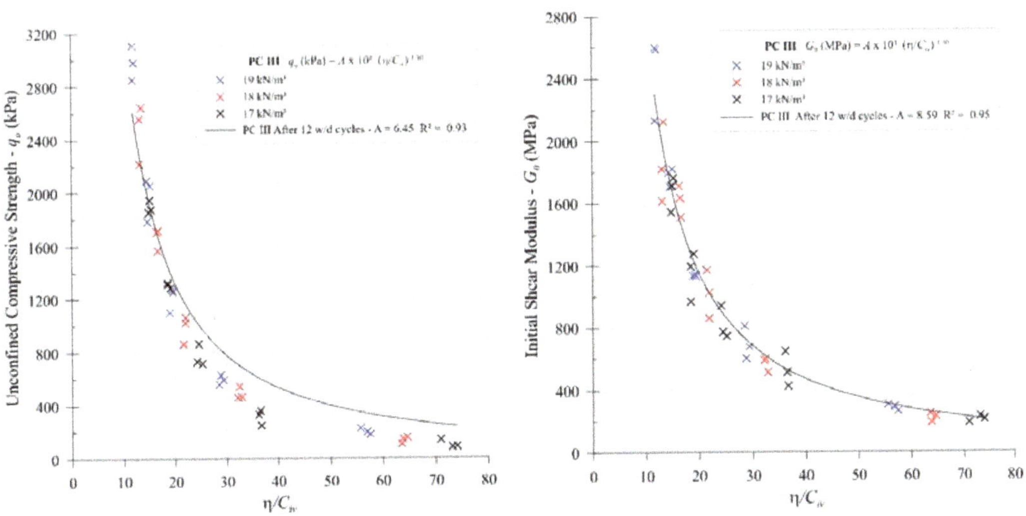

(**a**) (**b**)

Figure 7. Compacted filtered iron mining tailings treated with early strength Portland cement and cured for 2, 4 and 7 days: (**a**) q_u versus η/C_{iv} after 12 wet-dry cycles and (**b**) G_0 versus η/C_{iv} after 12 wetting–drying cycles.

On the other hand, focusing on the G_0 results in Figure 7b, it can be noted that after 12 wetting–drying cycles, G_0 is related to η/C_{iv} index through Equation (5). Such an equation has the same form as Equation (3), and the scalar of present equation is higher than the results of 90 days of curing at 23 °C. However, for q_u and G_0, results after 12 wetting–drying cycles are of the same order of magnitude as the results after 90 days of curing at 23 °C.

$$q_u(\text{kPa}) = 6.45 \times 10^4 \times \left[\frac{\eta}{C_{iv}}\right]^{-1.30} \tag{4}$$

$$G_0(\text{MPa}) = 8.59 \times 10^4 \times \left[\frac{\eta}{C_{iv}}\right]^{-1.30} \tag{5}$$

4.4. Triaxial

Figure 8 presents the stress–axial strain–volumetric strain curves of the standard consolidated drained triaxial tests of artificially cemented specimens of iron ore tailings moulded with γ_d of 17 kN/m^3 and 19 kN/m^3. All specimens have shown a quite stiff response at small axial strains (connected to the contraction of the material), followed by quite brittle behaviour (strong strain-softening response), and the tendency to dilation of the material. The brittleness and dilation tendency are gradually suppressed due to the increase of confining pressures.

The peak failure envelope leads to a peak angle of shearing resistance (φ'_{peak}) of about 34.1° for both dry unit weights and a peak cohesion intercept of (c'_{peak}) of 80.9 kPa for (γ_d) of 17 kN/m^3 and 157.2 kPa for (γ_d) of 19 kN/m^3. The increase of the degree of compaction at standard Proctor energy from 89% to 99% did not cause any change in φ'_{peak} but almost double c'_{peak}. On the other side, the critical state line reaches an angle of shearing resistance at a critical state (φ'_{cs}) of 36.3°.

Values of secant deformation modulus (E_{sec}), obtained at axial strains of 0.1%, 0.5% and 1.0% and for confining stresses ranging from 50 to 200 kPa, are presented in Table 3. Regarding the specimens prepared with γ_d = 17 kN/m^2, it can be seen in Table 3 that the higher modulus is E_{sec} = 816.1 MPa (for ε_a = 0.1% and confining pressure of 200 kPa), while for specimens prepared with γ_d = 19 kN/m^3, the higher modulus is E_{sec} = 2599.9 MPa; the latter (γ_d = 19 kN/m^2) being more than three times the secant modulus value at γ_d = 17 kN/m^3.

Table 3. Secant modulus (E_{sec}) of cement treated iron ore tailings (at distinct axial strains) considering dry unit weights of (a) γ_d = 17 kN/m^3 and (b) γ_d = 19 kN/m^3.

γ_d = 17 kN/m^3 & 3% PC III				**γ_d = 19 kN/m^3 & 3% PC III**			
		E_{sec} (MPa)				**E_{sec} (MPa)**	
Confining Pressure	ε_a (%) = 0.1	ε_a (%) = 0.5	ε_a (%) = 1.0	Confining Pressure	ε_a (%) = 0.1	ε_a (%) = 0.5	ε_a (%) = 1.0
50 kPa	714.7	441.6	355.1	50 kPa	1888.9	1412.8	378.6
100 kPa	740.2	605.3	524.2	100 kPa	2042.7	1652.9	812.5
200 kPa	816.1	500.1	526.6	200 kPa	2599.9	1808.6	965.3
	(a)				(b)		

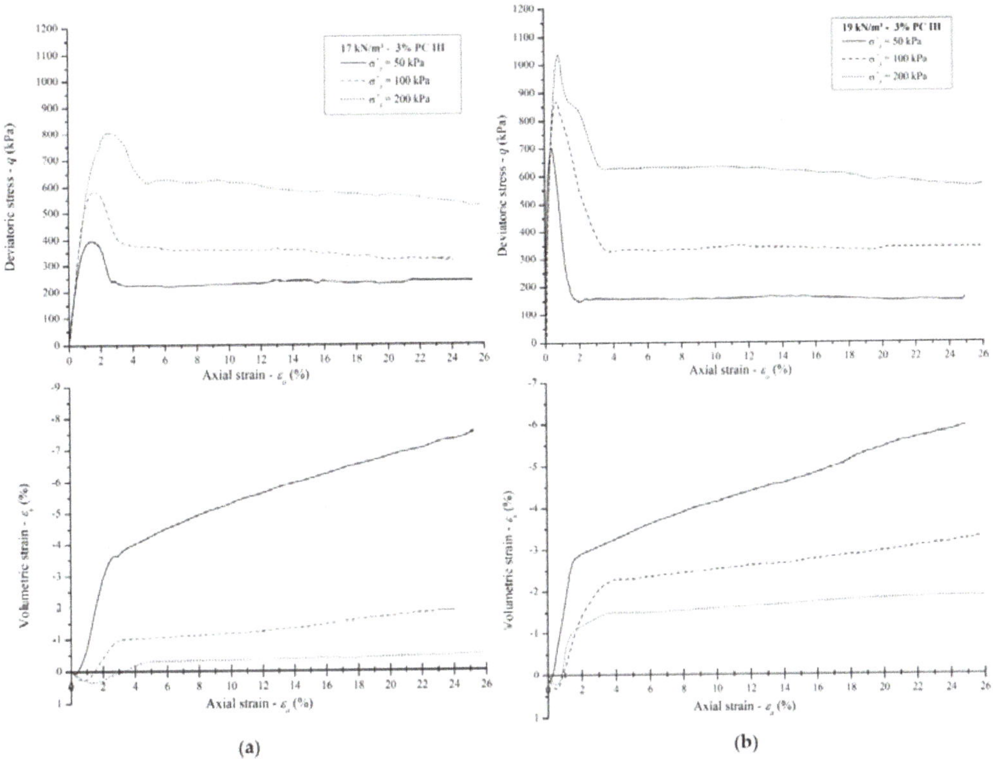

Figure 8. Stress–axial strain–volumetric strain curves for the consolidated drained triaxial tests of specimens moulded with (**a**) γ_d of 17 kN/m^3 and 3% of early strength Portland cement blended with iron tailings under confining pressures of 50, 100 and 200 kN/m^3 and (**b**) γ_d of 19 kN/m^3 and 3% of early strength Portland cement blended with iron tailings under confining pressures of 50, 100 and 200 kN/m^3.

5. Discussion

An original study with the objective of contributing to the understanding of the geomechanical behaviour of a new form of iron ore tailings disposal (stacking of compacted filtered ore tailings–Portland cement blends) was presented as an alternative method to the conventional tailings dam disposal. Adequate correlations between the η/C_{iv} index with q_u and G_0 through power functions were obtained (Figure 5). In artificially cemented soils the η/C_{iv} ratio is usually adjusted by a power (ξ) applied to the variable C_{iv} (defined by curve fitting) to make the rates of variation of η and $1/C_{iv}$ compatible [20]. The value of ξ determines the greater or lesser contribution of porosity or cement content in the mechanical response. According to Diambra et al. [34], its magnitude is directly associated with the properties of the soil matrix and usually approximates the inverse of the exponent of the power function ($\xi \approx 1/D$ or $1/E$ in Equations (2) and (3), respectively). In the present study, an assumed value of ξ = 1 allowed the best fit (R^2 > 0.92) for correlating the η/C_{iv} index with q_u, G_0 and durability.

Rios et al. [35], working with a residual soil of very low (or no) plasticity corresponding to a well graded silty sand and with three different grain size fractions of this same soil determined that, under conditions of similar mineralogy, the particle size distribution is the most relevant factor in the definition of the magnitude of ξ. The research concludes that soils with higher fines (silt) content, fine sand fraction, and better graded, with broader grain size distribution curves, reported lower power values ($\xi \approx 0.21$) compared to poorly graded and fine to coarse sandy soil fractions ($\xi \approx 1$). However, mineralogical composition (related to particle shape) is reported as the most decisive factor in the magnitude of ξ, reporting adjusted values of $\xi = 1$ and $\xi = 0.1$ when comparing two uniform sands (with similar particle size distribution) characterized by having majority quartz and mica phases, respectively. The preponderant contents of quartz and iron minerals (hematite and goethite) in the filtered iron ore tailings determine the value of the fitting power ($\xi = 1$). Values of $\xi = 1$ have been widely reported in the literature for the definition of dosage equations in soils of granular or frictional nature treated with Portland cement [36]. This value, equal to unity, determines an equivalent influence between porosity and cement volumetric content on q_u, G_0 and durability.

Adding cement is considered an effective procedure to prevent liquefaction of soils. In general, the behaviour pattern of filtered iron ore tailings–Portland cement blends is determined by brittle and strain softening behaviour at low confining stresses (mainly due to cementing agent bonds), which evolves to more gentle strain softening with peak strength occurring at higher axial strains as confining stress levels increase (Figure 8). This behaviour is analogous to that reported for a wide range of cemented sands tested at low confining stresses [37,38]. The larger the γ_d of the compacted specimens (lower η), the larger the peak deviator stress reached, the stiffer and more dilative the material, and the greater the post-peak drop in the deviator stress. On the other hand, volumetric strains are strongly dilatant at low stress levels. Some authors (e.g., Airey [39], Coop and Willson [40], Consoli et al. [38]), from the study of different artificially and naturally cemented sands, agree that at high confining stresses (higher than those investigated here) volumetric strains tend towards compression. Additionally, a cohesive behaviour (dominated by cement) at low confining stresses and/or high cement contents tends to evolve to a frictional behaviour (dominated by the sand matrix) at high confining stresses and/or low cement contents.

Figure 9 shows the deviatoric stress—axial strain—volumetric strain curves (for consolidated drained triaxial tests) of the uncemented filtered iron ore tailings and 3% early strength Portland cement mixed with the iron ore tailings, both compacted to a γ_d of 17 kN/m^3 and submitted to confining stresses of 200 kPa. The uncemented filtered iron ore tailings show strong contractive behaviour, confirming the relevance of compressibility of the filtered iron ore tailings and the possibility of uncontrolled positive pore-pressure generation if under undrained shear conditions, which would lead to loss of effective stresses and increased liquefaction potential of the tailings at relatively low confining stresses. In contrast, the occurrence of volumetric dilatational strains (generation of negative pore-pressures if under undrained shear conditions), at low stress levels, and high peak cohesion intercepts (c'_{peak}) reported in tailings treated with the addition of 3% cement would reduce the liquefaction potential of compacted filtered iron ore tailings piles.

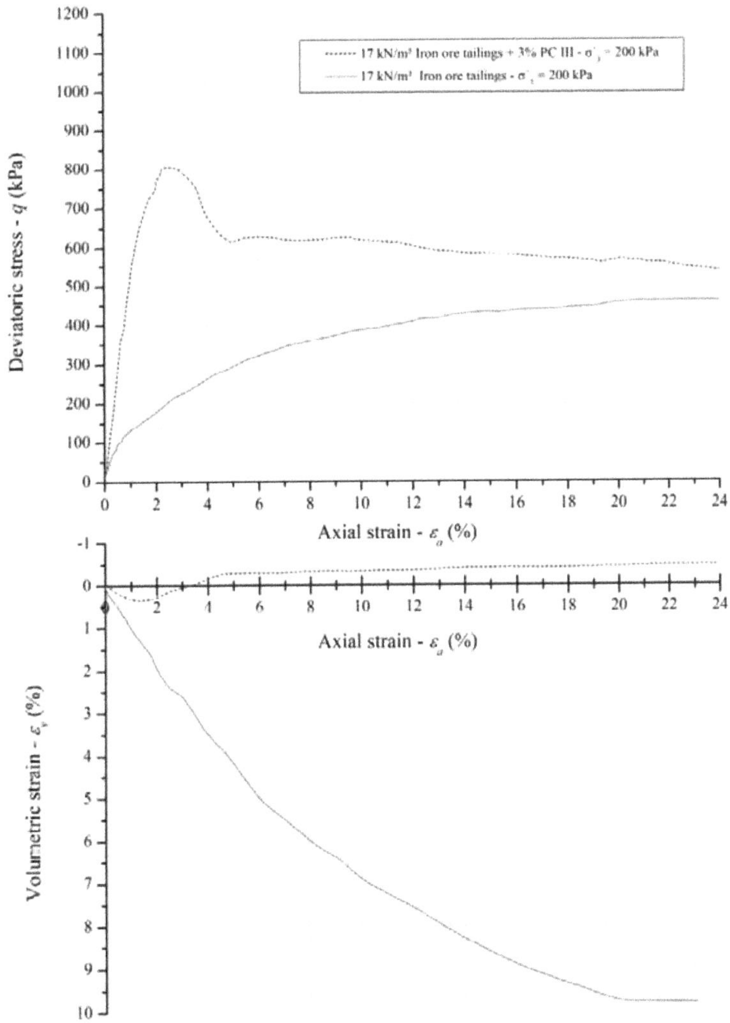

Figure 9. Stress–axial strain–volumetric strain curves for the consolidated drained triaxial tests of specimens moulded with filtered iron ore tailings and 3% early strength Portland cement mixed with iron ore tailings at a dry unit weight (γ_d) of 17 kN/m^3 under confining pressures of 200 kN/m^3.

6. Conclusions

An extensive laboratory testing program was carried out to investigate the effectiveness of using Portland cement and compaction energies to evaluate the engineering behaviour of filtered iron ore tailings. The observations and conclusions can be summarized as follows:

The employment of the porosity/cement index (η/C_{iv}) with the purpose of expressing the performance of iron ore tailings combined with the incorporation of Portland cement and densification through compaction, with curing periods varying from 2 to 90 days, can be considered successful. High coefficients of determination were obtained when q_u and G_0 results were correlated with this parameter. Based on the dosage equations established in present research for the studied iron ore tailings–Portland cement blends, there are several technical ways of reaching a q_u or a G_0 target value for a given project and the best solution might change from situation to situation depending on the time period available for curing, accessibility to equipment to reach a given porosity and cost of Portland cement.

The stress–strain response showed a strength peak for all the samples and a softening following a peak. Also, the increase in effective stress causes an expansive response in volumetric strain. The peak failure envelope leads to a peak angle of shearing resistance (φ'_{peak}) of about 34.1° for both dry unit weights and a peak cohesion intercept of (c'_{peak}) of 80.9-kPa for (γ_d) of 17-kN/m^3 and 157.2-kPa for (γ_d) of 19-kN/m^3. The increase of the degree of compaction at standard Proctor energy from 89% to 99% did not cause any change in φ'_{peak} but almost double c'_{peak}. On the other side, the critical state line reaches an angle of shearing resistance at critical state (φ'_{cs}) of 36.3°. The use of 3% of Portland cement for triaxial testing represents an intermediate amount of cement studied in this research.

The present work has been envisaged as a contribution to the behaviour of compacted iron ore tailings–Portland cement blends to be disposed by stacking. The influence of degree of compaction as well as the amount of Portland cement on strength and stiffness properties was evaluated. The blends studied herein were compacted at optimum moisture content. It might not be possible to do so in the field, especially during rainy seasons. Therefore, the influence of compaction moisture content in the mechanical behaviour requires further study. Another point requiring future research is the development of alternative sustainable binders for stabilization of stacking filtered tailings in order to have a less costly, greener engineering solution. It is also necessary to emphasize that the present study was constrained to the range of low to medium confining pressures, making it attractive to tailings disposal by stacking up to heights of 10–12 m. Other studies are necessary to evaluate changes in the behaviour of the material under higher stackings, when the confining pressure will be greater than the studied range. At last, the addition of a binder to the compacted filtered tailings reduces the volume of hydraulically carried out sediments, thus allowing smaller sedimentation structures downstream of the disposal structure (e.g., ponds and sedimentation dikes).

Author Contributions: Conceptualization, N.C.C.; methodology, N.C.C. and A.L.; validation, N.C.C., H.M.C. and E.B.M.; formal analysis, N.C.C. and H.C.S.F.; investigation, A.L., H.M.C. and E.B.M.; resources, N.C.C.; data curation, N.C.C., H.C.S.F. and E.B.M.; writing—original draft preparation, N.C.C.; writing—review and editing, N.C.C., J.C.V., J.P.S.S. and A.L.; supervision, N.C.C.; project administration, N.C.C.; funding acquisition, N.C.C. All authors have read and agreed to the published version of the manuscript.

Funding: The authors wish to make explicit their appreciation to VALE S.A. (IAP-001247 and IAP-001466) and CNPq (Brazilian Research Council) for the support to the research group.

Institutional Review Board Statement: Not applicable.

Informed Consent Statement: Not applicable.

Data Availability Statement: Some or all data, or models, used during the study are available from the corresponding author upon reasonable request.

Conflicts of Interest: The authors declare no conflict of interest. The funders had no role in the design of the study; in the collection, analyses, or interpretation of data; or in the decision to publish the results.

Abbreviations

A, C, D, E	scalars
B	Skempton's parameter
c'_{peak}	peak cohesion at effective stresses
C_{iv}	volumetric cement content
d	particle diameter
E_{sec}	secant modulus
G_0	initial shear modulus

IOT	iron ore tailings
PC	Portland cement
q_u	unconfined compressive strength
ξ	power function parameter
ε_a	axial strain
ε_v	volumetric strain
γ_d	dry unit weight
γ_s	unit weight of solids
φ'_{cs}	angle of shearing resistance at critical state
φ'_{peak}	peak angle of shearing resistance at effective stresses
η	porosity
η/C_{iv}	porosity/cement index

References

1. Li, W.; Coop, M.; Senetakis, K.; Schnaid, F. The mechanics of a silt-sized gold tailing. *Eng. Geol.* **2018**, *241*, 97–108. [CrossRef]
2. Kossoff, D.; Dubbin, W.; Alfredsson, M.; Edwards, S.; Macklin, M.; Hudson-Edwards, K. Mine tailings dams: Characteristics, failure, environmental impacts, and remediation. *Appl. Geochem.* **2014**, *51*, 229–245. [CrossRef]
3. Armstrong, M.; Langrené, N.; Petter, R.; Chen, W.; Petter, C. Accounting for tailings dam failures in the valuation of mining projects. *Resour. Policy* **2019**, *63*, 101461. [CrossRef]
4. Hu, L.; Wu, H.; Zhang, L.; Zhang, P.; Wen, Q. Geotechnical properties of mine tailings. *J. Mater. Civ. Eng.* **2017**, *29*, 04016220. [CrossRef]
5. World Mine Tailings Failures. State of World Mine Tailings Portfolio 2020. WMTF. 2020. Available online: https://worldminetailingsfailures.org/ (accessed on 15 November 2021).
6. United Nations Environment Programme. New Report Urges Global Action on Mining Pollution. UNEP. 2017. Available online: https://www.unep.org/news-and-stories/story/new-report-urges-global-action-mining-pollution (accessed on 15 November 2021).
7. Islam, K.; Murakami, S. Global-scale impact analysis of mine tailings dam failures: 1915–2020. *Glob. Environ. Chang.* **2021**, *70*, 102361. [CrossRef]
8. Lyu, Z.; Chai, J.; Xu, Z.; Qin, Y.; Cao, J. A comprehensive review on reasons for tailings dam failures based on case history. *Adv. Civ. Eng.* **2019**, *2019*, 4159306. [CrossRef]
9. Santamarina, J.C.; Torres-Cruz, L.A.; Bachus, R.C. Why coal ash and tailings dam disasters occur. *Science* **2019**, *364*, 526–528. [CrossRef] [PubMed]
10. Statista. Iron Ore Mine Production in Brazil from 2015 to 2020. Statista. 2021. Available online: https://www.statista.com/statistics/1026351/brazil-iron-ore-mine-production/ (accessed on 15 November 2021).
11. Vick, S.G. *Planning, Design, and Analysis of Tailings Dams*; Bitech: Vancouver, BC, Canada, 1990.
12. Consoli, N.C.; Sills, G.C. Soil formation from tailings: Comparison of predictions and field measurements. *Géotechnique* **2000**, *50*, 25–33. [CrossRef]
13. James, M.; Aubertin, M.; Wijewickreme, D.; Wilson, G.W. A laboratory investigation on the dynamic properties of tailings. *Can. Geotechnical J.* **2011**, *48*, 1587–1600. [CrossRef]
14. Jefferies, M.G.; Been, K. *Soil liquefaction: A critical state approach*; CRC Press: Boca Raton, FL, USA, 2015.
15. Villavicencio, G.; Breul, P.; Bacconnet, C.; Fourie, A.; Espinace, R. Liquefaction potential of sand tailings dams evaluated using probabilistic interpretation of estimated in-situ relative density. *Rev. De La Construcción* **2016**, *15*, 9–18. [CrossRef]
16. *ASTM D2487*; Standard practice for classification of soils for engineering purposes (Unified Soil Classification System); American Society for Testing and Materials: West Conshohocken, PA, USA, 2017.
17. *ASTM C150*; Standard specification for Portland cement; ASTM International: West Conshohocken, PA, USA, 2017.
18. Selig, E.; Ladd, R.S. Preparing test specimens using undercompaction. *Geotechnical Test. J.* **1978**, *1*, 16. [CrossRef]
19. *ASTM D7263*; Standard test methods for laboratory determination of density and unit weight of soil specimens; American Society for Testing and Materials: West Conshohocken, PA, USA, 2021.
20. Consoli, N.C.; da Fonseca, A.V.; Cruz, R.; Rios, S. Voids/cement ratio controlling tensile strength of cement treated soils. *J. Geotechnical Geoenvironmental Eng.* **2011**, *137*, 1126–1131. [CrossRef]
21. *ASTM C39*; Standard test method for compressive strength of cylindrical concrete specimens; American Society of Civil Engineers: West Conshohocken, PA, USA, 2018.
22. Mitchell, J.K. Soil improvement—State-of-the-art report. In Proceedings of the 10th International Conference on Soil Mechanics and Foundation Engineering, International Society of Soil Mechanics and Foundation Engineering, Stockholm, Sweden, 15–19 June 1981; pp. 509–565.
23. Consoli, N.C.; Foppa, D.; Festugato, L.; Heineck, K.S. Key parameters for strength control of artificially cemented soils. *J. Geotechnical Geoenvironmental Eng.* **2007**, *133*, 197–205. [CrossRef]
24. Consoli, N.C.; Ferreira, P.; Tang, C.-S.; Marques, S.F.V.; Festugato, L.; Corte, M.B. A unique relationship determining strength of silty/clayey soils–Portland cement mixes. *Soils Found.* **2016**, *56*, 1082–1088. [CrossRef]

25. *ASTM D2845*; Standard test method for laboratory determination of pulse velocities and ultrasonic elastic constants of rock; American Society for Testing and Materials: West Conshohocken, PA, USA, 2008.
26. Mitchell, J.K.; Soga, K. *Fundamentals of Soil Behavior*, 3rd ed.; John Wiley & Sons: Hoboken, NJ, USA, 2005.
27. *ASTM D559*; Standard test methods for wetting and drying compacted soil-cement mixtures; American Society for Testing and Materials: West Conshohocken, PA, USA, 2015.
28. Consoli, N.C.; Samaniego, R.A.Q.; González, L.E.; Bittar, E.J.; Cuisinier, O. Impact of severe climate conditions on loss of mass, strength, and stiffness of compacted fine-grained soils–Portland cement blends. *J. Mater. Civ. Eng.* **2018**, *30*, 04018174. [CrossRef]
29. *BS 1377*; Methods of test for soils for civil engineering purposes; British Standards: London, UK, 1990.
30. Clayton, C.R.I.; Khatrush, S.A. A new device for measuring local axial strains on triaxial specimens. *Géotechnique* **1986**, *36*, 593–597. [CrossRef]
31. Maswoswe, J.J. Stress Path Method for a Compacted Soil during Collapse Due to Wetting. Ph.D. Thesis, University of London, London, UK, 1985.
32. La Rochelle, P.; Leroueil, S.; Trak, B.; Blais-Leroux, L.; Tavenas, F. Observational approach to membrane and area corrections in triaxial tests. In Proceedings of the Advanced Triaxial Testing of Soil and Rock, Louisville, KY, USA, 19–20 June 1986; p. 715, (published by ASTM STP 977–1988).
33. Consoli, N.C.; Morales, D.P.; Saldanha, R.B. A new approach for stabilization of lateritic soil with Portland cement and sand: Strength and durability. *Acta Geotechnica* **2021**, *16*, 1473–1486. [CrossRef]
34. Diambra, A.; Ibraim, E.; Peccin, A.; Consoli, N.C.; Festugato, L. Theoretical derivation of artificially cemented granular soil strength. *J. Geotech. Geoenvironmental Eng.* **2017**, *143*, 04017003. [CrossRef]
35. Rios, S.; da Fonseca, A.V.; Consoli, N.C.; Floss, M.; Cristelo, N. Influence of grain size and mineralogy on the porosity/cement ratio. *Géotechnique Lett.* **2013**, *3*, 130–136. [CrossRef]
36. Diambra, A.; Festugato, L.; Ibraim, E.; da Silva, A.P.; Consoli, N.C. Modelling tensile/compressive strength ratio of artificially cemented clean sand. *Soils Found.* **2018**, *58*, 199–211. [CrossRef]
37. Cuccovillo, T.; Coop, M.R. On the mechanics of structured sands. *Géotechnique* **1999**, *49*, 741–760. [CrossRef]
38. Consoli, N.C.; Cruz, R.C.; Da Fonseca, A.V.; Coop, M.R. Influence of cement-voids ratio on stress-dilatancy behavior of artificially cemented sand. *J. Geotech. Geoenvironmental Eng.* **2012**, *138*, 100–109. [CrossRef]
39. Airey, D.W. Triaxial testing of naturally cemented carbonate soil. *J. Geotech. Eng.* **1993**, *119*, 1379–1398. [CrossRef]
40. Coop, M.R.; Willson, S.M. Behavior of hydrocarbon reservoir sands and sandstones. *J. Geotech. Geoenvironmental Eng.* **2003**, *129*, 1010–1019. [CrossRef]

MDPI
St. Alban-Anlage 66
4052 Basel
Switzerland
Tel. +41 61 683 77 34
Fax +41 61 302 89 18
www.mdpi.com

Applied Sciences Editorial Office
E-mail: applsci@mdpi.com
www.mdpi.com/journal/applsci

www.ingramcontent.com/pod-product-compliance
Lightning Source LLC
Chambersburg PA
CBHW040817250926
51850CB00007BA/1193